EAC Occasional Paper No. 11

When Valletta meets Faro. The reality of European archaeology in the 21st century

EAC Occasional Paper No. 11

When Valletta meets Faro
The reality of European archaeology in the 21st century

Proceedings of the International Conference
Lisbon, Portugal, 19–21 March 2015

Edited by Paulina Florjanowicz

EAC Occasional Paper No. 11

When Valletta meets Faro. The reality of European archaeology in the 21st century

Edited by Paulina Florjanowicz

Published by:
Europae Archaeologia Consilium (EAC), Association Internationale sans But Lucratif (AISBL),
Siège social/ Official address
rue des Brigades d'Irlande 1
5100 Namur
BELGIUM
www.e-a-c.org

© The individual authors 2016

The views expressed in this volume are those of the individual authors, and do not necessarily represent official policy, nor the opinion of EAC.

ISBN 978-963-9911-76-5

Brought to publication by Archaeolingua, Hungary
Managing editor: Elizabeth Jerem

Copy editing Paulina Florjanowicz
English language editing by Barbara M. Gostyńska
Layout and cover design by Rita Kovács

Printed by Aduprint Printing and Publishing Ltd, Hungary
Distribution by Archaeolingua, Hungary

Cover image:
'Heritage Day' in 2010 at a megalithic tomb of Wartberg Culture (3400–2800 BC)
© LWL-Archäologie für Westfalen, photo by Hermann Menne

Contents

	Introduction Paulina Florjanowicz	7
1	Trajectories towards a knowledge-producing contract archaeology Kristian Kristiansen	9
2	Challenging attitudes – delivering public benefit Adrian Olivier	13
3	From Valletta to Faro with a stopover in Brussels. International legal and policy background for archaeology or simply the understanding of heritage at the European level Paulina Florjanowicz	25

Session 1 Setting the scene

4	A survey of heritage management in Germany, with particular reference to Saxony-Anhalt Konstanze Geppert and Harald Meller	35
5	The organisation of Czech archaeology – a socialist legal system applied in a market economy Jan Mařík	47
6	Archaeological research in the Slovak Republic – positives and negatives Matej Ruttkay, Peter Bednár, Ivan Cheben and Branislav Kovár	53
7	French preventive archaeology: administrative organisation, role of the stakeholders and control procedures Bernard Randoin	59
8	A view from Turkey on the Valletta and Faro Conventions: effectiveness, problems and the state of affairs Mehmet Özdoğan and Zeynep Eres	65
9	Everything you always wanted to know about commercial archaeology in the Netherlands Marten Verbruggen	77

Session 2 Balancing stakeholders

10	Scotland and a 'national conversation' Rebecca H Jones	85
11	The General Directorate of Cultural Heritage's competencies in the context of safeguarding and promoting the Portuguese archaeological heritage Maria Catarina Coelho	89
12	Working for commercial clients: the practice of development-led archaeology in the UK Dominic Perring	95
13	Balancing stakeholders in the Netherlands. A plea for high-quality municipal archaeology Dieke Wesselingh	105
14	The legal basis and organisation of rescue archaeology in Poland Michał Grabowski	113
15	Preventive archaeology in Wallonia: perspectives Alain Guillot-Pingue	119

Session 3 Assuring quality

16 | Is everybody happy?
User satisfaction after ten years of quality management in development-led archaeology in Europe   125
Monique H. van den Dries

17 | Challenges and opportunities for disseminating archaeology in Portugal:
different scenarios, different problems   137
Ana Catarina Sousa

18 | From Valletta to Faro – avoiding a false dichotomy and
working towards implementing Faro in regard to archaeological heritage
(reflections from an Irish perspective)   157
Margaret Keane and Sean Kirwan

19 | Assuring quality: archaeological works on Irish national road schemes   167
Rónán Swan

20 | Archaeology as a tool for better understanding our recent history   175
Peep Pillak

21 | Archaeological sites: the need
for management and legislation improvements (some thoughts on the Albanian reality)   187
Ols Lafe

Résumés   191

Contributors   197

Introduction

Over the past decades, European archaeology has focused on different ways of researching and protecting sites in areas intended for construction and other forms of land development. This type of archaeology, which has become the predominant model of this scientific discipline, has been given different names all over Europe: for example preventive, rescue, commercial, contract, development-led.

Whichever term we use to describe it – it is worth discussing. Therefore, the European Archaeological Council chose it as the theme for its annual symposium held in Lisbon in March 2015. With this event, the EAC completed a triptych of debates on the true effects of the Valletta Convention on European archaeology. It started with the symposium *The Valletta Convention: Twenty Years After. Benefits, Problems, Challenges* held in Albania in 2013 (EAC Occasional Paper no. 9), followed by the conference *Setting the Agenda: Giving New Meaning to the European Archaeological Heritage* organised in the Netherlands in 2014 to discuss the priorities for future EAC actions (EAC Occasional Paper no. 10).

The idea behind the third symposium, this time organised in Portugal under the title *When Valletta meets Faro. The reality of European archaeology in the 21st century*, was not only to analyse the technical aspects of different legal systems in force but to define the ways to assure the lasting quality of the work by making the results more accessible. The symposium aimed to review the different ways of delivering preventive or rescue archaeology across Europe, and to look at the challenges and benefits of state and private (or commercial) archaeology.

The anticipated outcome was to gain a greater shared understanding of the benefits and challenges faced, and the various approaches taken by European States to provide well-informed advice to governments on the application or modification of policy.

The discussion was backed by the concept of integrating the approach of the Valletta Convention, which shaped preventive archaeology policies as we know them, with the concept of heritage communities contained in the Faro Convention, which determines the 21st-century holistic and participatory approach to heritage governance.

The symposium comprised three sessions outlined by the EAC Board as a consequence of experience from the two previous conferences. Each of the sessions included an opening lecture introducing the topic and was followed by five presentations of national case studies answering the same questions from different countries' perspectives. This allowed the participants to explore the variety of approaches and challenges of modern archaeology across Europe.

This volume, EAC Occasional Paper no. 11, has brought together nearly all of the presentations delivered at the Lisbon symposium along with additional contributions from experts who chaired the sessions and moderated the discussions, as well as those who joined the summarising group to sum up the results of the lively and inspiring debates.

Session 1: Setting the scene

The aim of this session was to introduce the theme of the symposium by presenting the legal and organisational framework for different preventive archaeology models applied across Europe. The range of available solutions is very wide – from strictly centralised schemes to full free-market models. Different countries developed their policies in different legal, social, and economic circumstances. The main questions that need to be addressed refer to assigning significance: Who chooses? How do we choose which sites warrant action? What is the appropriate action to take? And last but not least – who does the work?

Session 2: Balancing stakeholders

This session was designed to focus on the effects. Its aim was to critically analyse the practical outcomes of different rescue archaeology solutions that have been applied around Europe and to show ways of balancing everyone's expectations. One of the most important aspects is arbitrating the goals of the different stakeholders in the planning process. An important issue to tackle is whether the delivery model for preventive archaeology is still a scientific endeavour or whether it is just another pre-construction service.

Session 3: Assuring quality

The final session was meant as a practical reminder of the actual reason for undertaking preventive archaeology measures. It is important to acknowledge that current measures used for protecting archaeological heritage in the planning process are not taken for granted and that good relations with the public are essential. One of the greatest challenges of preventive archaeology is to determine why and how to monitor the quality of the research process and, last but not least – ensure lasting public benefit.

Overall, the volume covers 21 contributions from archaeologists throughout Europe. The scope of issues tackled is quite broad, from pure legal analysis to emotions unleashed with archaeological discoveries related to the tragic history of Europe in the 20th century. Wide geographical representation is provided by authors from a range of countries extending from Portugal to Estonia.

The overview of such a variety of archaeological experience shows huge differences, but there are some common points.

The articles show that we should not rely too heavily on legal acts, as although they provide the necessary framework, this is only part of the solution. These are the policies that go beyond the law and they are extremely important. Therefore, we must make sure that policies are written by experts and approved by politicians or decision makers and not the other way around. Effective policies balance aims of different stakeholders and focus on public benefit. It seems that success stories happen only where a solid policy exists and is actually being implemented.

It is equally necessary to realise the existence of various stakeholders in every archaeological undertaking. Their reasons may differ significantly but they are all important and require consideration.

Another conclusion from the papers presented in this volume is that assuring quality applies both to research results and to raising public awareness. Neither of these should ever be neglected. Adequate analysis and quality presentation of research results is the reason for launching any archaeological field project. And promoting the results to the public and engaging them is an investment for the future.

The 16th EAC Heritage Management Symposium in Lisbon would not have happened if it were not for all the support and encouragement from the members of the EAC Board. The hospitality and brilliant organisation of the conference provided by Maria Coelho, Filipa Neto, João Marques and others from the General Directorate of Cultural Heritage in Portugal as well as António Carvalho and his colleagues from the National Archaeological Museum made this event a memorable one. It was also successful thanks to the professional and dedicated speakers, chairs and a focused summarising group, as well as an active audience that did not hesitate to ask incisive questions.

I would like to acknowledge all the experts for their ideas, for sharing their experience and for their hard work and extreme patience. Your contributions have created this book. I would also like to thank Barbara M. Gostyńska for her diligent work and perseverance in the editing process and Marie-Jeanne Ghenne for translating the abstracts into French.

Paulina Florjanowicz
Scientific Coordinator of the 16th EAC Heritage Management Symposium and Editor of this volume

1 | Trajectories towards a knowledge-producing contract archaeology

Kristian Kristiansen

Abstract: In this presentation I trace some recent changes in the development of contract archaeology in Europe, and the debates following from it. Not least the economic crisis after 2008 made apparent the vulnerability of certain forms of contract archaeology, which raised demands for a more sustainable organisation of the field. Also the rise of 'Big Data' and demands for open access define new challenges for a better European integration of archaeological data and research. It is therefore time for a modernisation of international conventions and codes of conduct from professional associations to match these new realities.

Keywords: European contract archaeology, sustainable knowledge production, quality management, Big Data, Valletta Convention

Introduction

We can date the beginning of a serious and more systematic discussion of how to maintain good archaeological standards in contract archaeology in Europe to the book edited by Willem J. H. Willems and Monique H. van den Dries: *Quality Management in Archaeology*, published in 2007 (Willems & Van den Dries 2007). I used it as a starting point for my comparative contribution in *World Archaeology* in 2009 about contract archaeology in Europe (Kristiansen 2009). There had been earlier foundational texts such as the Valletta (or Malta) Convention of the Council of Europe from 1992 (Council of Europe 1992), and the European Association of Archaeologists/EAA Code of Practice from 1997, followed by the EAA Principles of Conduct for Contract Archaeology in 1998 (Link 1). Shortly after, in 1999, the European Archaeological Council was founded, to act as a forum for heads of national archaeological services in Europe (Link 2). Their annual meetings resulted in a series of books (or Occasional Papers), some of which deal with contract archaeology, such as *Large-scale excavations in Europe: Fieldwork strategies and scientific outcome*, from 2008 (Bofinger & Krausse 2012), and most recently *The Valletta Convention: Twenty Years after – Benefits, Problems, Challenges*, from 2014 (Van der Haas & Schut 2014). From 2012 we also have a systematic overview of north-west Europe in the book: *Development-led Archaeology in North-West Europe* (Bradley et al. 2012). Finally, Jean-Paul Demoule took stock of the historical development in Europe in his 2012 article: *Rescue archaeology – a European view* (Demoule 2012). It contains a historical presentation of the change from public to private commercial archaeology in Europe, with good regional coverage. He prefers the public system, such as the French or Scandinavian, but provides a fair presentation of the different developments in Europe. He also raises important questions as to ownership of heritage and its results, and raises the question of whether contract archaeology can live up to the more recent Faro Convention from 2005 on the Value of Cultural Heritage for Society (Council of Europe 2005) without public/state intervention.

The debate continues

In 2011 Monique van den Dries argued in an article titled: *The good, the bad and the ugly? Evaluating three models of implementing the Valetta Convention*, that the two dominant models, earlier termed socialist versus state models, and hybrids between them, can produce high quality archaeological knowledge. She proposed that:

- It depends largely on who decides and controls quality whether national, regional or local;
- As differences in organisation in Europe will persist, we must strive to create conditions for knowledge production irrespective of such differences.

She further stressed the role of stakeholders and the role of decision-making priorities by local authorities in the Netherlands. Monique van den Dries sees local priorities as more democratic, even if they sometimes overrule research interests at a national or international level. What she does not address, however, is the quality of priorities employed by local authorities. As demands are raised on academic qualifications of project leaders in many countries, we should similarly expect them to be raised on decision makers, as they form an essential link in the system.

She finally points out how geographical and economic differences in Europe impact on the organisation of contract archaeology, and asserts that the rapid infrastructure expansion in some European countries, such as the Netherlands, could never have been handled without the formation of a commercial archaeological sector.

In 2013 Mads Ravn responded to Monique van den Dries in an article titled: *It's about knowledge, not systems: a contribution to a complex discussion about good, bad and ugly production of archaeological knowledge* (Ravn 2013). He argues that state-agency-controlled organisation of contract archaeology, as practised in Scandinavia and France, produces better archaeological knowledge. His arguments can be summarised as follows:

- The capitalist model of private companies are more prone to economic crises, as exemplified in Ireland and the UK. Even if such crises also affect public institutions, they offer other benefits to outweigh them.
- Large museums offer the best organisational framework for contract archaeology due to their infrastructure, which provides larger research environments that can withstand economic fluctuations. They can also support the dissemination of results.

He raises an important problem: how can we ensure research-based academic quality in contract archaeology more generally? While he stresses the organisational framework, there are several other aspects to consider: from legislation and national guidelines that support publication, as in Sweden, to various ways of creating larger research environments and raising academic qualifications.

In Denmark PhDs are now recommended for archaeologists leading large rescue operations, whilst for those in Sweden, holding a PhD is also strongly recommended. A new graduate school for PhD students in rescue archaeology named GRASCA and based at the Linnaeus University in Kalmar has just been launched, financed by a research foundation named *The Knowledge Foundation*. In Denmark small museums have been urged by the National Cultural Heritage Agency to create joint organisations for contract archaeology to secure quality, and at the large Moesgård museum, where the archaeology department is also housed, the leadership position of contract archaeology is shared between the museum and the university to create synergies.

After the crisis – what did we learn?

In a book titled: *Archaeology and the global economic crisis. Multiple impacts, possible solutions*, the editors Nathan Schlanger and Kenneth Aitchison (2010) documented the devastating effects of the 2008 crisis on much of European contract archaeology. In some countries there was a nearly total collapse of the existing market, companies folded and thousands of archaeologists lost their jobs. They concluded that:

- Sustainable organisations must be yet another quality parameter.

Does this suggest that greater public involvement or interaction between contract archaeology, museums and academia can secure such sustainability? We are reminded of the arguments put forward by Ravn that existing public or semi-public organisations, such as museums, are more stable in the long run.

However, what about the sustainability of the lives of individual archaeologists?

An analysis of archaeological salaries in the UK shows a somewhat depressing picture (Everill 2007). In Japan a trend from public to private commercial archaeology has weakened the position of archaeology, according to Katsu Okumura (2013). Is contract archaeology mainly for junior archaeologists, a career starter or dead end? In larger companies, whether public or private, there seem to be better opportunities for career development, but this is another issue which deserves to be examined in a European-wide survey.

- The social and economic standing of contract archaeologists is therefore another quality parameter to be considered.

It is just as important as their academic standing for providing high quality. The task of providing a Europe-wide survey of archaeologists, their employment, salaries, etc. is covered by the project *Discovering the Archaeologists of Europe* (Link 3). From its webpage we learn that in the 21 countries covered by this project €1 billion is spent annually on professional archaeology, employing 24,740 archaeologists, and for Europe as a whole the figure is estimated at 33,000 people. In 12 of the participating countries salaries were below the national average. Fulltime employment had decreased since the crisis from 86% to 78%. Projects such as this form a foundation for strategies to improve conditions for archaeologists in Europe, devised both by the EAA and by national archaeological organisations.

Some methodological considerations

The road from principle to practice may sometimes result in rather diverging interpretations and practices, and this variety I tried to exemplify in my 2009 article (Kristiansen 2009). Whether one argues for a 'market'-based model based on tendering or a more controlled public model, sometimes with less tendering, there must be methodological standards in place to safeguard quality. Such quality instruments are often a mix of public and private control mechanisms some old, some new:

Some state agency/public sector quality control mechanisms:

- Legislation and national standards/guidelines;
- Permits for companies;
- National databases;
- Quality supervision of reports.

Some professional-/private-sector control mechanisms:

- Ethical and professional codes (EAA, Council of Europe, ICAHM);
- Professional organisations;
- Certification;
- Research agendas.

Most countries in Europe exhibit a mix of these quality mechanisms, some grounded in very old traditions, such as national archives/databases, while others are a direct result of the last 40–50 years of expanding rescue or contract archaeology, such as professional organisations, certification and associated professional codes of conduct.

However, the debate needs also to take stock of the digital revolution and the impact of Big Data on archaeological research, not least because rescue/contract archaeology has produced much of this new Big Data.

Quality and open access: public and academic responsibilities

A recent issue of *World Archaeology* discussed the new expanding role of 'open access' to data and publications for archaeology (Lake 2012). We can observe that:

- Big data and digital humanities are fast changing the conditions for knowledge production;
- Archaeology needs a quality standard for European formats and access to archaeological reports. Otherwise we cannot argue that results can be used to produce new knowledge for the common good.

Digital documentation supports such a move, but needs to be geared to modelling and analytical tools, and all data must be open-access in the future. This represents a global trend towards shared scientific databases/infrastructures, also funded by both EU and national research councils. But it demands the addition of a new set of rules/legislation for contract archaeology. A number of recent large-scale research projects have demonstrated the potential in systematising Europe- wide data from mostly contract archaeology, such as *The Later Prehistory of North-West Europe* (Bradley et al. 2015). We are also beginning to see joint European projects financed by the European Research Council, projects taking advantage of Big Data, such as Alistair Whittle's 'The Times of Their Lives: Towards Precise Narratives of Change for the European Neolithic through Formal Chronological Modelling' (Link 4), or Stephen Shennan's 'The Cultural Evolution of Neolithic Europe' (Link 5). We may conclude that 50 years of contract archaeology is now starting to exert a profound impact upon European archaeological research and knowledge production when summarised and analysed as Big Data. However, it demands sustained efforts on a European scale to support such a development.

Proposals to secure sustainable knowledge production in contract archaeology

In summarising the recent debates I make the following proposals to secure a more sustainable, long-term production of archaeological knowledge in contract archaeology:

- Processes of decision-making in tendering should be included in a system of quality management that comprises the whole production process of a contract/project. I list some essential elements of such a system.
- Stated research goals in all large projects, as relevant knowledge is defined by up-to-date research.
- Full interpretation of results, otherwise no new knowledge.
- Full publication of large projects to make results available for research and popular dissemination.
- Full digital access to results in a trans-European database or interlinked databases, providing Big Data.
- PhD required for leadership of large projects. Collaboration with universities.
- Sustainable organisations or networks of organisations.
- Sustainable salaries comparable with other academic institutions.

Some of these suggestions have already been implemented in several countries in Europe with very good results and can be seen as a code of conduct for the future. However, a systematic, Europe-wide implementation of the above proposals demands integrated, comparable documentation systems of the relevant parameters to be followed by a modernisation of current conventions and professional codes of conduct.

References

Bofinger, J. & Krausse, D. (eds) 2012: *Large-scale excavations in Europe: Fieldwork strategies and scientific outcome*, EAC Occasional Paper No 6.

Bradley, R., Haselgrove, C., Vander Linden, M. & Webley, L. (eds) 2012: *Development-led Archaeology in North-West Europe*, Oxbow Books, Oxford.

Bradley, R., Haselgrove, C., Vander Linden, M. & Webley, L. 2015: *The Later Prehistory of North-West Europe. The Evidence of Development-led Fieldwork*, Oxford University Press.

Council of Europe 1992: *European Convention on the Protection of the Archaeological Heritage* (revised). European Treaty Series 143. Council of Europe, Strasbourg, http://www.coe.int/sv/web/conventions/full-list/-/conventions/treaty/143 (accessed 21.12.2015).

Council of Europe 2005: *Framework Convention on the Value of Cultural Heritage for Society*. European Treaty Series 199. Council of Europe, Strasbourg, http://www.coe.int/t/dg4/cultureheritage/heritage/Identities/default_en.asp (accessed 21.12.2015).

Demoule, J.-P. 2012: Rescue Archaeology: A European View, *Annual Review of Anthropology* 41, 611–26.

Everill, P. 2007: British Commercial Archaeology: Antiquarians and Labourers; Developers and Diggers in: Y. Hamilakis & P. Duke (eds): *Archaeology and Capitalism. From Ethics to politics*, One World Archaeology 54, 119–36, Left Coast Press. Walnut Creek, California.

Kristiansen, K. 2009: Contract archaeology in Europe: an experiment in diversity, *World Archaeology* 41(4), 641–8.

Lake, M. (ed.) 2012: Open archaeology. Thematic issue of *World Archaeology* 44 (4).

Okumura, K. 2013: Ethics in Commercial Archaeology: Japan. In: C. Smith (ed.), *Encyclopedia of Global Archaeology*, Springer, New York.

Ravn, M. 2013: It's about knowledge not systems: a contribution to a complex discussion of good, bad and ugly production of archaeological knowledge, *World Archaeology* 45(4), 642–52.

Schlanger, N. & Aitchison, K. (eds) 2010: *Archaeology and the global economic crisis. Multiple impacts, possible solutions*, Culture Lab Editions, Tervuren.

Van den Dries, M. 2011: The good, the bad and the ugly? Evaluating three models of implementing the Valetta Convention. *World Archaeology* 43(4): 594–604.

Van der Haas, V. & Schut, P. A. C. (eds.) 2014: *The Valetta Convention. Twenty Years After – Benefits, Problems, Challenges*, EAC Occasional Paper No. 9, Archaeolingua, Budapest.

Willems, W. & Van den Dries, M. (eds) 2007: *Quality Management in Archaeology*. Oxbow Books, Oxford.

Websites

Link 1: e-a-a.org/codes.htm (accessed 21.12.2015).

Link 2: http://european-archaeological-council.org/ (accessed 21.12.2015).

Link 3: http://www.discovering-archaeologists.eu/blog_index.html (accessed 24.12.2015).

Link 4: http://totl.eu/ (accessed 24.12.2015).

Link 5: http://www.ucl.ac.uk/archaeology/research/directory/euroevol_shennan (accessed 28.12.2015).

2 | Challenging attitudes – delivering public benefit

Adrian Olivier

Abstract: Much of the work that archaeologists undertake today draws on public funds and public financing and is carried out in the name of the public. Past decades have seen a real increase in the level of public awareness of, and interest in, archaeology; however, much of this communication is top down and one-way. Public benefit is easy to claim, but much more difficult to define or demonstrate in practice. Approaches to delivering public benefit are changing, but there remains little understanding of, or articulation with, what the public (or publics) want from archaeologists. If archaeology is to survive and prosper, archaeologists must learn better how to fulfil a public role by engaging with communities as co-creators placing the past at the service of the public so that it is relevant and useful in the context of their daily lives.

Keywords: public benefit, preventive archaeology, heritage values.

Introduction

In one of his last and most perceptive publications about archaeological heritage management, Willem Willems explored the impact and some of the consequences of the Valetta Convention on the practice of archaeology in Europe. In particular, he challenged the orthodoxy of preservation in situ (Willems 2014). In doing this, Willem offered a fundamental critique of archaeological heritage management and its associated practices in the field of preventive archaeology (regardless of specific national models whether centralised or free market). As Willem intended, this critique has very significant implications for the essential function and purpose of archaeological practice that will reverberate across the discipline for many years to come. One of his other key themes concerned the central and critical role of research in all aspects of archaeological practice – whether in a commercial/contract or in a more conventionally academic context. This is a topic that was very, very close to Willem's heart, and I will explore some of the issues related to this in another locus (EAA 2015 Glasgow Annual Meeting Session CA26: The Role of Research in Heritage Management Heritage Management & Research: the dynamics of dialogue).

Willem also considered the need for archaeologists to demonstrate the public benefit of their work and was optimistic about developments here over the past 20 years (Willems 2014, 151). The widespread implementation of the Valletta Convention has clearly led to a significant increase in the costs of archaeological interventions (and of public expenditure on archaeology), and Willem considered that archaeologists have been forced to justify and legitimise these increased costs by focusing on the public benefit of their work through better communication with the public. As Willem noted (loc. cit.), there has been a real increase in the level of public awareness of, and interest in, archaeology. In this context, the situation has undoubtedly improved somewhat since the Convention came into force in 1995 and gives cause for some celebration. It is less clear, however, whether this increase in awareness and interest actually reflects public benefit or whether, as Willem surmised, public benefit has indeed truly become a central theme for archaeologists today.

Much of the work that archaeologists undertake today draws on public funds and public financing (directly or indirectly, as taxpayers or as shareholders or stakeholders) and is carried out in the name of the public (i.e. for the public good or for the public benefit). In a heritage management context, there has been considerable discussion and debate around the nature and purpose of preventive archaeology as it is practised in different European countries (for a survey of European practice: Bozóki-Ernyey 2004; for examples of different aspects of the debate: Demoule 2002a; Thomas 2002; Demoule 2002b; Van den Dries 2011; Demoule 2012, 6189). Regardless of the merits or otherwise of different models of preventive archaeology, the essential function of preventive archaeology can be defined as serving a wider public benefit by safeguarding, one way or another, archaeological values through the management of the impact of change on the historic environment (Wilkins 2013).

Nonetheless, public benefit is easy to claim, but much more difficult to define or demonstrate in practice. Continuing debate (e.g. Goskar 2012) and a growing body of literature on this topic (e.g. Little 2002; 2012), much of it focused on the role of the archaeologist in society (e.g. Richardson 2014, 4), shows that archaeologists are still struggling to understand and come to terms with concepts of public benefit. It remains difficult for many archaeologists to demonstrate the actual and lasting public worth (and value) of what they do in a way that reaches beyond either straightforward public communication or the provision of raw material (information) for quasi-educational and essentially passive education (entertainment), i.e. 'info-tainment' (Olivier 2016).

Archaeologists generally have a very clear and strong belief about why their activities should be relevant, either to the other disciplines that they encounter in the course of their work, or to the public at large (Little 2012, 403). Over the years this has been articulated in different ways for different contexts (e.g. Clark 2006; English Heritage 2007; and in the USA, the National Parks Service brochure: 25 Simple things you can do to promote the public benefits of archaeology). However, this belief system is to a great extent self-justifying and self-fulfilling. Too often it is given expression only through a process of top-down and one-way communication where information flows outwards from archaeologists to their audience.

Different approaches to assessing the public value of heritage are being developed that move away from the usual preoccupations with intrinsic, instrumental, and institutional values to ones that shift attention to broader concepts of public value focusing more on the relationship between heritage outcomes and the requirements of people expressed as a use value (Accenture 2006; Accenture & National Trust 2006). However, in much of the work to date outcomes (public expectations) are usually defined by 'expert' agencies acting on behalf of the public, rather than directly by the public or as a result of public consultation, so these approaches remain essentially top-down and 'expert'-led (op. cit. 13). If new approaches to defining public benefit are to realise their promise as a means of expressing public value and public benefit that is meaningful to the public, then they will have to be more firmly grounded in a realistic understanding of public attitudes and needs (below).

Until recent years, only limited attention has been devoted to finding out and recognising what the public actually think is relevant, and to incorporating alternative public perspectives into 'professional' archaeological activities. To do this requires a two-way traffic between the archaeologist and the public as an essential foundation to building genuine public engagement. This must take us far beyond defining the ways in which archaeology can contribute to society (Little 2012, 403) – an essentially expert and elitist perspective – to acquiring a much better understanding of what society wants from archaeology and from its archaeologists (cf. Agendakulturarv 2004).

The evolution and growth in recent years both of Public Archaeology and of Community Archaeology, each now with its own specialist literature and journals (Public Archaeology established in 2000, and the Journal of Community Archaeology and Heritage established more recently in 2014), shows that, in some quarters at least, thinking about the relationship between archaeologists and the public has begun to extend beyond simple outreach to include attempts to engage the public directly with archaeological practice and process. However, as Richardson notes, in the UK context at least, where archaeology as a professional discipline seeks to maintain its professional expert status, community and visitor participation give the semblance of community involvement, but often remain completely subservient to professional archaeological expertise (Richardson 2014,3).

There have been many advances in community archaeology, and many, many excellent projects worldwide that incorporate genuine community engagement and involvement in all aspects of their work. Examples of good practice have been presented at numerous conference sessions (e.g. at annual meetings of the European Association of Archaeologists and the Society for American Archaeology), in specialist journals (above), and in other publications (e.g. Little 2002; Merriman 2004; Skeates, McDavid & Carman 2012; Thomas & Lea 2014, etc.). However, despite all this work, such practice is by no means as embedded in the discipline as it should be if archaeologists are to achieve the degree of engagement and relevance that they sometimes espouse, and the meaningful public/political support that they certainly desire.

This disparity between apparent (and ever-increasing) public interest in archaeology, and the lack of public engagement in, and active support for, the matter of archaeology has been identified repeatedly over the last 25 years. As long ago as 1989 Merriman observed that:

> Although the value of archaeology in the abstract is affirmed by a large majority of the public, for most people it is seen to have little relevance to their lives, and it is this lack of perceived relevance which leads to lack of interest and understanding of the subject. (1989, 73)

A decade later Merriman's axiom still carried its original force, and the continuing lack of hard statistical evidence about the nature and level of public support and interest in archaeology was emphasised by Schadla-Hall who, in a since oft-quoted editorial of the European Journal of Archaeology, called for serious and sustained research into public attitudes towards archaeology (Schadla-Hall 1999, 151). Ascherson followed by suggesting in the editorial of the first issue of Public Archaeology that professional archaeologists were beginning to overcome an apparent indifference to what local inhabitants or visitors thought about their work and at least were starting to care about public perceptions, even if they didn't know about them – a hiatus that Public Archaeology was intended to fill (Ascherson 2000, 4). However, in 2004, Schadla-Hall was still lamenting that 'the vast majority of the public has no interest or direct contact with what members of the archaeological profession consider to be their subject' and that 'the development of archaeology as an academic subject across the world in the last two hundred years has left most of humanity untouched and unworried' (Schadla-Hall 2004, 255).

Today in 2015, the situation essentially remains little changed. Despite apparently high levels of visceral interest in archaeology in the press or on television, the continued financial pressures of recession and related austerity programmes in most countries of Europe have generally resulted in significant changes in the priorities of governments and agencies with regard to heritage and archaeology, accompanied by a steady reduction in the provision of public archaeological services. These pressures have had, and will continue to have, serious impacts on how archaeology is practised in future. The apparently widespread lack of interest in the consequences of this for the archaeological environment or on heritage at large, and the absence of coherent public political support to counter this (except in very specific and usually quite local contexts), ought to be surprising but actually simply affirms the perceptions (summarised above) that archaeologists have yet to make a convincing case to the public of the relevance of what they do for society at large – and this despite the ever-increasing focus of attention on programmes of public outreach and engagement (Orser 2001, 464).

It is also certainly possible to over-estimate the assumed level of public interest. The few detailed studies amenable to rigorous analysis that have been carried out (e.g. Heritage Council 2007) appear to show that although people attach a (theoretical) importance to heritage and place a high value on its protection, the public continues to have a poor understanding of what is meant by 'heritage' (as defined by heritage organisations). This reinforces the discordance between public and 'professional' perspectives. Only a modest proportion of people develop an active interest in heritage, but this is conditioned by existing conceptions (and in some views, misconceptions) of heritage and the manner in which it is experienced (op. cit. 745). Large numbers profess an interest in heritage and archaeology but there is still precious little evidence about what this really means and the degree to which this translates into a meaningful (and understandable) connection between people and their past. A significant core of people readily admits to having little or no interest in heritage at all. Is the answer to this conundrum simply to build better, stronger, or more convincing arguments in favour of archaeology? The evidence (what there is) suggests that we have to go much further than this, and understand what the public (or different publics) actually think about, and what they want from, archaeology if we are to engage them in a real two-way dialogue.

In trying to achieve greater public interest and awareness (and public support), heritage professionals and archaeologists therefore need to speak to people who may be neither well-versed in archaeology nor particularly interested in it (Orser 2001, 464). Making projects more public-friendly is of course entirely laudable, and of considerable value in itself, but this is only the starting point for engagement, and cannot and should not be seen as an end in itself that will deliver public benefit per se – rather it is merely a part of a much more complicated and long drawn-out process. Archaeologists therefore need to be more, much more, than good communicators and ambassadors of their subject if we are to reshape the essential nature of our relationship with society and the public at large.

However successful the Valletta Convention may have been in raising levels of public awareness and interest in archaeology, and despite all the energy, effort, and resources devoted to outreach, coupled with the very significant and continuing advances in understanding and knowledge achieved in the course of the last 50 years, one thing is startlingly clear. Regardless of the degree to which the public 'consume' archaeological 'product' (often with considerable appetite) the relevance and significance of archaeology outside its own enclosed and rather internalised world has hardly advanced at all indeed in many respects it is regressing.

The work of heritage managers and archaeologists is defined by political decisions, and there is a growing recognition that the practice of archaeology is 'political' in the sense of its being public, serving the best interests of the public (and democratic society) by 'enabling and encouraging people to draw on the power of their history and heritage to shape their lives and surroundings' (Agendakulturarv 2004, 7). The newly formed European Association of Archaeologists Working Group in Public Archaeology is devoting a session at the 2015 Annual Meeting in Glasgow to the topic of Making (Public) Archaeology more Political (Session CA15). The intention is to explore the political aspects of Public Archaeology, and especially how archaeologists affect politics and wider society the role of politics in archaeology and of archaeology in politics (Link 1). The key principle outlined by the session organisers is that Public Archaeology was born out of a critique of traditional ways of doing archaeology (loc. cit.), and in this context it is well beyond time that archaeologists should begin to examine all the different facets of their political relationship with society and the public, and follow, for example, the Swedish Operation Heritage Policy Statement by redefining the essential features of heritage management (Agendakulturarv 2004). Only by doing this will we be able to understand how it might be possible to transform the present, generally stagnant, relationship between archaeologists and people into something more active and dynamic that has the potential to be genuinely transformational by engaging positively with the public to develop mutual and complimentary interests.

Such an approach is not without its dangers not the least of which is that different publics and different communities will engage with their archaeological heritage in different ways. Nevertheless, if today's archaeologist is to be genuinely more reflexive and responsive to public attitudes and needs as a matter of general practice (rather than through case studies – however exemplary), this will require a fundamental rethinking of what currently pass for existing archaeological orthodoxies. The many practical and intellectual challenges that will be encountered in this process are well-known and well-debated (e.g. Schadla-Hall 2004; Richardson 2014; Thomas & Lea 2014). Unless archaeologists face and rise to these challenges, learning from the hard won experience of others, they will never win the public over to their cause however well-developed their communication skills might be.

To achieve this requires working with the public (or 'publics') actively and responsively, listening and responding to public interests responsibly in what has been aptly described as a 'carefully choreographed dance between archaeological expertise and public co-curation and creation' (Richardson 2014, 11). If this can be achieved we will move away from the top-down expert role so deeply embedded, not only in archaeological management structures, but also in the attitudes of so many archaeologists, and begin to foster real and genuinely co-operative and equal partnership with the public to explore and understand our archaeological heritage in all its myriad facets.

Instrumental framework

The shift outlined above from offering the public passive access to their past through communication, to an active involvement and engagement through participation is also reflected in the evolution of international instruments related to archaeology (and the wider heritage). Article 2 of the 1990 ICOMOS Charter for the Protection and Management of the Archaeological Heritage (the Lausanne Charter) calls for active participation by the general public as part of the development of wider policies for the protection of the archaeological heritage (although focusing mainly on the provision of information to the public as a component of integrated protection). Article 6 also emphasises the need actively to seek and encourage local commitment and participation as a means of promoting the maintenance of the archaeological heritage.

The European Convention on the Protection of the Archaeological Heritage (Revised) – the Valletta Convention (Council of Europe 1992) references the need to develop public awareness of the value of the archaeological heritage for understanding the past, and to promote public access to important elements of the archaeological heritage (Article 9.i & 9.ii), but otherwise focuses almost entirely on scientific and technical values (i.e. matters of professional concern current at the time of drafting). The explanatory report to the convention does echo the Lausanne Charter by noting the increasing demand by members of the public to have access to their past (1992b, 2) – but again more as passive recipients of professional expertise (Olivier 2016).

The European Landscape Convention – the Florence Convention (Council of Europe 2000) and the Framework Convention on the Value of Cultural Heritage for Society – the Faro Convention (Council of Europe 2005) take a more integrative approach that moves away from intrinsic 'scientific' heritage values to having a greater concern with social issues. The Florence Convention emphasises the role of civil society both in contributing to understanding of landscapes and in participating actively in landscape policies and decision making (Article 5), as well as focusing specifically on awareness raising (Article 6A) and training and education (Article 6B). The Faro Convention goes further, and begins to put some flesh on the aspirations of the Lausanne Charter (although in a necessarily wider cultural heritage context) by putting people and human values at the centre of an enlarged and cross-disciplinary concept of cultural heritage (Article 1b). The whole approach of the Faro Convention is one of inclusivity. In addition to articulating access to cultural heritage as a fundamental human right, like the Florence Convention, the Faro Convention specifically supports public participation in cultural heritage activities and decision-making (Articles 4 & 5).

Although socially attractive, the integrated people-oriented approach to archaeology and heritage exemplified by the Faro Convention, and to a lesser extent by the Florence Convention, is harder for government administrations to understand or to instrumentalise and operationalise than earlier, more traditionally structured instruments such as the Valletta Convention. The latter focus more on expert and professional concerns rather than trying to grapple, as the Faro Convention does, with the complex concept of multiple heritage communities each with multiple values. The more traditionally oriented conventions are therefore more straightforward to implement and more amenable to monitor for impact, although, as experience shows, this is not without its own difficulties (Olivier 2014). The advantage of the new approaches exemplified by the Faro Convention, and given practical meaning by a new generation of public and community archaeologists, is that it can provide a powerful link between archaeological practice and social cohesion (i.e. public benefit), a link that (as set out above) by and large has yet to be firmly established either in the minds of many practising archaeologists or as reflected in the changing attitude of the general public.

Certainly this approach can be challenged, especially the Faro concept that everyone has a personal right to benefit from, and contribute to, his or her cultural heritage, whilst respecting the cultural heritage of others (Article 4). The multiple and sometimes conflicting values tied up, for example, with different aspects of social and ethnic identity, or between different groups (heritage communities?) with different intellectual and/or economic interests, make this very difficult for national, regional, or local administrations to realise in any practical sense. A possible solution may lie in changing the way heritage (and archaeology) is taught, so that the matter of archaeology is made to be relevant to the interests and daily concerns of culturally diverse and mixed populations. Only by making archaeology useful in today's world will we be able to bridge this gap between archaeologists and people (Díaz-Andreu 2016). It may therefore be important for archaeologists to focus less on cultural identity as it is reflected in the archaeological record, and more on the expression of sense of place, which helps to define how people feel about their relationship with the physical world. In this context, landscape becomes a key component of the identity that shapes communities. Reacting positively to such social drivers will require a further shift in the way that archaeologists think, to one in which they can understand all the different values that contribute to sense of place, rather than using their own expert knowledge and professional standpoint to define it. The discipline of archaeology is certainly broad enough to accommodate such an approach, although it remains to be seen whether archaeologists as a 'heritage community' are mature enough and bold enough to move wholeheartedly in this direction.

The Florence Convention offers a framework to involve people in the identification and definition of landscapes (and inter alia their landscape heritage), and in this way may provide a useful bridge between the more traditional and expert-based values of the Valletta

Convention, and the more outward facing and socially inclusive values of the Faro Convention. However, the Florence Convention also includes the requirement to establish procedures for the participation of the general public (together with local and regional authorities and other interested parties) in the implementation of landscape policies and related decision-making. This is much more difficult to achieve in practice (Goodchild 2007), although case studies show that it can be successful in specific and local contexts (e.g. Bruns 2012; 38-43; Golobič 2007). I have previously expressed concerns about how such participation can be operationalised by administrations at a general level (Olivier 2016), but am now moving towards a position where I think that administrative and process-based solutions may, in fact, be unnecessary – the answer might lie rather in replicating the good practice demonstrated by case studies and applying it more widely in local and specific contexts – this undoubtedly represents a practical and perhaps a more realistic way of delivering widespread public participation on the ground.

The European Union

Article 3.3 of the Treaty of Lisbon, which entered into force on 1 December 2009 requires the EU 'to ensure that Europe's cultural heritage is safeguarded and enhanced' (EU 2007a), and in pursuit of this goal the EU can carry out actions to support, coordinate or supplement Member States' actions in the fields of culture and education. Article 167.2 of the Treaty on the Functioning of the European Union specifies that the EU will support and supplement the actions of member states in improving 'the knowledge and dissemination of the culture and history of the European peoples and in conservation and safeguarding of cultural heritage of European significance'. Importantly, the Treaty also requires the EU to 'take cultural aspects into account in its action under other provisions of the Treaties, in particular in order to respect and to promote the diversity of its cultures' (Article 167.4). The EU therefore has an explicit interest in cultural heritage as a key factor that contributes to and helps to define a common European heritage at the same time as respecting national and regional cultural heritage diversity.

The European Agenda for Culture (EU 2007b) recognised cultural heritage not just as a source of knowledge and identity, but as a 'valuable resource for economic growth, employment and social cohesion' that is also a 'driver for cultural and creative industries' (loc. cit. 2). It firmly positioned cultural heritage as a shared resource and a 'common good' and identified in particular the need to improve the evidence base for the analysis of the economic and social impact of cultural heritage. The Agenda recognised the impacts of decreasing public budgets on traditional cultural activities, and emphasised the need to adapt management and practice to involve a broader range of stakeholders through a more integrated and outward facing approach to heritage activities as a focus for participative community interaction and social integration.

Critically, the Agenda for Culture established heritage as a priority in the EU's work plans for culture, and since then political interest in cultural heritage in the EU has grown significantly. This culminated recently in the Namur Declaration made at the 6th Conference of Ministers responsible for cultural heritage, meeting in Namur on 2224 April 2015. The declaration reaffirms the importance of cultural heritage as a key component in European identity and focuses on four priorities:

- the contribution of heritage to quality of life and the environment;
- the contribution of heritage to Europe's attractiveness and prosperity, based on the expression of its identities and cultural diversity;
- education and life-long training;
- participatory governance in the heritage field.

As do the Florence and Faro conventions, the EU now places considerable emphasis on the different social values of the cultural heritage. It has also recognised the need to understand the economic and social impacts of cultural heritage on society and strongly promotes the (theoretical) concept of participation (including public participation) in heritage governance. However, in identifying specific actions to support these priorities considerable weight is placed on the development of guidelines related to Heritage and citizenship, Heritage and societies, Heritage and the economy, Heritage and knowledge, Heritage and territorial governance, and Heritage and sustainable development. Guidelines, of course, have much merit and utility, but they somehow reflect a rather old-fashioned and perhaps more top-down and professional, expert-led approach than may be desirable if the objective is genuinely to engage all stakeholders, including civil society, in the development of a shared and unifying approach to cultural heritage management.

The Amersfoort Agenda

All these developments at a pan-European level reflect an increasing awareness amongst archaeologists and heritage managers that the Valletta Convention is very much an artefact of its time that mirrors the scientific and professional values of the drafting group and the professional attitudes and concerns of the early 1990s. It is apparent today that attitudes and approaches have changed and evolved in reaction to changing circumstances. The preoccupations of archaeologists in the late 80s and early 90s are less directly relevant to the work of today's archaeologist than they would have been 25 years ago. Of course there is still a great deal of lasting value contained in the Valletta Convention; some of the key issues (e.g. Article 6, dealing with the financing of archaeological research and conservation) have been comprehensively (and usually successfully) addressed in many European countries (although sometimes with serious and unlooked for consequences). Nevertheless, there is also much that remains to be delivered in terms of meeting some of the other aspirations of the Convention, for example, Article 10 on the illicit circulation of elements of the archaeological heritage (Olivier & Van Lindt 2014). The social and economic context of archaeology, and especially of preventive archaeology, is very different today to the position in 1992, and, as observed above, practising archaeologists must overcome some different and very real challenges if their discipline is to survive and thrive in the 21st century.

Archaeologists in general, and the Board of the EAC in particular, have for some years been very conscious of the need to refresh the implicit agenda for archaeology contained in the Valetta Convention, and to bring it up to date in a real 21st-century context. For several years, the EAC's annual symposium has focused on different facets of the role and meaning of the archaeological heritage in Europe. In 2011 the annual symposium discussed the social significance of heritage and the need to understand the concept of Europe through its cultural diversity (Callebaut et al. 2011); in 2012 the symposium explored different perspectives on public awareness, participation, and protection in archaeological heritage management (Lagerlöf 2013); in 2013 the significance of the Valletta Convention and its positive and negative effects were considered (Van der Haas & Schut 2014); and the symposium of 2014 was devoted to developing a new strategic agenda for archaeology built on the foundations of the Valletta Convention but moving the debate forward to encompass critical forward-facing issues – the Amersfoort Agenda (Schut et al. 2015).

The Amersfoort Agenda is a vision document that builds on the firm foundations of the Valletta Convention and takes forward its key principles in the spirit of the Faro Convention. The Agenda focuses on three contemporary themes that confront important issues facing archaeological heritage management today:

- **Embedding archaeology in society**
  Stimulating society's involvement in archaeology and at the same time encouraging archaeology's involvement in society by linking it to the challenges of today's world; interacting with and understanding the needs and expectations of society, and integrating archaeology into education for children and young people.
- **Dare to choose**
  Facing up to the many choices confronting archaeologists today; being transparent about choices that have been made, understanding the consequences of those choices, and accepting that choices may be constrained by the values of other disciplines and stakeholders that lie beyond the traditional boundaries of archaeological 'scientific' concern.
- **Managing the sources of European history**
  Using new (digital) technologies to provide better and wider access to archaeological information that can be shared with other disciplines and the public to create added value and benefit.

The Amersfoort Agenda therefore encapsulates and gives coherent expression to many of the issues related to public benefit (and the role of archaeologists) summarised briefly in this paper. It is also entirely congruent with the direction of travel set out in the Namur Declaration (CoE 2015), although perhaps the Amersfoort Agenda adopts a rather more people-oriented and bottom-up approach. The EAC is currently developing an Action Plan that will set out priorities and appropriate actions to translate the aspirations of the Agenda into practice in a real-world context. This will assist EAC members in following through the agenda both at a strategic level, but also, more importantly, in the context of specific actions to underpin and implement the main themes and agenda items of the Amersfoort Agenda. It is intended that the Action Plan will provide a practical 'road map' that can be adapted to changing needs, priorities, capacity, and resources to implement the Agenda. In this way it is hoped that archaeological heritage managers will explore and develop practical solutions to address some of these problems, and that they will therefore play an important role in making the practice of (preventive) archaeology more relevant to the needs and desires of civil society, and in this way deliver clear, recognisable public benefit.

Discussion

Heritage management framework

By and large, archaeologists have overcome many of the problems that beset the profession 25 years ago. Most countries have a functioning system for the protection, conservation, and management of the archaeological heritage (although under significant and continuing economic pressures at present). Archaeology has successfully been integrated into the spatial planning process (although this is now under threat in some countries). Standards of work (including research) are more or less consistent across Europe and are generally high (although there is always room for improvement). There has been a real shift in preventive archaeology from data production to knowledge building. Knowledge production (including scientific analysis and publication) has now reached an unprecedented level, and the impacts of continuing advances in digital technology and communication on all aspects of archaeological work will continue to be profound.

Some significant challenges remain. Legal systems (and legal constraints) to prevent illicit destruction of the archaeological resource and the associated illicit trade in antiquities are generally ineffective (despite the success of some high-profile individual cases). The volume of material produced by archaeologists (records and artefacts) continues to grow exponentially and, with only a few exceptions, facilities to organise, care for, and store this archive are inadequate. The gap between academic research (universities) and preventive archaeology is still sometimes too wide, and in many countries some of the doctrines of protection, preservation, conservation, and management, which underpin preventive archaeology, may require review and revision to reflect changing circumstances.

One fundamental tenet of archaeological practice, and in particular of heritage managers engaged in preventive archaeology, is that the archaeological heritage is a unique, finite, and non-renewable resource (e.g. Lausanne Charter Article 2 and echoed in many national legislations), although this view (and the management decisions that flow from it) are coming under increasing challenge (e.g. Carman 1996, 78; Holtorf 2005, 130–49; Pace 2012, 2778). Under most legal systems in Europe, the archaeological resource is protected and managed one way or another for its intrinsic values (significance, rarity, etc.). Generally, these protection systems often ignore the non-intrinsic, societal, and personal values that people assign to the archaeological heritage, although there are exceptions that incorporate the broader approaches set out in the Burra Charter (Australia ICOMOS Burra Charter 2013) and encapsulated in modern principles of conservation (e.g. Drury & McPherson 2008). There is also often a discrepancy between idealised and sometimes theoretical legal and administrative frameworks for heritage protection, however modern they might appear, and their practical application, which is often significantly under-resourced.

For all countries in Europe, society and societal attitudes have changed significantly over the last 100 years, although parts of the 'scientific' archaeological community may not always have kept pace with social changes as much as they have with technological advances in their discipline. Perhaps the archaeological community at large needs to come to better terms with the ongoing and inevitable loss of knowledge resulting both from man-made interventions and natural processes acting on the historic environment. We know considerably more about the past today than we ever did, and the natural corollary of this is to question the extent to which we may still feel the need to investigate and analyse everything in the historic environment that is at risk simply because (theoretically) it is unique and irreplaceable. Indeed, understanding the nature of the different choices that confront archaeologists and developing the ability, where appropriate, to make flexible, pragmatic, and open responses to those choices is one of the key elements of the Amersfoort Agenda.

In too many instances, the product of preventive archaeology is defined as the production of the 'scientific' results of fieldwork (usually as an academic publication) and this is taken to represent the 'gain' in, and contribution to the sum of 'scientific' knowledge (and understanding) achieved by the archaeologist. There is, of course, no need whatsoever to gainsay the necessity of applying the highest practical and academic standards to the fieldwork, analysis, and research that is undertaken in the course of preventive archaeology. However, the product of this work ('scientific' knowledge) is all too often locked within a closed (and self-justifying) information system. A system that adds to the existing specialised pool of knowledge that underpins and feeds the continuing process of heritage management, but that is of limited interest except to other experts: specialists and academics. Archaeologists are habituated to producing 'scientific' and 'academic' results (for other archaeologists) and their professional practice focuses rather more on self-referential processes and maintaining standards than looking at either the wider function of archaeology in society, or the potential impacts that their work can have on society.

There is indeed a growing and welcome trend to make the increase in our understanding of the past derived from preventive archaeology more publicly accessible and publicly available. However, as noted above, this is usually a one-way and top-down process that does not necessarily contribute to the active engagement of the public with their past. It is legitimate to ask whether the detailed information produced by archaeologists is actually useful to society at large. Without a more active dialogue, it is difficult to see how existing product (however attractively packaged) can be used to achieve identifiable (and quantifiable) public and social outcomes.

Public engagement, public benefit, and social outcomes

Most archaeological work, and all preventive archaeology, is carried out in the name of the public, for the public benefit, and is paid for (one way or another) by the public, but often with little, or no explicit public participation and involvement in the decisions that are made, in the activities that are carried out, or in creating the products and outcomes of this work. Do the public want this work? Are they prepared to pay for it? Do they even care? There appears to be little or no real, demonstrable public (political) support for the role and function of archaeologists in society, and despite high levels of activity and public interest, archaeology seems to remain as irrelevant to society today as it was in the past.

In a structural context, and at a theoretical level, the trajectory to public engagement and participation has been clearly signposted by a shift to more socially aware practice by the evolution of international heritage instruments to incorporate public values, and even in a political context by the European Union – keen to use a revised concept of heritage to promulgate a vision of a shared, democratic, and participatory European heritage. It is equally clear that a significant number of archaeologists and some administrations in Europe (and world-wide) are already grappling with many of these issues, and in specific circumstances are successfully delivering real public engagement and public benefit (e.g. Operation Heritage in Sweden).

The social aspirations and principles that have been set out in recent heritage conventions and other international instruments provide a useful operational structure (albeit one that is more theoretical than practical). There have also been repeated calls from inside and outside the profession by 'experts' and by politicians for heritage managers to move beyond the technical aspects of heritage management and conservation to 'drawing out local skills, knowledge and experience of place rather than dictating what is of cultural significance' (Lammy 2006, 69). However, the counterpoint to this is that most, if not all, heritage management regulatory processes, procedures, and related decisions (across Europe) are built on 'professional' assessments of the implicit and intrinsic 'scientific' values of the archaeological heritage, usually with very limited or no articulation with the public, and often without incorporating public or social values into the process. The issue is whether it is possible for practising archaeologists to accept and integrate these public and social values into their work at a general level, and then actually to incorporate them into day-to-day practice in specific national and administrative contexts.

The wide range of issues associated with public participation and public benefit reprised in this paper have been discussed by archaeologists for (too) many years and it is clear that there are no easy answers or simple, straightforward solutions that can be adopted 'out of the box'. The issues are complex, and achieving real, meaningful, and sustainable outcomes will involve challenging a number of current orthodoxies, however uncomfortable these may be. These include:

- agreeing what the appropriate roles for government, agencies, voluntary bodies, communities, and private individuals are in making decisions about archaeology;
- understanding what the appropriate balance is between the role of archaeologist as 'expert' defining heritage values for other people to consume, and as 'facilitator' enabling other people's perceptions of heritage values;
- recognising and understanding better how society values heritage and being able to incorporate other perceptions into our professional belief system;
- finding ways to integrate public engagement skills into the repertoire of all archaeologists so that delivering real identifiable public benefit is built into archaeological practice and process.

Models of preventive archaeology and responses

Debates about the different models of preventive archaeology in Europe (above) are unhelpful. As demonstrated elsewhere and at the EAC annual symposium 2015: *When Valletta meets Faro. The reality of European archaeology in the 21st century*, each model has benefits and disbenefits, advantages and disadvantages, whether the system is state-funded or developer-funded, centralised or dispersed, commercial or subsidised, competitive or monopolistic, regulated or deregulated, or indeed any combination of these characteristics. In such discussions, attention is too often focused on the role (and responsibilities) of the state, the existence and nature of an archaeological 'market' (or not), and the essential role of the heritage manager and archaeologist as the guardian and protector of the past notwithstanding the fact that in many countries a great many decisions related to heritage that impact significantly on the heritage are increasingly made with little (or even no) reference to heritage 'expertise'.

The somewhat prosaic reality is that these questions may be of little interest except to students of heritage management and may be irrelevant to the discussion at hand. Heritage managers and archaeologists at large are only very rarely in a position to exercise any choice or influence about the sort of model that operates in their country. Structures for preventive archaeology, even when successful and well-resourced, are at best tolerated in a political and economic context, and at worst have to operate in a more or less hostile environment defined by changing political imperatives (or whims). The real issue is not about the role of the expert in making decisions or participating in decisions; nor is it about how different systems and procedures for preventive archaeology operate in different countries. It is about the nature of the outputs that derive from preventive archaeology (and their quality) and the uses to which they are put that is to say the product and the outcomes of preventive archaeology. Unless preventive archaeologists turn from their current preoccupations and give much greater attention to these factors and to their role in facilitating public and community engagement, they will find themselves and their practice increasingly marginalised by administrations (and society) that have other priorities.

Many problems beset the archaeological profession across Europe today: reduced funding, hostile social attitudes, fragmentation, lowering quality of life, to name a few. The natural response of archaeologists is to band together in a pan-European context to create a larger, more powerful, more coherent, and more sustainable body to address these challenges. Such a grouping (the European Association of Archaeologists) of course has inestimable value in very many respects, but no matter how large such an association can become (possibly 4, 6, or perhaps even as many as 10,000 potential members), the fundamental reality is that it will only ever be (in European terms) a comparatively small and restricted special interest group. If current practices continue unchecked, producing specialised knowledge that no one wants (apart from other archaeologists), material that no one can afford to store, and doing a job that no one cares about, archaeology will remain marginalised only serving the interest of its own relatively small constituency with very limited relevance or influence. Until archaeologists can deliver outputs and outcomes that are directly relevant to the broader interests of the public and society at large, we will continue to exist on the fringes of other policy arenas, largely ignored by the public (and their politicians).

Conclusion

Archaeologists must learn how best to fulfil a public role, moving from expert intellectual owner and/or guardian of the past, and adapt to the role of a facilitator or mediator who places the past (and knowledge about the past) at the service of the public. In trying to develop public participation in the production of archaeological knowledge and multiple-voiced, participatory approaches to heritage issues (Richardson 2014,11), the archaeologist will have to confront issues of intellectual ownership of the past that may prove both challenging and sometimes uncomfortable. To do this, archaeologists need real public support and public opinion on their side in contexts other than 'rescue' campaigns (e.g. the Temple of Mithras, the Rose Theatre, the Newport Ship). Archaeologists will not achieve this if, as is so often the case, they continue to patronise the public with their own values, without taking the trouble to find out either what it is that the public (or publics) actually want or, without helping the public, contribute as co-creators to build a broader-based understanding of the past. Ultimately, in a democratic society, if broad sections of the public can be engaged fully in the archaeological process, then the politicians and administrators will inevitably follow their lead.

Many archaeologists today work extremely hard to put a value or premium – economic or otherwise – on the archaeological heritage, but the arguments remain unconvincing both to politicians who control the environment in which we operate and to the public for whom archaeology is little more than a passing and passive interest. At a professional level, demonstrating public benefit means more than simply justifying 'scientific' and 'academic' outputs, showing a return (intellectual or fiscal) on investment in archaeological works, or sharing results with the public – none of these will ensure lasting public benefit. Delivering true public benefit means taking the natural product of archaeological work (knowledge for research) and *transforming* the results into something that is interesting, meaningful, relevant, and above all *useful* to communities and to the public in the context of their daily lives. It means taking public values into account in archaeological work and including public values in decision-making (participation).

The key is not to try to change the attitudes of society directly – this is almost certainly bound to fail – but more realistically to change the attitudes and approaches of archaeologists so that they are more inclusive and aligned to the needs of society. If this can be managed, then there is a chance, at least, that the aspirations of recent international instruments will be met, and that society at large will begin to appreciate the true value of archaeology and the contribution that it can make not just in a rarefied intellectual environment but to the daily life of people at large. It is time to stop talking about the theory of public benefit and time to try to achieve it in practice.

References

Accenture. 2006: Capturing the public value of heritage: looking beyond the numbers in K. Clark (ed.): *Capturing the Public Value of Heritage: the proceedings of the London conference 25–26 January 2006*. English Heritage, 19–22.

Accenture & National Trust. 2006: *Demonstrating the Public Value of Heritage*. National Trust.

Agendakulturarv. 2004: *Putting People First: Operation Heritage Policy Statement*. http://agendakulturarv.raa.se/opencms/export/agendakulturarv/sidor/in_english/in_english.html (accessed 08.09.2015).

Ascherson, N. 2000: Editorial. *Public Archaeology*. 1 (1), 1–4.

Australia ICOMOS. 2013: *The Burra Charter: the Australia ICOMOS Charter for Places of Cultural Significance 2013*. Australia ICOMOS Inc, Australia.

Bozóki-Ernyey, K. (ed.) 2004: *European Preventive Archaeology: Papers of the EPAC Meeting, Vilnius 2004*. National Office of Cultural Heritage, Hungary & Council of Europe.

Bruns, D. 2012: Landscape, towns and peri-urban and suburban areas, in Council of Europe *Landscape facets: Reflections and proposals for the implementation of the European Landscape Convention*, 9–52.

Callebaut, D., Mařík, J. & Maříková-Kobková, J. (eds) 2011: *Heritage reinvents Europe: proceedings of the international conference; Ename, Belgium, 17–19 March 2011*, EAC Occassional Paper No. 7, Namur.

Carman, J. 1996: *Valuing Ancient Things: Archaeology and Law*. Leicester University Press.

Clark, K. 2006: *Capturing the Public Value of Heritage: the proceedings of the London conference 25–26 January 2006*. English Heritage.

Council of Europe. 1992a: *European Convention on the Protection of the Archaeological Heritage (revised)*. European Treaty Series 143. Council of Europe, Strasbourg.

Council of Europe. 1992b: *European Convention on the Protection of the Archaeological Heritage (revised), Explanatory Report*. Council of Europe, Strasbourg.

Council of Europe. 2000: *European Landscape Convention*. European Treaty Series 176. Council of Europe, Strasbourg.

Council of Europe. 2005a: *Framework Convention on the Value of Cultural Heritage for Society*. European Treaty Series 199. Council of Europe, Strasbourg.

Council of Europe. 2005b: *Framework Convention on the Value of Cultural Heritage for Society, Explanatory Report*. Council of Europe, Strasbourg.

Council of Europe. 2015: *Namur Declaration: Cultural heritage in the 21st century for living better together. Towards a common strategy for Europe*. 6th Conference of Ministers responsible for Cultural Heritage (22–24 April 2015).

Demoule, J. P. 2002a: Rescue archaeology: the French way. *Public Archaeology* 2: 170–77.

Demoule, J. P. 2002b: Answer to Roger Thomas. *Public Archaeology* 2: 239–40.

Demoule, J.P. 2012: Rescue Archaeology: A European View. *Annual Review of Anthropology*, 41, 611–26.

Díaz-Andreu, M. 2016 (forthcoming): *Interacting with heritage: social inclusion and archaeology in Barcelona*. Paper presented at the Heritage Values Network Workshop: Heritage Values and the Public. Barcelona 19–20 February, 2015, Barcelona.

Drury, P. & McPherson, A. 2008: *Conservation principles: Policies and guidance for the sustainable management of the historic environment*. English Heritage, London.

English Heritage. 2007: *Valuing our heritage: the case for future investment in the historic environment*. English Heritage, Heritage Link, Heritage Lottery Fund, The National Trust.

European Union. 2007a: *Treaty of Lisbon: amending the Treaty on European Union and the Treaty Establishing the European Community*. Official Journal of the European Union (2007/C 306/1).

European Union. 2007b: *Resolution of the Council of 16 November 2007 on a European Agenda for Culture*.

Official Journal of the European Union (2007/C 287/01).

European Union. 2012: *Consolidated version of the Treaty on the Functioning of the*

European Union. Official Journal of the European Union (2012/C 346/47).

Golobič, M. 2007: Recommendations for their improvement – the example of the Alpine region, in *Proceedings of the fourth meeting of the Council of Europe Workshops for the implementation of the European Landscape convention*, European Spatial Planning & Landscape, No. 83, 87–96.

Goodchild, P. 2007: The skills of training the public for participation in decision-making processes, in *Proceedings of the fourth meeting of the Council of Europe Workshops for the implementation of the European Landscape convention*, European Spatial Planning & Landscape, No. 83, 211–17.

Goskar, T. 2012: On Commercial Archaeology and Public Benefit. http://pastthinking.com/2012/12/06/on-commercial-archaeology-and-public-benefit/ (accessed 08.09.2015).

Heritage Council. 2007: *Valuing Heritage in Ireland: A report by Keith Simpson and Associates, Lansdowne Market Research, Optimize Consultants, and the Heritage Council.*

Holtorf, C. 2005: *From Stonehenge to Las Vegas: Archaeology as Popular Culture.* Rowman Altamira.

ICOMOS. 1990: *Charter for the Protection and Management of the Archaeological Heritage.*

Lagerlöf, A. (ed.) 2013: *Who cares? Perspectives on public awareness, participation and protection in archaeological heritage management: proceedings of the international conference; Cité des Sciences, Paris, France, 15–17 March 2012*, EAC Occasional Paper No. 8, Namur.

Lammy, D. 2006: *Community, identity, and Heritage*, in K. Clark (ed.): *Capturing the Public Value of Heritage: the proceedings of the London conference 25-26 January 2006.* English Heritage, 65–9.

Little, B. L. 2002: *Public benefits of archaeology.* Gainesville: University Press of Florida.

Little, B. L. 2012: Public Benefits of Public Archaeology in R. Skeates, C. McDavid & J. Carman (eds): *The Oxford Handbook of Public Archaeology (Oxford Handbooks)*, Oxford University Press, 396–413.

Merriman, N. J. 1989: Museums and archaeology: the public view, in E. Southworth (ed.): *Public Service or Private Indulgence?* Liverpool: Society of Museum Archaeologists. 10–25.

Merriman, N. (ed.) 2004: *Public Archaeology*, London: Routledge.

Moshenska, G. 2009: Beyond the viewing platform: Excavations and audiences, *Archaeological Review from Cambridge*, 24(1), 39–53.

National Parks Service. 1995: *25 Simple things you can do to promote the public benefits of archaeology.*

Olivier, A.C.H. 2014: The Valletta Convention: twenty years after a convenient time, in V. Van der Haas & P. Schut (eds): *The Valletta Convention: Twenty years after – Benefits, Problems, and challenges*. EAC Occasional Paper No. 9.

Olivier, A.C.H. 2016 (forthcoming): *Communities of Interest – Values: whose values?* Paper presented at the Heritage Values Network Workshop: Heritage Values and the Public. Barcelona, 19–20 February 2015. Public Archaeology.

Olivier, A.C.H & Van Lindt, P. 2014: Valletta Convention perspectives: an EAC survey in V. Van der Haas & P. Schut (eds): *The Valletta Convention: Twenty years after – Benefits, Problems, and challenges*. EAC Occasional Paper No. 9.

Orser, C.E. 2001: Review: Little, B. (2002). Public benefits of archaeology. *Journal of Field Archaeology*, Vol. 28, 464–66.

Pace, A. 2012: From Heritage to Stewardship: defining the sustainable care of archaeological places, in R. Skeates, C. McDavid & J. Carman (eds): *The Oxford Handbook of Public Archaeology*, 275–95.

Richardson, L-J. 2014: Understanding Archaeological Authority in a Digital Context, *Internet Archaeology* 38. http://dx.doi.org/10.11141/ia.38.1 (accessed 08.09.2015).

Schadla-Hall, T. 1999: Public archaeology, *European Journal of Archaeology*, 2, 147–58.

Schadla-Hall, T. 2004: The comforts of unreason, in N. Merriman (ed.) *Public Archaeology*, London: Routledge, 255–71.

Schut, P., Scharff, D., & de Wit, L. C (eds) 2015: *Setting the Agenda: Giving New Meaning to the European Archaeological Heritage*. EAC Occasional Paper No. 10. Amersfoort.

Skeates, R. McDavid, C. & Carman, J. (eds) 2012: *The Oxford Handbook of Public Archaeology (Oxford Handbooks)*, Oxford University Press.

Southworth, E (ed.) 1989: *Public service or private indulgence?* Society of Museum Archaeologists: The Museum Archaeologist v13 Conference proceedings, Lincoln 1987.

Thomas, R. 2002: Comment: rescue archaeology: the French way, by Jean-Paul Demoule. *Public Archaeology* 2, 236–38.

Thomas, S. & Lea, J. (eds) 2014: *Public Participation in Archaeology*, Boydell Press.

Van den Dries, M. 2011: The good, the bad and the ugly? Evaluating three models of implementing the Valletta Convention. *World Archaeology* 43(4), 594–604.

Van der Haas, V. M. & Schut, P. A. C. (eds) 2014: *The Valletta Convention: twenty years after; benefits, problems, challenges*, EAC Occasional Paper No. 9, Brussels.

Wilkins, B. 2013: Response Posted to Goskar 2012. http://pastthinking.com/2012/12/06/on-commercial-archaeology-and-public-benefit/ (accessed 08.09.2015).

Willems, W. J. H. 2014: Malta and its consequences: a mixed blessing, in V.M. Van der Haas & P.A.C. Schut (eds): *The Valletta Convention: Twenty Years After – Benefits, Problems, Challenges* EAC Occasional Paper No. 9, 151–6.

Websites

Link 1: http://eaaglasgow2015.com/session/making-public-archaeology-political/ (accessed 08.09.2015).

3 | From Valletta to Faro with a stopover in Brussels. International legal and policy background for archaeology or simply the understanding of heritage at the European level

Paulina Florjanowicz

Abstract: Contemporary archaeology is more linked with 'real life' than any other part of cultural heritage. Land development, transport infrastructure, environment protection, agriculture – all these areas have a direct impact on the archaeological heritage and put it at risk. In order to neutralise this risk, different legal measures and policies have been introduced both at national and European level. This is an attempt to present the latter from an archaeologist's perspective. The most widely known is of course the *European Convention on the Protection of the Archaeological Heritage* (Council of Europe 1992); however, the *Framework Convention on the Value of Cultural Heritage for Society* (Council of Europe 2005) is just as important. These two Council of Europe conventions are quite different, thus illustrating the evolution of the approach to heritage. But even though they differ significantly, they are still more complimentary than contradictory.

When discussing the international policy and legislation relating to archaeology, one must not forget the European Union. According to Article 3.3 of the Treaty of Lisbon, '[The Union] shall respect its rich cultural and linguistic diversity, and shall ensure that Europe's cultural heritage is safeguarded and enhanced'. At the same time, the Treaty stipulates that culture is a policy area where the Union only supports member states, which excludes any harmonisation of national laws. However, archaeology, being linked with so many other fields of people's activity, is constantly affected by EU laws. Hitherto, the social and economic potential of this category of heritage was being ignored, thus increasing the threats. That is why various attempts were made over the past few years to change the European Union's understanding of cultural heritage and its role in Europe. Recent achievements, such as adoption of two EU Council conclusions in 2014 and the European Parliament resolution of 8 September 2015, directly recognising the positive aspects of cultural heritage for the European community, pave the way for changes which might have enormous consequences for archaeology as well – if they are good or bad depends largely on the archaeologists themselves. That is why it is so important to understand the processes which are now taking place.

Keywords: Council of Europe, European Union, European Parliament, international legal framework, stakeholders, integrated approach

Point of departure

Archaeology is arguably more linked with 'real life' than any other part of cultural heritage. Land development, urban planning, transport infrastructure, environment protection, agriculture – all these areas have a direct impact on the archaeological heritage and put it at risk. It has always been a challenge for archaeologists, especially those dealing with rescue archaeology, that they operate on the frontline, having to deal with all types of stakeholders who do not have much understanding of traditional heritage values. For instance, the definition of heritage still binding in Polish law states that heritage is protected for its 'scientific, historic and artistic values' (Act 2003, Art. 3.1). Obviously, this set of values has always been difficult to explain to an investor whose primary concerns are time and money, or to a citizen who is eagerly waiting for a new highway to be opened. In order to neutralise these risks, different legal measures and policies have been introduced both at national and European level to protect heritage. I will try to present the latter from the archaeologist's perspective.

Council of Europe

The most widely known instrument is of course the European Convention on the Protection of the Archaeological Heritage (Council of Europe 1992); however, the Framework Convention on the Value of Cultural Heritage for Society (Council of Europe 2005) is just as important, which I will try to demonstrate. The Valletta Convention (Council of Europe 1992) is commonly considered as a way of protecting archaeological heritage and allowing scientific research, and as a convention that aims to secure professional standards in archaeology (which is increasingly linked with the construction process), while the Faro Convention (Council of Europe 2005), is a community-oriented document giving the heritage

> *Council of Europe Framework Convention on the Value of Cultural Heritage for Society*
> *Article 2 – Definitions*
> *For the purposes of this Convention,*
>
> a **cultural heritage is a group of resources inherited from the past** which people identify, independently of ownership, as a reflection and expression of their constantly evolving values, beliefs, knowledge and traditions. **It includes all aspects of the environment resulting from the interaction between people and places through time**;
>
> b a heritage community consists of people who value specific aspects of cultural heritage which they wish, **within the framework of public action**, to sustain and transmit to future generations.

Figure 3.1: Definitions in the Framework Convention on the Value of Cultural Heritage for Society (Council of Europe 2005, Art. 2). Bold type by the author.

(including archaeology) to the people and – as some dare to say letting them decide about it. But is this really the case?

Both these conventions are about incorporating archaeology into real life. The first one is about protecting archaeology from human-generated threats, and the second one is about bringing the heritage back to the people. So, they are not contradictory but complementary; logically one follows the other. In my opinion, the Faro recommendations definitely do not override those inscribed in the Valletta Convention. As one of the authors of the Faro Convention, Daniel Thérond, once said: the earlier conventions focused on seeking the answer to the question of how to protect the cultural heritage, while the Faro convention makes us ask a new question: Why do we protect it? (Thérond 2007).

I think that one convention follows and supplements the other. The first one, the Valletta Convention, recommends solutions that require immediate implementation in order to save the archaeological heritage in a situation of increased construction activity. The second one, the Faro Convention, provides solutions to allow cultural heritage protection in the long-term perspective. And none of these measures would have worked if they had been applied in reverse order.

Besides, if we take a closer look at Valletta, it is obvious that the approach it recommends is not that different from the one Faro promotes. Already in the first article it states that 'The aim of this (revised) Convention is to protect the archaeological heritage as a source of the European collective memory and as an instrument for historical and scientific study' (Council of Europe 1992, Art. 1). This shows that the reference to the social value of heritage the intangible value actually comes first. The Convention also refers to dissemination of scientific information. Article 7 states that 'For the purpose of facilitating the study of, and dissemination of knowledge about, archaeological discoveries, each Party undertakes … to take all practical measures to ensure the drafting, following archaeological operations, of a publishable scientific summary record before the necessary comprehensive publication of specialised studies' (ibidem, Art. 7). This call for publishing a summary of research results might seem minimalist, considering the different stakeholders' needs for access to information, but one must remember that the Valletta Convention came before the age of the World Wide Web, not to mention social media.

Finally, it is also worth mentioning that the Valletta Convention refers directly to the question of raising public awareness. In its Article 9, the Convention calls for each Party 'To conduct educational actions with a view to rousing and developing an awareness in public opinion of the value of the archaeological heritage for understanding the past and of the threats to this heritage, and to promote public access to important elements of its archaeological heritage, especially sites, and encourage the display to the public of suitable selections of archaeological objects' (ibidem, Art. 7). This is not yet a heritage community as defined in Faro, but the rights of the public to have access to heritage are already recognised. Of course, to what extent the Valletta Convention is actually incorporated into national regulations and further on – into daily practice is another matter, but some good examples are already presented in this book (see in this volume: Jones; Swan; Wesselingh).

The evidence demonstrating that the Faro Convention is a natural consequence of the earlier Council of Europe conventions (including Valletta) is already apparent in its preamble, where it refers directly to all of them (Council of Europe 2005). Furthermore, the Faro Convention is not at all about giving away the responsibility for decision-making to the community, society or any group of non-professionals. Article 1 mentions the right to participate (only!) in cultural life and recognises responsibility towards promoting cultural diversity. The crucial part of the Faro Convention are, however, the definitions of cultural heritage and heritage communities, which are very broad and very inclusive (Figure 3.1).

Further provisions of the Faro Convention in many ways prove that individuals and communities will benefit from the cultural heritage and equally have responsibilities towards it; however, it does not indicate anywhere that non-professionals should make binding decisions regarding the quality or scope of protection or research. On the contrary, Article 6b states that 'No provision of this Convention shall be interpreted so as to … affect more favourable provisions concerning cultural heritage and environment contained in other

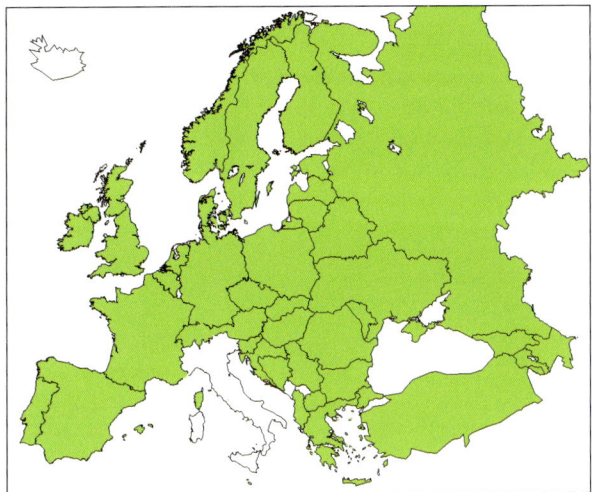

Figure 3.2: Countries which have ratified the Valletta Convention (© Paulina Florjanowicz).

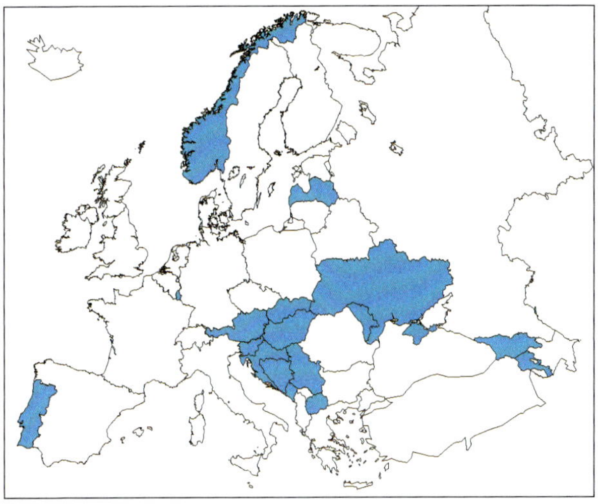

Figure 3.3: Countries which have ratified the Faro Convention (© Paulina Florjanowicz).

national or international legal instruments' (ibidem, Art. 6). It also states that economic policies should not affect the values and integrity of cultural heritage and promotes an integrated approach (ibidem, Art. 8 & 10).

The question therefore is not whether the Faro Convention limits the Valletta Convention as it does not but to what extent the provisions of the two conventions are actually implemented in countries that ratified them. This question remains open, especially given that according to the Council of Europe, whereas the Valletta Convention has already been ratified by nearly 40 member states (Link 1), only 17 member states have so far ratified the Faro Convention (Link 2), (Figures 3.2 and 3.3).

The Council of Europe has recently recognised the value of all its heritage-related conventions once again. The so-called Namur Declaration, adopted at the Conference of Ministers responsible for heritage, held in Namur, Belgium, from 22 to 24 April 2015 in the context of the Belgian Chairmanship of the Committee of Ministers of the Council of Europe, calls for the need to develop a common European strategy for heritage (Council of Europe 2015). It also encourages the member states who have not yet done so to sign and ratify the four heritage-related conventions, including Valletta and Faro.

The European Union

In theory, implementation of the two abovementioned Council of Europe conventions in national regulations and policies should solve the problem and firmly protect cultural heritage, including archaeological heritage. But what about the European Union? It is quite obvious to any EU citizen that we are affected by EU laws and regulations all the time. The question is whether this relates to cultural heritage as well.

The legal basis for the EU's existence are two treaties, both amended several times: the Treaty on the European Union (TEU, originally the Maastricht Treaty, EU 2012a), and the Treaty on the Functioning of the European Union (TFEU, originally called the Treaty of Rome, EU 2012b). Their latest, consolidated version is the so-called Lisbon Treaty, which entered into force in 2009 (EU 2007).

The EU can only act within the limits of the competences conferred on it by these treaties, and where the treaties do not confer competences on the EU they remain with the member states (Article 5.2 TEU). Any EU action must comply with the principle of subsidiarity, which means that every problem should be solved at the lowest possible level, starting from local, through regional to national and finally European. So the Union shall act only if and insofar as the objectives of the proposed action cannot be sufficiently achieved by the member states at national, regional or local level (Article 5.3 TEU).

The EU also has the power to issue the following types of legal acts, which have different impacts on the member states (Article 288 TFEU):

- **Decisions:** Legislative acts of the EU which are binding upon those to whom they are addressed. If a decision has no addressees, it binds everyone.
- **Directives:** Legislative acts of the EU which require member states to achieve a particular result without dictating the means of achieving that result. Directives must be transposed into national law using domestic legislation.
- **Regulations:** Legislative acts of the EU which are directly applicable in member states without the need for national implementing legislation.

In these circumstances, it would seem most adequate for the EU to ratify all of the heritage-related Council of Europe conventions, including Valletta and Faro, and thus automatically make them binding for any legal act of the EU. It seems especially so given that the Lisbon Treaty includes an article that was missing in the previous versions of the treaties: according to Art. 3.3 of the consolidated version of the Treaty of Lisbon, '[The Union] shall respect its rich cultural and linguistic diversity, and shall ensure that Europe's cultural heritage is safeguarded and enhanced' (EU 2007).

Unfortunately though (or perhaps fortunately), at the same time, the Treaty stipulates that culture is a policy area where the Union only supports, complements or supplements the actions of the member states, which excludes the harmonisation of national laws and regulations in this field (Article 167 TFEU). Needless to say, given this legal framework, ratification of these Council of Europe conventions is simply beyond the EU's competence. Yet, EU policies and laws still affect cultural heritage in many ways, and archaeology is not an exception (Ronchi & Nypan 2006; Guštin & Nypan 2010). An obvious question to be asked is what can be done, given these legal circumstances: on the one hand the EU is supposed to ensure that the cultural heritage is safeguarded, on the other hand it can take no legal action in this area.

This issue was discussed for a number of years but the turning point, when it was brought up at an official EU forum, was in 2010, in Bruges, during the Belgian presidency of the Council of the European Union. The so-called Bruges Declaration calls on the EU to recognise the cross-sector character of cultural heritage, its value and potential, and advises close cooperation at EU level in heritage policy-making (Bruges Declaration 2010).

Soon afterwards, two independent actions were initiated as a follow-up of this declaration: The Reflection Group 'EU and Cultural Heritage' (RG), composed of national governmental experts on heritage policy, was setup (Report 2012) and Heritage Alliance 3.3 – a representation of the non-governmental sector was formed (Link 3). Both groups soon started close cooperation and acted within their capacity. The RG closely cooperated with successive EU Council presidencies to include cultural heritage issues on their agenda, and Heritage Alliance actively lobbied the European Commission to increase their interest in cultural heritage as an asset. The following EU Council presidencies actively contributed to the RG's work in the first years of its activity: Belgium, Poland and Lithuania (Vilnius Declaration 2013); also France was very active and chaired the RG in 2014. This led to the turning point in 2014 during the Greek and Italian presidencies of the EU Council (also very active in the work of the RG). That year, for the first time, the EU Council adopted Council conclusions relating directly to cultural heritage – and they did it twice!

EU Council conclusions of 2014

EU Council conclusions are not legally binding on EU member states; they are political statements by the EU Council that facilitate cooperation between member states which may involve changes in practices or the law at national level. Conclusions also set out the direction of policies to be pursued by the European Commission. They shape the policies for the EU and member states.

Figure 3.4. Discussions which led to the final shape of the first ever EU Council Conclusions on the potential of cultural heritage took place in the new Acropolis Museum in Athens. The museum is a great example of the potential of archaeological heritage (© Paulina Florjanowicz).

> *Old approaches sought to protect heritage by isolating it from daily life. New approaches focus on making it fully part of the local community. Sites are given a second life and meaning that speak to contemporary needs and concerns.*

Figure 3.5: Quote from the EC Communication *Towards an integrated approach to cultural heritage in Europe* (EU 2014c, 5).

On 21 May 2014, the EU Council, presided over by Greece at that time, adopted the first ever Council conclusions referring directly to cultural heritage: *Council conclusions on cultural heritage as a strategic resource for a sustainable Europe* (EU 2014a), (Figure 3.4). This important document may be considered as the EU's official reply to the Faro Convention; it also explains WHY heritage is important but puts it in the context of EU priorities, which are social and economic development. It refers directly to Article 3.3 of the Lisbon Treaty, and includes a definition of cultural heritage, inspired by the one in the Faro Convention. Furthermore, the conclusions recommend mainstreaming cultural heritage in national and EU policies for different sectors and encourage investment in cultural heritage, also by means of EU funds. At the same time, which is of utmost importance, the conclusions stress that cultural heritage is a non-renewable asset and that it is unique. They also call on member states to enhance the role of cultural heritage in sustainable development (urban and rural planning, rehabilitation projects).

The EU therefore regards cultural heritage as a valuable asset due to its potential for social and economic development. In order to apply this type of recognition to archaeological heritage, one must value it not only for the fact that it tells us about the past, because that is not enough anymore, but because knowledge about the past strengthens the community thus enforcing social capital, inspiring economic development, etc. Even though the document does not mention archaeology explicitly, it does mention sites and links between heritage and agricultural or maritime policies, so obviously it refers to this type of heritage asset as well.

On 25 November 2015, the EU Council, led at the time by Italy, recognised the potential of cultural heritage once again and adopted the *Council conclusions on participatory governance of cultural heritage* (EU 2014b). This document recognises heritage as a shared resource and aims to reduce the risk of its misuse and at the same time to increase the social and economic benefits resulting from its exploitation. It invites member states to develop multi-level and multi-stakeholder frameworks for heritage management and recommends cross-cutting policies enabling cultural heritage to contribute to different areas. It also promotes evidence-based research to make even stronger arguments for the benefit of cultural heritage.

This is an immediate continuation of the Greek Council conclusions and it goes a step further. This document already recommends concrete solutions for dealing with the heritage by different stakeholders in EU member states, recommends cooperation and a cross-sectoral approach. One could argue that it is nothing new from the archaeology perspective, but actually it puts archaeology in a privileged position, as it is one of the few heritage disciplines that has dealt with all types of stakeholders for decades. Whether this dialogue has always been successful or not is another matter, yet still archaeology can provide some evidence-based case studies, which are of crucial value, as there are not many of them available across Europe.

EU follow-up on cultural heritage policy

2014, being an extremely important year for heritage policy development at the European level, brought two further important documents. In response to both EU Council Conclusions, the European Commission issued on 22 July 2014 an official communication addressed to the European Parliament, the Council, the European Economic and Social Committee and the Committee of the Regions entitled *Towards an integrated approach to cultural heritage in Europe* (EU 2014c). The major difference is that this time it is the European Commission expressing its views on how it sees cultural heritage and its place on the EU's agenda, whereas the Council conclusions are adopted by the EU Council composed of member states governments' representatives. The approach presented in the communication does not differ from the content of the conclusions but it is a solid declaration that the EU shall now prioritise heritage in its actions and will support it at least until 2020. It now regards cultural heritage not only as an asset for all but also as a responsibility for all. The ultimate aim is to make Europe a laboratory for heritage-based innovation. It recognises conservation (one can assume this term includes archaeological research as well) as a process concerning the entire cultural landscape, not merely an isolated site, and that it is becoming increasingly people-centred (Figure 3.5).

The communication has been further reviewed by the European Parliament's Committee on Culture and Education, which published a draft report on 3 March 2015, open for consultation (EP 2015a). The draft version did not include any direct reference either to the Valletta Convention or to the archaeological heritage, but in the course of consultations which the EAC, following the Lisbon symposium, strongly encouraged its members to participate in this situation has changed. The final report, presented at a plenary sitting on 24 June 2015, already includes several important references to archaeology (EP 2015b). It mentions the Valletta Convention as a source for internationally recognised standards for archaeological work, and it asks that a policy framework be set out for the historic environment, including archaeology. Finally, the report acknowledges that many archaeological sites, especially underwater sites, are still at risk of despoliation by organised relic hunters.

Recently, on 8 September 2015, the European Parliament adopted a resolution towards an integrated approach to cultural heritage for Europe, which includes all of the provisions mentioned above, also with regard to archaeology (EP 2015c). Again, a resolution of the European Parliament is not a binding act of law for the member states, but it suggests a political desire to perform in a given area. It allows the European institutions to provide guidelines for coordination of national legislations or administrative practices in a non-binding manner, i.e. without any legal obligations for the addressees member states and/or citizens. Even though it is not binding, such a resolution might and should be used as a very important argument for strengthening the position of cultural heritage in any legal or policy negotiations and/or consultations with national governments. The European Parliament resolutions, once adopted, are also forwarded to national parliaments of the member states.

Another important policy document that will strongly affect the way the EU regards heritage is the next Work Plan for Culture in which cultural heritage, for the first time, is one of four priority areas (EU 2014d). The Work Plan foresees three actions in priority area B – Cultural Heritage:

1. Setting up an OMC (Open Method of Coordination) group on participatory governance of cultural heritage, with the aim of identifying innovative approaches to the multilevel governance of tangible, intangible and digital heritage, which involve the public sector, private stakeholders and civil society. The experts in this OMC group will map and compare public policies at national and regional level to identify good practices and prepare recommendations, also in cooperation with existing heritage networks. The group has already been set up and has started work, and the report, in the form of a handbook, is expected in late 2016.
2. Setting up another OMC group on skills, training and knowledge transfer: traditional and emerging heritage professions. This OMC group, which will be operational during 2017–2018, will focus on the transmission of traditional skills and know-how and on emerging professions, including in the context of the digital shift.
3. A study by the European Commission on risk assessment and prevention for safeguarding cultural heritage from the effects of natural disasters and threats caused by human action. The study, to be prepared during 2016, will include mapping of the existing strategies and practices at national level. Over-exploitation, pollution, unsustainable development, conflict areas and natural catastrophes (fire, floods, earthquakes) are among the factors to be considered.

Obviously, all three actions under the new Work Plan for Culture cover issues which are also important for archaeological heritage management. At this point it is only the archaeologists themselves and their contacts with their respective national authorities that can assure they are involved in these tasks. As the case of the European Parliament resolution showed, it is possible to put archaeology on the agenda.

Conclusions

Having gone through all the recent developments in the European-level approach towards cultural heritage, the following conclusions may be drawn which shape the framework for the future protection of the archaeological heritage:

- There are sufficient legal measures to protect archaeological heritage internationally and nationally;
- The EU has its limitations both in terms of what it can and cannot do in reference to heritage (rule of subsidiarity);
- Concrete decisions regarding cultural heritage are exclusively in the hands of national governments (either at national level, or adopted by the EU Council or European Parliament – both bodies comprising national representatives);
- An integrated approach to heritage, valuing it mostly for its social and economic potential, is a fact and it is not likely to change. It is highly recommended that it be applied in policy-making both at EU and member state level. At the same time, cultural heritage is recognised as a unique asset;
- Cultural heritage has not been valued so highly at EU level ever before, and it is up to us how this is used for the benefit of archaeology;
- For archaeology this means that the understanding of the value of heritage has changed and both parties should acknowledge it (archaeologists because it is not just about research anymore, and decision-makers because they cannot deny its importance). Still, cultural heritage is protected for a reason and constitutes a valuable, yet non-renewable, asset – no one questions this;
- Participatory governance of cultural heritage is the approach of the future. It does not mean sharing the decision-making process with all stakeholders, but it does mean prioritising the public benefit.

Personally, I think archaeology has huge potential in this context, even though it is not mentioned explicitly in most of the abovementioned documents, besides the Valletta Convention and the recent European Parliament resolution. A participative approach to heritage can be very beneficial and effective in protecting archaeological heritage. The reason for this is that archaeology has the most interesting story to tell: a real one, about real life and real people.

Acknowledgements

I would like to thank the EAC Board for inviting and encouraging me to tackle this issue, and especially Leonard de Wit, who provided me with some valuable comments on the first draft of this text. I would also like to thank all EAC members that lobbied their MEPs to include archaeology in the resolution of the European Parliament.

References

Act 2003: Act on the protection of monuments and the guardianship of monuments, Dz. U. no. 162, item 1568, (Polish).

Bruges Declaration 2010: Declaration of Intent closing the Belgian EU Council Presidency Conference, *Cultural heritage: a resource for Europe. The benefits of interaction*, Bruges, http://www.kunstenenerfgoed.be/sites/default/files/uploads/pdf/Declaration-of-Bruges2010-eng.pdf (accessed 17.11.2015).

Council of Europe 1992: *European Convention on the Protection of the Archaeological Heritage* (revised). European Treaty Series 143. Council of Europe, Strasbourg.

Council of Europe 2005: *Framework Convention on the Value of Cultural Heritage for Society*. European Treaty Series 199. Council of Europe, Strasbourg.

Council of Europe 2015: Namur Declaration closing the 6th Conference of Ministers responsible for cultural heritage (2224 April 2015), under the Belgian Chairmanship of the Committee of Ministers of the Council of Europe, *Cultural heritage in the 21st century for living better together. Towards a common strategy for Europe*, Namur, https://www.coe.int/t/dg4/cultureheritage/heritage/6thConfCulturalHeritage/Namur-Declaration-final.pdf (accessed 15.10.2015).

EP 2015a: *Draft Report. Towards an integrated approach to cultural heritage for Europe*, (2014/2149(INI)), rapporteur: Mircea Diaconu, Committee on Culture and Education, European Parliament, 3 March 2015, http://www.europarl.europa.eu/sides/getDoc.do?pubRef=-//EP//NONSGML+COMPARL+PE-546.783+01+DOC+PDF+V0//EN&language=EN (accessed 27.09.2015).

EP 2015b: *REPORT. Towards an integrated approach to cultural heritage for Europe*, (2014/2149(INI)), rapporteur: Mircea Diaconu, Committee on Culture and Education, European Parliament, 24 June 2015, http://www.europarl.europa.eu/sides/getDoc.do?pubRef=-//EP//NONSGML+REPORT+A8-2015-0207+0+DOC+PDF+V0//EN (accessed 27.09.2015).

EP 2015c: European Parliament resolution of 8 September 2015 towards an integrated approach to cultural heritage for Europe (2014/2149(INI)), http://www.europarl.europa.eu/sides/getDoc.do?pubRef=-//EP//NONSGML+TA+P8-TA-2015-0293+0+DOC+PDF+V0//EN (accessed 27.09.2015).

EU 2007: Treaty of Lisbon: amending the Treaty on European Union and the Treaty Establishing the European Community. Official Journal of the European Union (2007/C 306/1), http://eur-lex.europa.eu/legal-content/EN/TXT/?uri=CELEX:12007L/TXT (accessed 01.10.2015).

EU 2012a: Consolidated version of the Treaty on European Union, Official Journal of the European Union (2012/ C 326/13), http://eur-lex.europa.eu/legal-content/PL/TXT/?uri=celex:12012M/TXT (accessed 01.10.2015).

EU 2012b: Consolidated version of the Treaty on the functioning of the European Union. Official Journal of the European Union (2012/C 326/47), http://eur-lex.europa.eu/legal-content/EN/TXT/?uri=celex:12012E/TXT (accessed 01.10.2015).

EU 2014a: Council conclusions on cultural heritage as a strategic resource for a sustainable Europe, http://www.gr2014.eu/sites/default/files/conclusion%20cultural%20heritage.pdf (accessed 27.09.2015).

EU 2014b: Council conclusions on participatory governance of cultural heritage, http://data.consilium.europa.eu/doc/document/ST-15320-2014-INIT/en/pdf (accessed 27.09.2015).

EU 2014c: Communication from the Commission to the European Parliament, the Council, the European Economic and Social Committee and the Committee of the Regions *Towards an integrated approach to cultural heritage in Europe*, 22 July 2014, http://eur-lex.europa.eu/legal-content/EN/TXT/?uri=COM:2014:477:FIN (accessed 27.09.2015).

EU 2014d: Conclusions of the Council and of the Representatives of the Governments of the Member States, meeting within the Council, on a Work Plan for Culture (20152018), http://data.consilium.europa.eu/doc/document/ST-16094-2014-INIT/en/pdf (accessed 01.06.2015).

Guštin, M. & Nypan, T. 2010 (eds): *Cultural heritage and legal aspects in Europe*, Annales Mediterranea, Koper, http://ehhf.eu/sites/default/files/Cultural%20heritage%20and%20legal%20aspects%20in%20Europe%20BOOK.pdf (accessed 27.09.2015).

Report 2012: Report from the Polish chairmanship of the Reflection Group "EU and Cultural Heritage" in 2012, National Heritage Board of Poland, http://www.msz.gov.pl/resource/98119d63-d134-43ad-8a5a-2a012972437f:JCR (accessed 27.09.2012).

Ronchi, A.M. & Nypan, T. 2006 (eds): *European legislation and cultural heritage. A growing challenge for sustainable cultural heritage management and use*, European Working Group on EU directives and cultural heritage, Milan, http://ehhf.eu/sites/default/files/EU_DIRECTIVES_BOOK_2006.pdf (accessed 27.09.2015).

Thérond, D. 2007: *The Faro Convention and new approaches of cultural heritage in a changing society*. Paper presented during Heritage and Society Seminar held by the Portuguese Presidency in the EU Council, 5–6 December 2007, Lisbon.

Vilnius Declaration 2013: Final statement of the Lithuanian EU Presidency Conference, *Cultural heritage and the EU-2020 strategy – towards an integrated approach*, Vilnius, http://www.europanostra.org/UPLOADS/FILS/Final-statement-Vilnius-en.pdf (accessed 27.09.2015)

Websites

Link 1 http://www.coe.int/t/dg4/cultureheritage/heritage/Archeologie/default_en.asp (accessed 27.09.2015).

Link 2 http://www.coe.int/t/dg4/cultureheritage/heritage/Identities/parties_en.asp (accessed 27.09.2015).

Link 3 http://www.europeanheritagealliance.eu (accessed 27.09.2015).

Session 1

Setting the scene

Private archaeological companies also receive significant government contracts.
Excavations at Bratislava Castle (from the Slovak case study, see Ruttkay et al.)
© Archaeological Institute in Nitra

4 | A survey of heritage management in Germany, with particular reference to Saxony-Anhalt

Konstanze Geppert and *Harald Meller*

Abstract: The Valletta Convention of 1992 is embodied in the federal law of Germany (Art. 36, §4 of the Constitution of Saxony-Anhalt). Since the cultural sector comes under the jurisdiction of individual states in Germany, each of the 16 states of the German Federation (the so-called *Bundesländer*) has its own cultural heritage law, but they all have structural similarities. The legal and organisational framework for the preventive archaeology model of Saxony-Anhalt is laid down in its Law on the Protection of Historic Monuments, which will be exemplified here by reference to prominent finds from Saxony-Anhalt.

The primary principle is the preservation of monuments in the unusually rich archaeological landscape of Saxony-Anhalt. Archaeological sites as records of human history are non-renewable resources, which means that every excavation is in fact a process of destruction. The State Office for Heritage Management and Archaeology, and especially the department of Archaeological Conservation, fulfils the duties set out in the Law on Protection of Historic Monuments with respect to archaeological monuments. Its central tasks include the preservation and protection of the physical substance of the archaeological monuments, as well as recording them, documenting them scientifically and studying them. To complete these tasks, various methods (field surveys, preliminary investigations in advance of planned building activities, aerial photography, geophysical prospection, LiDAR scans, among others) are used to record systematically the physical substance of the monuments. Archaeological conservation by the state has, in our opinion, several advantages, which are discussed and contrasted with other heritage management models. The state's sophisticated work of archaeological conservation is in many respects the first stage of the scientific study and evaluation of archaeological finds and sites, whilst at the same time forming the basis for communicating and explaining them to the public. The financial burden of the documentation comes under the rule of the so-called 'cost-by-cause principle' (*Verursacherprinzip*). This means that the documentation of an archaeological site is funded by the developer who causes its destruction, up to a maximum of 20% of the whole planned investment.

A main focus of Archaeological Conservation Department's work is providing expert assistance in planning permission processes of every kind and offering supervision and execution of rescue excavations. These directly involve experts in various natural science disciplines, including archaeobotany, archaeozoology, and soil science, and specialists in various epochs are represented on the staff of the department itself. This is the only way to gain an understanding of the broader issues in environmental archaeology. By engaging these kinds of experts, continuing study of the archaeological monuments is also given increased attention. This often happens in collaboration with external and international partners.

Keywords: Valletta Convention, heritage management in Germany, Law on Protection of Historic Monuments Saxony-Anhalt, cost-by-cause principle, ownership of finds

The European Convention on the Protection of the Archaeological Heritage, known as the Valletta or Malta Convention, of 16 January 1992 (referred to below as the Valletta Convention) was ratified in Germany on 9 October 2002 (Art. 59, §2 of the Basic Law of the Federal Republic of Germany) and thereby became part of German federal law. Article 6 of the Convention was, and is, of great importance, since it deals how the study and preservation of the archaeological heritage are to be financed. Paragraph 2 lays an obligation on each country to make resources available for this purpose.

According to Article 6, §2, these resources are to be spent on preliminary surveys, scientific documentation and, in the case of monuments below ground, excavation, as well as comprehensive publication and cataloguing of the finds. However, these costs ought not to be borne by the tax-payer if they arise in the pursuit of private profit (Hönes 2005, 755).

Despite the adoption into federal law of the Valletta Convention, these provisions have no direct effect on the practice of heritage management; according to ruling 7 B 64/10 of 13 December 2010 by the Federal Administrative Court, they amount only to operational objectives. To clarify this, let us look again, more closely, at Article 6 of the Convention. It states that in the case of public or private developments, measures are to be put in place by the contract partners to ensure that the public does not bear any of the costs of surveys, excavation and documentation, which are

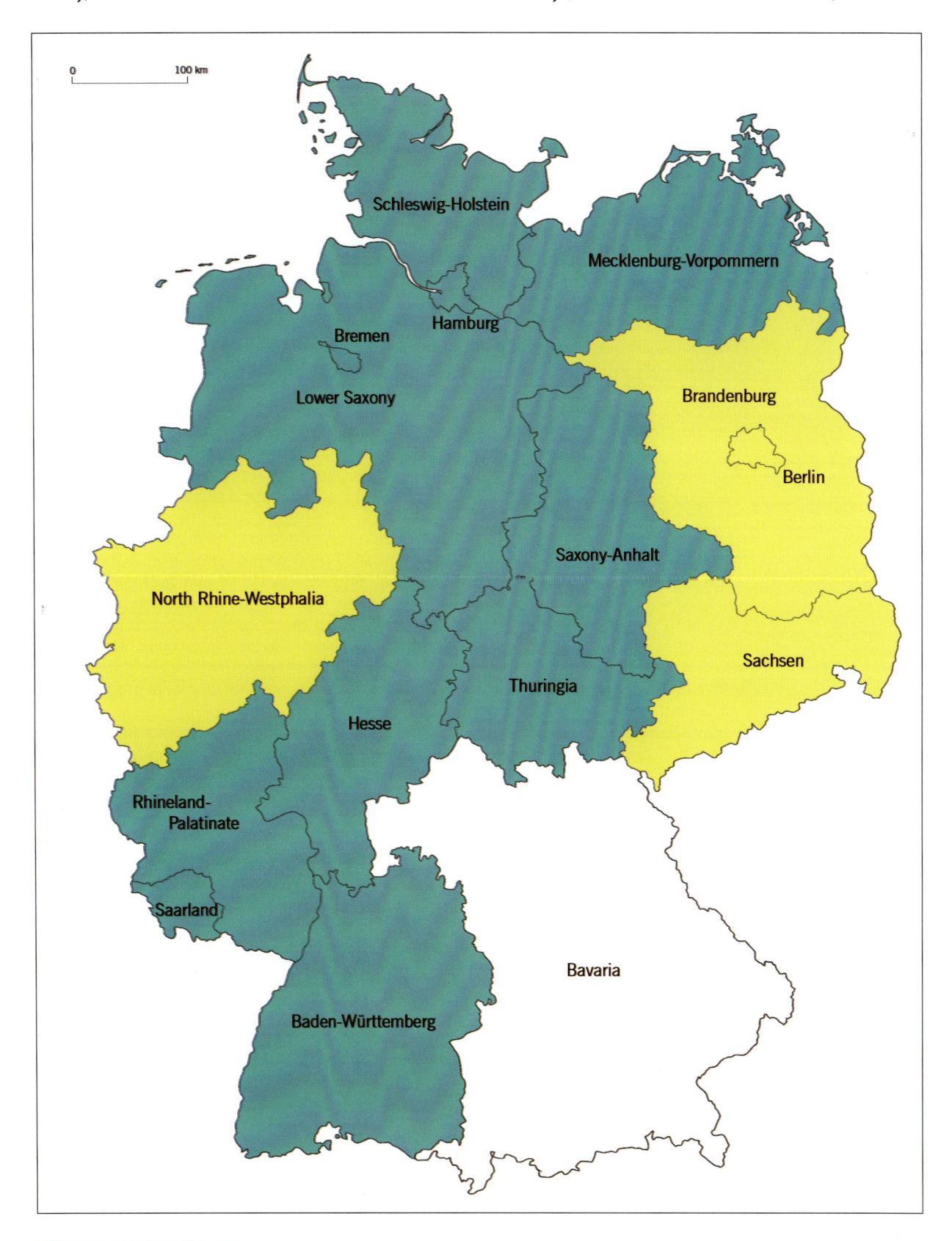

Figure 4.1. Map showing the different situations in the individual Bundesländer as regards legal provision for state ownership of archaeological finds on their Discovery: white – no treasure-trove law; blue – 'large' or extended treasure-trove law; yellow – 'umbrella' treasure-trove law (A. Reinholdt: State Office for Heritage Management Saxony-Anhalt, based on a map from http://d-maps.com/carte.php?num_car=17879&lang=de).

to be imposed on the project developer. The wording of the provision means that no direct obligation on the project developer to bear the costs of any archaeological documentation can be derived from the Valletta Convention (Kemper 2015, 4).

Although the German *Bundesländer* have sovereignty in cultural matters, they are still bound to harmonise their laws with the Basic Law of the Federation. Individual states therefore still have an obligation to implement the Valletta Convention in state law (Kemper 2015, 3). As a result, there are 16 different, although structurally similar, heritage protection laws. The duty of the project developer to bear the costs, as envisaged by the Valletta Convention (the cost-by-cause principle), is variously interpreted by these different state laws (see Gumprecht 2003, 33–34).

The purpose of this paper is to outline the most significant differences as regards the ownership of any archaeological finds discovered and the regulations on the allocation of costs, focusing on the example of Saxony-Anhalt and comparing it with other *Bundesländer*.

Ownership of finds

Section 984 of the German Civil Code regulates the question of the ownership of finds throughout the whole German Federation. Following the principle established in Roman Law on treasure trove, the finder and the owner of the land on which the find is discovered each receive half of its value. In order to safeguard scientific research, to ensure that the find can be put on public display, and lastly, to avoid the danger of items of cultural importance leaving the country, some *Bundesländer* have enshrined the state's claim to ownership of archaeological finds, defined as treasure trove, in their respective heritage protection laws. Basically, state entitlement to treasure trove can be divided into three categories according to how far-reaching the regulations are. While so-called 'small', or restricted, treasure trove law adjudges all finds to be the sole property of the state which originate from state-run excavations or from designated excavation areas, 'large', or extended, treasure trove law entitles the state to ownership of any finds with special scientific significance or value. Finally, 'umbrella' provisions give the state ownership of all archaeological finds, regardless of where they are found or of their value (Otten 2008, 31f.). The various legal provisions of the *Bundesländer* are shown in Figure 4.1.

According to Section 12 of the Law on Protection of Historic Monuments of Saxony-Anhalt, the state is entitled to claim as treasure trove any moveable items of cultural heritage whose owner can no longer be identified, provided they are discovered either in the course of state-run excavations or in designated excavation areas, or that they are of outstanding scientific value. Where appropriate, a finder who has fulfilled his or her obligation to surrender the find will receive a financial reward in keeping with its scientific value.

The importance of archaeological finds becoming the property of the state was strikingly illustrated by the events surrounding the discovery of the Nebra Hoard (Figure 4.2). The sky disc and the associated objects, which have since been recognised as some of the most important finds ever discovered in central Europe, were illegally excavated by treasure hunters in the territory of Saxony-Anhalt and subsequently put on the market by dealers in stolen goods. (Figures 4.3–4.6). In 2002 the finds were recovered as a result of a raid by Basel

Figure 4.2. Reconstruction of the context as it was when uncovered, based on criminal investigations and the statements of the finders (Juraj Lipták, Munich).

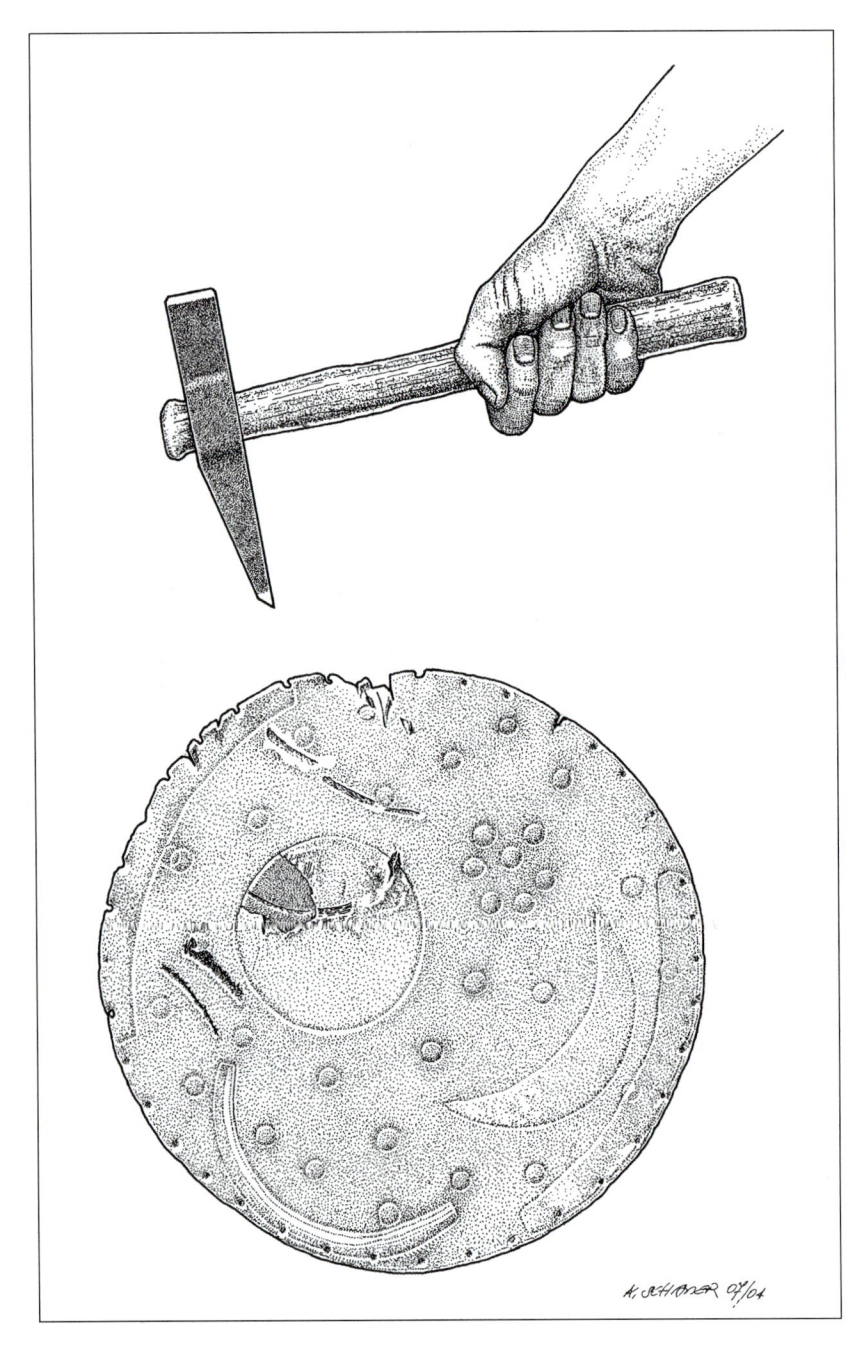

Figure 4.3. Reconstruction of the excavation tool on the basis of the damage to the disc as it stood upright in the ground. The hammer-shaped pick was subsequently submitted to the court by the finder and thus confirmed the reconstruction (Karol Schauer, Salzburg)

Figure 4.4. Examination of the find site by the State Office for Heritage Management and Archaeology of Saxony-Anhalt in 2002. The semi-circular dark patch in the right area of the excavation shows the illegal excavation in section (Juraj Lipták, Munich)

Figure 4.5. Photo taken by the stolen-goods dealer, showing the sky disc once cleaned (Unknown photographer).

Figure 4.6. The swords, the axes and the chisel have not yet been properly cleaned. The photograph was presumably taken soon after the illegal excavation (Unknown photographer).

police and shortly afterwards returned to Saxony-Anhalt and placed in the possession of the State Museum of Prehistory. The circumstances and history of their discovery and their original find-site were established by police investigations, verified in the course of protracted court proceedings and confirmed by archaeological and scientific studies (Meller 2010, 24-35). Preserving the sky disc and deciphering its meaning subsequently became matters of public and scientific interest, culminating in the inclusion of the disc in the UNESCO Memory of the World Register in 2013. Following this recognition, recent studies have continued to emphasise the archaeological importance of the find, indicating that cultural and scientific studies of the sky disc are far from finished (see Wunderlich 2014; Lockhoff & Pernicka 2014; Meller 2014; Meller 2015).

The importance of state ownership of finds in safeguarding such discoveries as the Nebra sky disc for the benefit of both the public and of scientific research, cannot be emphasised often enough, as this example illustrates.

The hoard was discovered in the course of illegal treasure hunting and not in the context of state-run excavations or within a designated excavation area. Had it been found in such a context, Section 12 of the Law on the Protection of Historic Monuments of Saxony-Anhalt would have ensured that it became the property of the state. However, because Saxony-Anhalt also has in place extended legal entitlement to treasure trove, the find still belonged to the state on the grounds of its outstanding scientific value, regardless of where and how it had been discovered. The same extended provisions also specify the duty to surrender items of cultural heritage, and the reward for their discovery. Since, however, the finders of the sky disc did not fulfil their duty to surrender the item, they were not entitled to any reward. On the contrary, as illegal treasure hunters, they were sentenced under Section 246 of the German Penal Code for breaking the treasure trove regulations and for misappropriation. The dealers also received suspended prison sentences of various lengths for handling stolen goods, or being accessories to the crime, under Section 259 of the German Penal Code.

Cost-by-cause principle

Not all of the 16 *Bundesländer* have incorporated regulations on liability for costs in their heritage protection laws. The most recent amendment relating to the cost-by-cause principle came into effect in Germany with the redrafting of the heritage protection law of the State of North Rhine-Westphalia on 27 June 2015, after previous administrative practice, over a period of almost two years, had been declared unlawful on the basis of a judgement by the Münster Higher Administrative Court (Kemper 2015, 1f.). Apart from the states of Bavaria and Baden-Württemberg and the city states of Bremen and Berlin, all *Bundesländer* have incorporated into their heritage protection laws regulations which explicitly impose costs on the development initiator (Kemper 2015, 14). In accordance with German constitutional law, the burden of costs for the developer must be within reasonable limits. Apart from Rhineland-Palatinate and Saxony-Anhalt, none of the other states have clearly defined what is 'reasonable' for the investor. In Rhineland-Palatinate, the developer is only liable for costs if the total value of a building project is over €500,000 and the investor's share of the costs is usually 1% of their total. In Saxony-Anhalt, the limits of what is reasonable when allocating costs have been determined by a judgement by the Higher Administrative Court of Saxony-Anhalt (Judgement 2 L 292/08 of 16 June 2010). Costs imposed must be within 10% and 20% of the total investment value and are usually 15%. A later judgement (2 L 154/10 of 20 August 2012) further obliges the investor to pay costs incurred in the process of conducting preliminary studies to establish whether the application can be approved. This initial documentation serves to confirm whether or not any action is required on the part of the heritage management authority.

In principle, the chargeable costs are to cover both excavation and documentation. With respect to documentation, however, there is no unanimity amongst the *Bundesländer* as to what is covered by the cost-by-cause principle. As a rule, cataloguing the finds after the excavation and their scientific publication are not included by law. Only in Schleswig-Holstein has the heritage protection law, updated in 2014, been reformulated so that the investor now bears the costs of publication of the excavation results (Section 14 of the Law on Protection of Historic Monuments of Schleswig-Holstein). It is, of course, possible to negotiate with the investor with regard to the presentation of the excavation results in publications or small exhibitions and come to a contractual agreement. Generally, however, it becomes apparent in many cases that scientific appraisal will not be possible, or cannot take place in the context of state- or privately funded heritage management.

Apart from the above-mentioned examples, where there is clear formulation of how the cost-by-cause principle works, in states without a legally binding cost-by-cause principle costs can be allocated in the context of the approval procedures for construction projects.

New tracks over old paths and the oldest nuclear family documented anywhere in the world so far

Saxony-Anhalt, situated in the heart of Germany, is unusually rich in its archaeological heritage. The basis for studying these extraordinarily rich sources are the records made in the course of archaeological excavations, including both numerous rescue excavations and the less frequent ones conducted solely for research purposes. The importance of rescue excavations in advance of infrastructure projects or mining operations can be illustrated from two prime examples from Saxony-Anhalt. In the context of the project to construct the new high-speed rail link from Erfurt to Halle and Leipzig (Figure 4.7) it was possible

Figure 4.7. View of the route of the Erfurt-Leipzig/Halle high-speed rail link
(G. Pie: State Office for Heritage Management Saxony-Anhalt).

to trace a hollow way, over 300 m long, dating from the Middle Bronze Age, which ran along the line of the railway embankment in the Oechlitz district of Mücheln, and, interestingly, coincided with it exactly (Figure 4.8). Wheel tracks from heavy waggons, their wheels 1.10–1.20 m apart, show where an overland route for the transport of goods and people once ran; it can be dated based on bronze objects recovered from the cart tracks to c. 1500 BC (Zich 2015, 98). A small section of the track was block-lifted and can now be seen in the permanent exhibition of the State Museum of Prehistory in Halle (Figure 4.9).

In 2005, in advance of gravel-mining activities in Eulau, in the district of Burgenland, archaeological features were uncovered which included four multiple burials in close proximity to each other. Three of the four graves were surrounded by circular ditches, with a diameter of around 6 m, representing the remains of what had once been burial mounds already documented several years earlier by aerial photography (Figure 4.10). The four graves held a total of 13 individuals – men, women and children – who were dated, on the basis of the characteristic funerary rites, grave goods and radiocarbon dating, to the Corded Ware culture (Haak et al. 2010, 54). The graves showed no evidence of having been disturbed and the individuals had been carefully laid in them, turned to face each other. The anthropological study of the individuals revealed important information about them. They were either newborn babies and children of up to 10 years of age or adults of 30 years and older. Surprisingly, there were no

Figure 4.8. The Middle Bronze Age hollow way near Oechlitz follows the line of the future railway track. Two cart-tracks can be seen in the sections (Juraj Lipták, Munich).

Figure 4.9. A section of the hollow way in the permanent exhibition of the State Museum of Prehistory in Halle (Juraj Lipták, Munich).

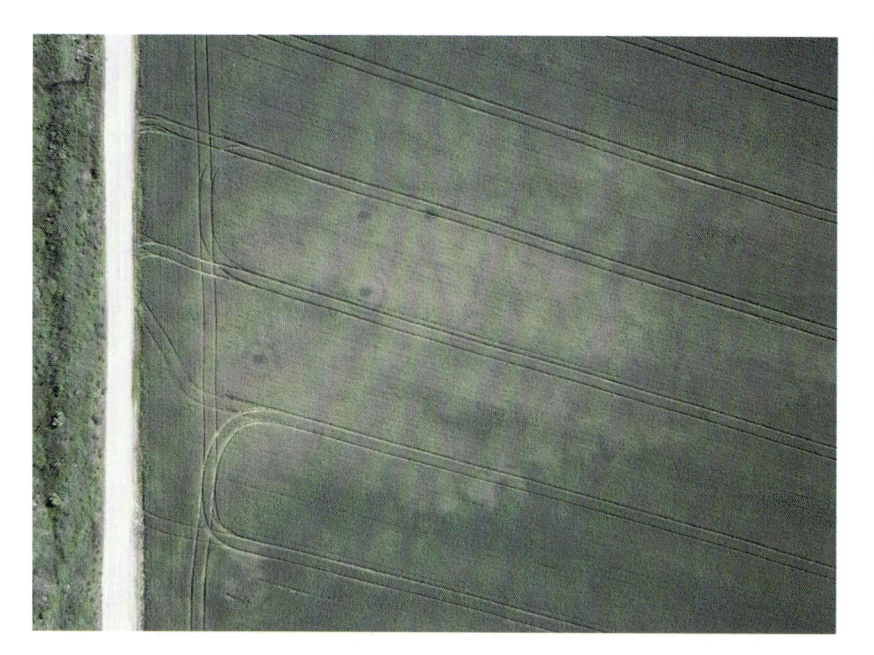

Figure 4.10. The gravel pits around the Eulau graves had already shown up in aerial photographs (R. Schwarz: State Office for Heritage Management Saxony-Anhalt).

adolescents or young adults amongst them. In the case of five individuals there was evidence of violent death, including skull fractures, thought to have been caused by the blows of stone axes. One death had been caused by an arrow shot, as evidenced by a flint arrowhead in the lumbar vertebrae of individual 5 from grave 90. Wounds typically incurred in warding off blows were also apparent, for example on the lower arms and metacarpus areas (Meyer et al. 2009, 420). Further information about the individuals in the graves could be gleaned using palaeopathological techniques. The kinship relations between the men, women and children were investigated, as were their places of origin. While strontium isotope analysis proved that the men and children were of local origin, the origin of the women turned out to be somewhere in the Harz region, 60 km away (Haak et al. 2008, 18229). Grave 99, which, like the other graves was block-lifted and examined under laboratory conditions, contained the greatest surprise for the interdisciplinary research team (Figures 4.11 and

Figure 4.11 Grave 99 from Eulau in the Burgenland district, containing the oldest nuclear family so far known in the world. Grave 99 is one of the three blocks displayed in the State Museum of Prehistory (see Fig. 4.13; Juraj Lipták, Munich).

Figure 4.12. Block-lifting of the Eulau graves, in order to examine them under laboratory conditions and then to exhibit them in the State Museum of Prehistory in Halle (A. Hörentrup: State Office for Heritage Management Saxony-Anhalt).

4.12). Examination of mitochondrial DNA proved that this was the earliest hitherto-known nuclear family in the world, consisting of a mother, father and two children (Haak et al. 2008, 18227). The final piece of the jigsaw, reconstructing the situation which led to the deaths of the buried community around 4,600 years ago, was provided by the archaeologists themselves. The two transverse arrowheads found in the lumbar vertebrae and ribcage of the adult female from grave 90, for example, were typical of the Schönfeld culture. The skull fractures are too narrow for Corded Ware culture axes, but the broad axes of the Schönfeld culture fit perfectly into the wounds left by the blows (Muhl et al. 2010, 135). The Schönfeld culture, unlike the Corded Ware culture, originated in the north of Saxony-Anhalt, the area to which the origin of the buried women points. The Eulau graves, now forming a central display in the State Museum for Prehistory (Figure 4.13), reflect, in all probability and in a particularly poignant way, the conflict between two Neolithic cultures, in which the women must have represented a particularly powerful motive for this 4,600-year-old crime.

Figure 4.13. Thanks to the latest restoration techniques, three of the four original Eulau graves now form a central display in the State Museum of Prehistory (Juraj Lipták, Munich).

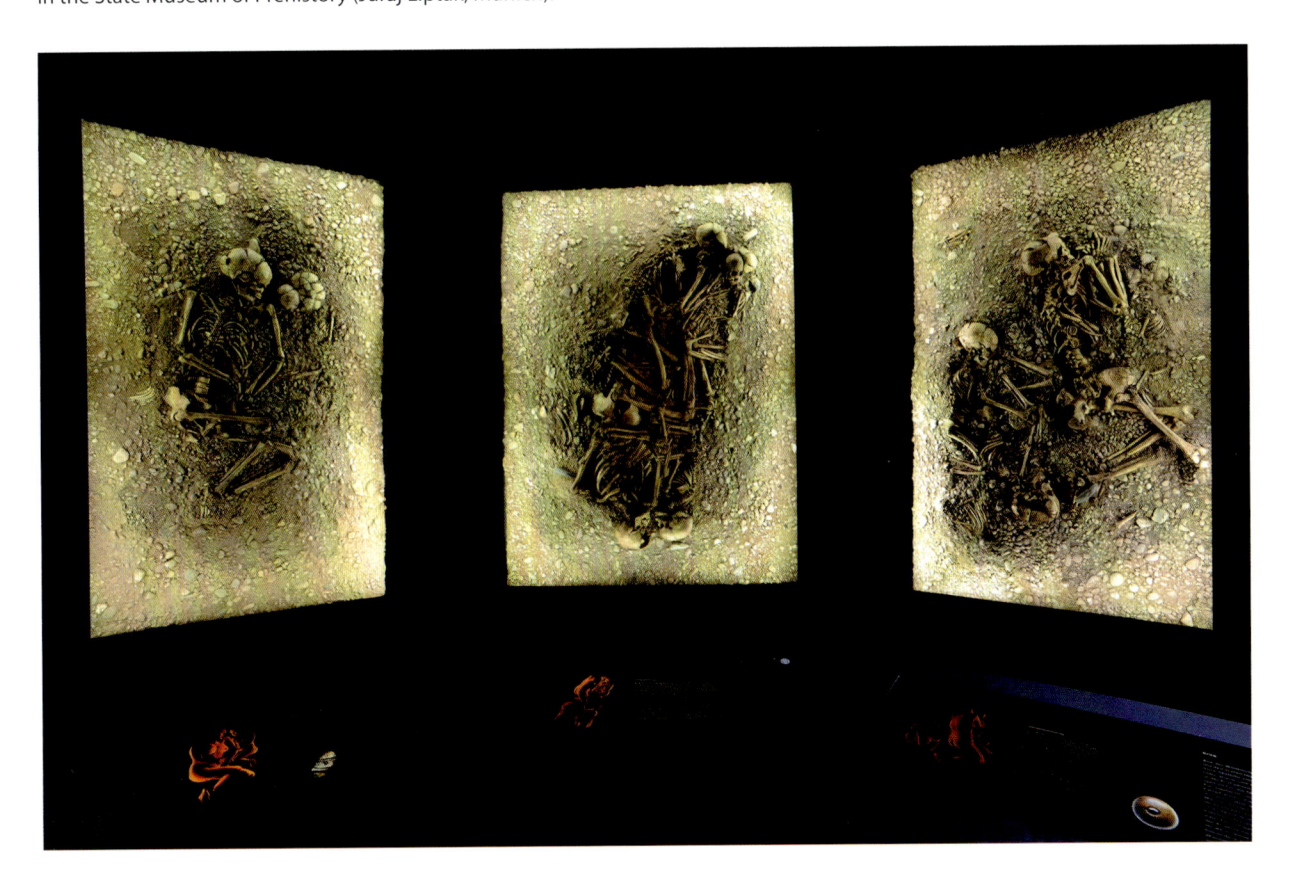

The insights derived from these two rescue excavations at Oechlitz and Eulau are just two examples of many, illustrating the importance of documenting archaeological remains before public or private construction or mining projects are undertaken. In Saxony-Anhalt, the cost-by-cause principle provides the necessary legal framework for this to be undertaken.

Private versus state-run excavation

Heritage protection and management always have to contend with the sometimes divergent interests of science, economy and publicity. Since as long ago as the first half of the 1980s, some states have been using the cost-by-cause principle in order to involve contractors in the financial responsibility for documenting archaeological remains during a building or mining project. Incorporating the Valletta Convention into the heritage protection laws of more *Bundesländer*, in the form of the cost-by-cause principle, along with large infrastructure projects in the states of former East Germany in the 1990s, led to an increased requirement for surveys and excavations to be carried out and, overall, to far greater rescue-excavation activity. For reasons to do with constitutional law, heritage management offices have no monopoly over archaeological activities. However, since surveying – i.e. ascertaining whether and what type of heritage sites are present in a given area – and excavation are duties of the heritage management authorities throughout Germany, these authorities do have an important role to play in the relevant planning approval procedures. The planning authorities are advised by the heritage management authorities through submissions and expert opinions on whether archaeological documentation is necessary and what its extent should be. Any necessary heritage management measures identified must be given due weight in the decisions of the planning authority. Increased demand for archaeological documentation measures can only be met by intensification of state and/or private archaeological heritage management activities. The extent of privatisation of archaeological heritage management varies widely amongst the individual *Bundesländer*.

The contribution of private excavation firms to archaeological heritage management has been much discussed in the past. Although it was initially suggested that the participation of private firms would mean complementary 'cost-saving' archaeological activity taking the burden off the heritage management authorities, this has not proved to be the case in reality. On the one hand, excavations by private firms have to be supervised by the state, requiring input by personnel from the responsible authorities, while on the other hand, the restoration of finds, the necessary cataloguing, and scientific appraisal in the form of publications or exhibitions still falls to the state bodies (Oebbecke 1997, 24; Tellenbach 1998, 241). No provision is generally made by the contracted excavation firms for evaluating the data, which means extra problems for those who do carry out this work, for example as a result of different software, different excavation methods, etc. Whilst the excavation firms are obliged to adhere to standards prescribed by the heritage management authorities, these usually relate only to the excavation techniques, in an attempt to ensure their uniformity. Questions of content, however, are not taken into account, although an appraisal of larger cultural landscapes requires a uniform approach to the documentation of finds to ensure comparability (Planck 1994, 68; Tellenbach 1998, 240–41). Normally it is the state department which is the repository of wider local knowledge, and enjoys collaborative partnerships with such bodies as universities, foundations and research institutes. Extensive excavation experience and expertise in using the latest documentation methods are important prerequisites for employees of both private excavation companies and state archaeological heritage departments. If the state heritage management authorities were to limit themselves to advising and supervising private archaeological firms, without undertaking any research activity of their own by mounting their own excavations, valuable knowledge and expertise in documentation methods, acquired over a long period, would be lost to state archaeological management, and the training and career development of future generations of qualified archaeologists would be in the hands only of the universities and private firms. The upholding of professional standards would no longer be guaranteed (Oebbecke 1997, 28).

Difficulties also arise in calculating how much needs to be done to document a site by rescue excavation. For example, a private firm will find it hard to justify an increase to the sum agreed with the contractor for documentation if unexpected finds or features are discovered. If the excavation is being undertaken by the state authority, there is the possibility of agreeing with an investor that any money which turns out not to be required will be refunded at the end of the excavation (Tellenbach 1998, 240). With a private excavation firm, which no doubt also has the interests of archaeological science at heart, economic considerations and the profit motive must nevertheless come first. The primary duty in relation to items of cultural heritage, i.e. their preservation, must of necessity take second place. When there is open market competition, moreover, equivalent working conditions and fair pay cannot always be taken for granted. On the other hand, the investor has no interest in the product which the excavation firm is offering. Yet interest in the product is the precondition for every market. The cultural-historical knowledge which can be gained in the context of a rescue excavation normally takes second place, for the investor, to such interests as cost-minimisation and a speedy progress of the construction work. This conflict of interests is best resolved in our opinion by the advantages of state heritage management, since this offers the best preconditions for the smooth and, above all, uniform collaboration of specialists in surveying, excavating, cataloguing, presenting the finds and publication of the research results. Beyond that, the union of the state office with a museum, which is in fact the structure in many *Bundesländer*, is yet another advantage to collect and present the archaeological finds to the public (Horn 2003, 43).

Uniform and binding standards for the work of private excavation firms throughout Germany do not yet exist (Andrikopoulou-Strack 2007, 16). This is because large-scale, supra-regional archaeological heritage management standards are becoming more and more challenging. An attempt to formulate Germany-wide norms, the excavation standards published by the Society of State Archaeologists (*Verband der Landesarchäologen*) and, on the Europe-wide level, the *The Standard and Guide to Best Practice in Archaeological Archiving in Europe* (Perrin et al. 2014) produced by the EU project ARCHES (Archaeological Resources in Cultural Heritage, a European Standard), must both be seen as recommendations rather than obligations.

References

Andrikopoulou-Strack, J.-N. 2007: Archaeology and heritage management in Germany, in W. J. H. Willems & M. H. van den Dries (eds): *Quality Management in Archaeology*, Oxford, 13–21.

Gumprecht, A. 2003: Der gesetzliche Rahmen für die Aufgaben der Bodendenkmalpflege in der Bundesrepublik Deutschland, in Verband der Landesarchäologen in der Bundesrepublik Deutschland (ed.): *Archäologische Denkmalpflege in Deutschland. Standort, Aufgabe, Ziel*, Stuttgart, 30–38.

Haak, W., Brandt, G., de Jong, H. N., Meyer, C., Ganslmeier, R., Heyd, V., Hawkesworth, C., Pike, A. W. G., Meller, H. & Alt, K. W. 2008: Ancient DNA, Strontium isotopes, and osteological analyses shed light on social and kinship organization of the Later Stone Age. *Proceedings of the National Academy of Science of the United States of America* 105, 18226–31.

Haak, W., Brandt, G., Meyer, C., de Jong, H. N., Ganslmeier, R., Pike, A. W. G., Meller, H. & Alt, K. W. 2010: Die schnurkeramischen Familiengräber von Eulau – ein außergewöhnlicher Fund und seine interdisziplinäre Bewertung, in H. Meller & K. W. Alt (eds): *Anthropologie, Isotopie und DNA. 2. Mitteldeutscher Archäologentag vom 08. bis 10. Oktober 2009 in Halle (Saale)*. Halle (Saale), 53–62.

Hönes, E.-R. 2005: Das Europäische Übereinkommen zum Schutz des archäologischen Erbes vom 16.1.1992. *Natur und Recht* 12, 751–57.

Horn, H.-G. 2003: Die Organisation der Bodendenkmalpflege, in Verband der Landesarchäologen in der Bundesrepublik Deutschland (ed.): *Archäologische Denkmalpflege in Deutschland. Standort, Aufgabe, Ziel*, Stuttgart, 39–44.

Kemper, T. 2015: Kostentragungspflicht und Zumutbarkeit für Verursacher im novellierten Denkmalschutzgesetz von Nordrhein-Westfalen im Vergleich mit den übrigen Bundesländern. *Archäologische Informationen* 38, Early View, http://www.dguf.de/fileadmin/AI/ArchInf-EV_Kemper.pdf (accessed 23.07.2015).

Lockhoff, N. & Pernicka E. 2014: Archaeometallurgical investigations of Early Bronze Age gold artefacts from central Germany including gold from the Nebra Hoard, in H. Meller, R. Risch & E. Pernicka (eds): *Metalle der Macht – Frühes Gold und Silber. 6. Mitteldeutscher Archäologentag vom 17. bis 19. Oktober 2013 in Halle (Saale)*. Tagungen des Landesmuseums für Vorgeschichte Halle 11. Halle (Saale), 223–35.

Meller, H. 2010: Nebra: Vom Logos zum Mythos – Biographie eines Himmelsbildes, in H. Meller & F. Bertemes (eds): *Der Griff nach den Sternen. Wie Europas Eliten zu Macht und Reichtum kamen. Internationales Symposium in Halle (Saale) 16.–21. Februar 2005*. Tagungen des Landesmuseums für Vorgeschichte Halle (Saale) 5. Halle (Saale), 23–73.

Meller, H. 2014: Die neolithischen und bronzezeitlichen Goldfunde Mitteldeutschlands – Eine Übersicht, in H. Meller, R. Risch & E. Pernicka (eds): *Metalle der Macht – Frühes Gold und Silber. 6. Mitteldeutscher Archäologentag vom 17. bis 19. Oktober 2013 in Halle (Saale)*. Tagungen des Landesmuseums für Vorgeschichte Halle 11. Halle (Saale), 611–716.

Meller, H. 2015: Zwischen Logos und Mythos. Zum Eigensinn der Himmelsscheibe von Nebra, in H. P. Hahn (ed.): *Vom Eigensinn der Dinge. Für eine neue Perspektive auf die Welt des Materiellen*, Berlin, 177–97.

Meyer, C., Brandt, G., Haak, W., Ganslmeier, R. A., Meller, H. & Alt, K. W. 2009: The Eulau eulogy: Bioarchaeological interpretation of lethal violence in Corded Ware multiple burials from Saxony-Anhalt, Germany. *Journal of Anthropological Archaeology* 28, 412–23.

Muhl, A., Meller, H. & Heckenhahn, K. 2010: *Tatort Eulau. Ein 4500 Jahre altes Verbrechen wird aufgeklärt*, Stuttgart.

Oebbecke, J. 1997: (Teil-)Privatisierung der Bodendenkmalpflege – kurzlebige Mode oder notwendige Modernisierung?, in J. Oebbecke (ed.): *Privatisierung in der Bodendenkmalpflege*, Baden-Baden, 13–31.

Otten, T. 2008: *Archäologie im Fokus – Von wissenschaftlichen Ausgrabungen und illegalen Raubgrabungen*. Schriftenreihe des Deutschen Nationalkomitees für Denkmalschutz 53. Bonn.

Perrin, K., Brown, D. H., Lange, G., Bibby, D., Carlsson, A., Degraeve, A., Kuna, M., Larsson, Y., Pálsdóttir, S. U., Stoll-Tucker, B., Dunning, C. & Rogalla von Bieberstein, A. 2014: *Archäologische Archivierung in Europa: ein Handbuch*. EAC Guidelines 1, Namur, http://archaeologydataservice.ac.uk/arches/Wiki.jsp?page=The%20Standard%20and%20Guide%20to%20Best%20Practice%20in%20Archaeological%20Archiving%20in%20Europe (accessed 29.07.2015).

Planck, D. 1994: Zum Einsatz privater Grabungsfirmen in der Archäologischen Denkmalpflege, in Verband der Landesarchäologen in der Bundesrepublik Deutschland (ed.): *Archäologische Denkmalpflege und Grabungsfirmen. Kolloquium im Rahmen der Jahrestagung 1993. Bruchsal, 10.13. Mai 1993*. Stuttgart, 67–69.

Tellenbach, M. 1998: Firmenarchäologie – Landesarchäologie. Unterschiedliche Leitbilder

– unterschiedliche Ergebnisse. *Archäologische Informationen* 21/2, 239–43.

Wunderlich, C.-H. 2014: Wie golden war die Himmelsscheibe von Nebra? Gedanken zur ursprünglichen Farbe der Goldauflagen, in H. Meller, R. Risch & E. Pernicka (eds): *Metalle der Macht – Frühes Gold und Silber. 6. Mitteldeutscher Archäologentag vom 17. bis 19. Oktober 2013 in Halle (Saale)*. Tagungen des Landesmuseums für Vorgeschichte Halle 11. Halle (Saale), 223–35.

Zich, B. 2015: Meilenstein der Mobilität, in H. Meller (ed.): *Glutgeboren. Mittelbronzezeit bis Eisenzeit*. Begleithefte zur Dauerausstellung im Landesmuseum für Vorgeschichte Halle 5. Halle (Saale), 96–100.

Websites

http://www.landesarchaeologen.de/fileadmin/Dokumente/Dokumente_Kommissionen/Dokumente_Grabungstechniker/grabungsstandards_april_06.pdf (accessed 29.07.2015).

Legal acts

European Convention on the Protection of the Archaeological Heritage (Revised), Valletta, 16 January, 1992.

Denkmalschutzgesetz des Landes Sachsen-Anhalt vom 21. Oktober 1991 (GVBl. LSA S. 368; Glied.-Nr: 2242.1), zuletzt geändert durch Artikel 2 des Dritten Investitionserleichterungsgesetzes vom 20. Dezember 2005 (GVBl. LSA S. 769, 801).

5 | The organisation of Czech archaeology – a socialist legal system applied in a market economy

Jan Mařík

Abstract: The first legal measures for the protection of archaeological finds in Bohemia and Moravia (historical regions of the Czech Republic) were taken already in the first half of the 19th century. Effective regulation, however, arrived only with the state decree issued in 1941. The current law came into force in 1987. The fundamental political as well as social transformations that occurred in the Czech Republic two years later brought much higher demands for rescue archaeology. Even though the law was created under the conditions of Real Socialism with a centralised and state-controlled economy, it is still, after more than 25 years, valid and applied in a democratic state and free market. Adaptation of the law to new social as well as economic conditions has mostly taken place with the approval of all involved parties. A series of regulations has been adopted that are more or less generally respected; however, their real enforceability relies more on moral appeal than on the letter of the law.

Keywords: legal acts, Czech Republic, archaeological heritage protection, political transformation

Introduction – pre-Second World War foundations

The foundations of the current archaeological heritage care system were laid in 1919 when, shortly after the establishment of the independent state of Czechoslovakia, the State Archaeological Institute was formed in Prague. This new state institution was subordinate to the Ministry of Education. Its major aim was to conduct systematic archaeological fieldwork, focusing mainly on more extensive excavations that were beyond the scope of regional museums. Nevertheless, the established practice of archaeological excavations conducted by various museum societies, private researchers as well as collectors was still upheld. The State Archaeological Institute was to be the leading authority which set the standards for scientific work. Moreover, the Institute was also expected to address the education of amateur archaeologists cooperating with archaeological departments in individual regional museums (Niederle 1919).

In the interwar period, however, this concept remained partly unfulfilled. The development of the new institute was significantly restricted in those days by the two following factors: the shortage of funds preventing employment of a sufficient number of specialists, and the absence of laws defining not only the rules for conducting archaeological fieldwork and treatment of archaeological finds but also the position of the State Archaeological Institute.

Even though several extensive archaeological fieldwork projects were successfully launched during this period, an overall knowledge about archaeological finds in the whole of Czechoslovakia could only be grasped mainly thanks to the activities of numerous amateur archaeologists and museum collaborators and, last but not least, thanks to various news articles in the daily press. Based on these fragmented and widely scattered sources of information, an archive was gradually built up at the State Archaeological Institute, subsequently becoming the most extensive professional (archaeological) archive in the Czech Republic (Rataj et al. 2003).

The Second World War – State decree No. 274/1941

Even though the rather low scientific level of numerous archaeological excavations had been increasingly brought into focus since the 1930s (Sklenář 2011, 47), a law that would consolidate the approaches of all entities dealing with archaeological finds was not approved during the entire interwar period. The most vociferous opponents of any regulations for conducting archaeological fieldwork were mainly private collectors, who also quite often used to excavate by themselves. State decree No. 274/1941 represented, in this respect, the principal turning point, for it provided, among other things, the first legal definition of an archaeological find and largely clarified the position of the State Archaeological Institute. The institute was the only organisation legally entitled to conduct archaeological fieldwork and was, moreover, appointed the supreme adjudicator in issues of care for archaeological monuments. Museum organisations that had traditionally conducted archaeological excavations could continue their activities only with the approval of the State Archaeological Institute, and they had to employ professionally educated specialists (archaeologists, etc.). The decree also specified that the owner of archaeological finds obtained in the course of archaeological fieldwork was the state, i.e. the Protectorate of Bohemia and Moravia.

The reasons behind the approval of this state decree are still not fully understood. From the historical evidence, we can conclude that the Protectorate government probably had the same intentions in passing this decree as the Dutch authorities had had, when they introduced similar legislation in 1940 (Willems 1997). Thus, the State Archaeological Institute, which was not completely controlled by the Nazis, unlike the department of archaeology of the German university in Prague, obtained legal means allowing a certain degree of supervision of various interventions carried out by the occupying power and its organisations such as, for example, the Ahnenerbe (Vencl 2002).

Post-Second World War development – Act No. 20/1987 on state landmark conservation

The provisions of State decree No. 274/1941 regarding archaeology were more or less completely adopted in a new law on the care of monuments, passed in 1958 (Act No. 22/1958 on cultural landmarks). Other significant changes did not appear until Act No. 20/1987 on state landmark conservation, which is still in force. As far as its general meaning is concerned, this law was regarded very modern at the time of its constitution. This is corroborated, among other things, by the fact that it included a series of points that even featured several requirements stipulated in the Valletta Treaty, agreed five years later in 1992 (Mařík & Prášek 2014).

However, the authors of the law could not have predicted the major political as well as economic transformations that occurred in the Czech Republic after 1989 – the fall of the Communist regime. Thus, paradoxically, a law created under the conditions of a totalitarian state that intentionally suppressed all private civil as well as business activities is still in force and has been valid for more than 20 years in a market economy and democratic society. Despite a series of attempts to establish a new legal norm, only several partial amendments to the law, predominantly of a technical nature, have been implemented. Even though the law has been progressively adjusted to the new social conditions, its limitations have gradually become visible, the main ones being that it affords weak control and sanction measures that make any enforceability extremely difficult.

Institutes of Archaeology

In 1953 the State Archaeological Institute lost the label 'state', became the Institute of Archaeology and was incorporated into the newly-established Academy of Sciences that centralised the majority of non-university research institutions. According to the 1987 Act, the institute has, in some respects, retained the position of state administrative authority. All information regarding archaeological fieldwork, from the moment of reporting a planned construction project that could threaten archaeological finds, to the starting date of the excavation and the final excavation report, is submitted to the Institute of Archaeology. The Academy of Sciences has also been granted new legal powers: it is the only institution with the authority to submit proposals for designating an archaeological site or a significant find as a cultural monument, and it has the power of veto in the process of obtaining a licence for conducting archaeological fieldwork.

The original detached departments of the Prague Institute of Archaeology were gradually transformed into individual Institutes of Archaeology in Brno (1983) and in Nitra, Slovakia (1953). Currently, two Institutes of Archaeology are active in the Czech Republic; the geographical scope of their respective authority is based on the historical borders of Bohemia (the Institute in Prague) and Moravia and Silesia (the Institute in Brno). Even though this spatial division is based on good reasons, it has given rise to a series of discrepancies in the practical implementation of legal requirements. This poses obstacles mainly in cases where a unified course of action by both institutes should be expected. Probably the most significant example of this is the absence of a joint information system for recording archaeological interventions and their results.

Licensing and licensed organisations

Besides the Institutes of Archaeology, other organisations and individuals are also entitled to conduct archaeological fieldwork on the authorisation (granting of a licence) issued by the Ministry of Culture of the Czech Republic. In order to obtain this licence the applicant has to employ at least one individual with a university master's degree in the field of archaeology, with a minimum of two years' excavation experience. Moreover, the applicant also has to meet other conditions, such as providing suitable space for temporary storage of archaeological finds and other equipment that is, however, not further specified in the Act.

The licence to conduct archaeological fieldwork can be issued by the Ministry of Culture of the Czech Republic only with the approval of the Czech Academy of Sciences. The approval of the Czech Academy of Sciences represents one of the most powerful regulatory measures that can influence the authorisation. In its decisions, the Czech Academy of Sciences primarily examines two factors: the scientific intent of the organisation (mainly in the case of university departments) and whether there is a need for another licensed organisation in the system of archaeological monument care. New licences are, therefore, issued mainly for regions where building activities and other interventions threatening the archaeological heritage are less well covered. Another significant aspect is the organisation's legal status because, at least according to the law, conducting archaeological fieldwork should be a non-profit-making activity and, thus, the licence is issued only for non-profit organisations.

When granted a licence, the successful applicant has to, moreover, make an agreement with the Czech Academy of Sciences specifying the conditions and extent of the archaeological fieldwork. This agreement usually designates a specific geographic area (district, region) where the licensed organisation is entitled to conduct excavations. Furthermore, the agreement specifies further obligations of the licensed organisation that are only generally described in the Act: mainly the requirement to submit excavation reports that are archived in the Institutes of Archaeology of the Czech Academy of Sciences.

Figure 5.1: Organisations licensed since 1988.

Currently, the right of veto exercised by the Czech Academy of Sciences is severely and frequently criticised, as it is supposedly obstructing free competition and the free market. The strong and authoritative position of the Czech Academy of Sciences represents one of the characteristic examples of the antiquated socialist legislation. The Act's authors wanted, in the first place, to create a system ensuring high-quality care for archaeological heritage. Even though the Act embodied the 'polluters pay' principle, it cannot be deemed to have given rise to the development of 'contract archaeology'. Only the expenses of conducting archaeological fieldwork were to be paid for, and that, in effect, meant that the finances were just transferred among entities established by the state. As a matter of fact, the Act's authors could not have foreseen the possibility that anything other than a state organisation would be authorised to carry out such activities. This approach also influenced the fact that the Act included only a minimum of supervisory mechanisms and essentially no sanctions that could be applied against the licensed organisations. Thus, termination of the agreement regarding the conditions of conducting archaeological fieldwork issued by the Czech Academy of Sciences and the consequent revocation of the licence represent the only real sanctions. In practice, however, these terminations occur very rarely and only in the case of long-term and repeated violation of the agreement on the part of the licensed organisation. Moreover, it is worth mentioning here that there is almost no immediate sanction that can be applied to penalise poorly conducted archaeological excavations.

The relatively rapid rise in private archaeological companies that occurred in the 1990s was connected with a significant increase in building activities, whose needs the state organisations were not able to meet. The emergence of private companies that filled the gap in the market represented a logical solution to this state of affairs (Figure 5.1). Thus, private firms have gradually become an integral part of the system of care for the archaeological heritage. In the last years, their annual share in the volume of conducted archaeological excavations has reached approximately 15–20%. However, the majority of excavations are still conducted by regional museums (Figure 5.2). Altogether, 110 licensed organisations exist in the Czech Republic, of which 15 organisations are private.

Conducting archaeological fieldwork

At the beginning of the 1990s, the majority of systematic archaeological excavations were concluded, with the exception of several long-term research projects under the guidance of university departments and Institutes of Archaeology of the Czech Academy of Sciences. Instead, the attention of most archaeological departments of regional organisations became almost completely absorbed by intensive building activities. If we compare the annual number of archaeological investigations conducted at the end of the 1980s with the current state of affairs, the increase is fourfold. Such a transformation of the social environment could not have been foreseen by the Act's authors. Moreover, the Act was relatively lenient in terms of prescribing methods to be used while conducting rescue archaeological excavations, with no strictly defined terms, rights and obligations and, last but not least, with a minimum of sanctions. Thus, relatively extensive possibilities for the Act's circumvention appeared,

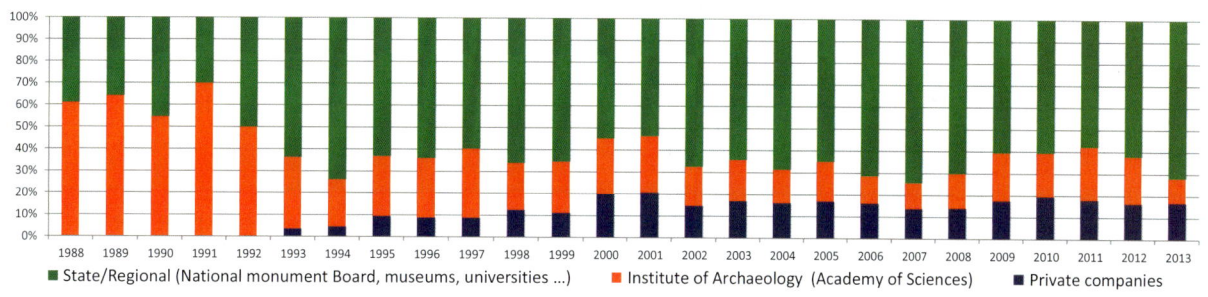

Figure 5.2: Participation of licensed organisations in rescue excavations.

which have, in extreme cases, led to the deliberate destruction of archaeological sites.

An archaeological excavation is initiated by a notice released, according to the Act, by the entity whose activities in an area with archaeological finds could threaten them in their original setting. The notice should be delivered to the relevant Institute of Archaeology of the Czech Academy of Sciences. Even though definition of the term 'area with archaeological finds' is not included in the Act, a relatively extensive reading is applied in practice: it represents an area where occurrence of archaeological finds cannot be completely excluded, such as in the case of opencast mines (which do not ostensibly represent an 'area with archaeological finds').

Due to the rather limited resources of the Institutes of Archaeology of the Czech Academy of Sciences, notices are transferred to licensed organisations active in the given regions. Based on the notice, the Institute of Archaeology, or any other licensed organisation, can conclude a contract with the builder (proprietor or leaseholder) in order to conduct a rescue excavation. The licensed organisation is obliged to report the launch of the rescue excavation to the Institute of Archaeology and, subsequently, also to deliver a final excavation report there.

In practice, however, this system is not strictly adhered to. Often the builders prefer direct communication with regional archaeological institutions and report their intentions directly to the local licensed organisation. Even though this practice, in fact, violates the law, it does not necessarily lead to damage of the archaeological heritage. On the other hand, problems can occur when the law is violated by a licensed organisation that ceases to report the launch of excavations or to deliver final excavation reports. Thus, a whole range of information of fundamental importance, not only for scientific research but also, for example, for landscape / urban planning, can be left concealed or, in worse cases, be completely lost. There are hundreds of cases when the Institutes of Archaeology of the Czech Academy of Sciences have only obtained final reports about excavations whose launch was not reported, or they have learnt about excavations from annual reports presented by the licensed organisations or, in worse cases, from the media.

To rectify this rather unsatisfactory state of affairs, internet portals registering all ongoing archaeological fieldwork have been created at the Institutes of Archaeology of the Czech Academy of Sciences in Prague and Brno. The Internet Database of Archaeological Fieldwork (IDAW, Link 1) registers notices regarding building and other activities conducted in areas with archaeological finds that have been submitted to either the Institute of Archaeology of the Czech Academy of Sciences in Prague or any other licensed organisation. Subsequently, the database follows the entire course of the archaeological fieldwork. Key data include launching and concluding dates of the fieldwork and information on the related excavation report. Each entry obtains a unique five-digit identifier, which is eventually also used for designation of the excavation report.

Circumstances accompanying the launch of the database also clearly illustrate the limitations of the currently valid Act. To persuade the majority of licensed organisations to voluntarily use the database took almost one and a half years, for such a requirement (to use an on-line register) is not stipulated by the law. Ultimately, the practical advantages of the database, such as simplification of communication with the Institute of Archaeology and easier access to information, outweighed initial distrust. However, only the database administered by the Institute of Archaeology of the Czech Academy of Sciences in Prague has been successfully put into practice, i.e. only in the region of historical Bohemia over 90% of licensed organisations are using the database. In contrast, the Institute of Archaeology of the Czech Academy of Sciences in Brno, covering the regions of Moravia and Silesia, was not so successful. Thus, the following data refer only to the region of Bohemia, where almost 70–80% of all archaeological fieldwork in the Czech Republic is conducted.

Currently, the IDAW is used by almost 300 registered users. The users include individuals from 61 organisations authorised to conduct archaeological fieldwork (who actively register and edit entries in the database), members of the state administration, students of archaeology and amateurs interested in archaeology, who are only entitled to view the database.

As expected, the database has provided better access to information not only about activities threatening the archaeological heritage but also about ongoing archaeological fieldwork. Based on previous experiences, it seemed obvious that information on these activities was often kept in those regions where the excavations were conducted. This assumption was corroborated shortly after the launch of the database in 2010. Prior to this, the Institute of Archaeology of the Czech Academy of Sciences in Prague had, on average, annually obtained 2,400 excavation reports. By 2011, the number of interventions recorded in the database had reached 7,000 and remained unchanged until 2013 (Figure 5.3). In the next year (2014), a significant increase was recorded; this development can probably be connected with the recovery of the construction business following the previous years of economic crisis. When the database was launched, the annual average number of delivered reports reached 4,200. Even though approximately three-quarters of these reports represented information on watching briefs at building sites where no archaeological finds were discovered, from the legal point of view they are archaeological excavations with negative results.

Funding of archaeological fieldwork

If the state-controlled socialist economy is taken into consideration, it seems at least strange that the Act's original version of 1987 already included the 'polluters pay' principle. According to the Act, the costs of rescue archaeology should be covered by the investor, with

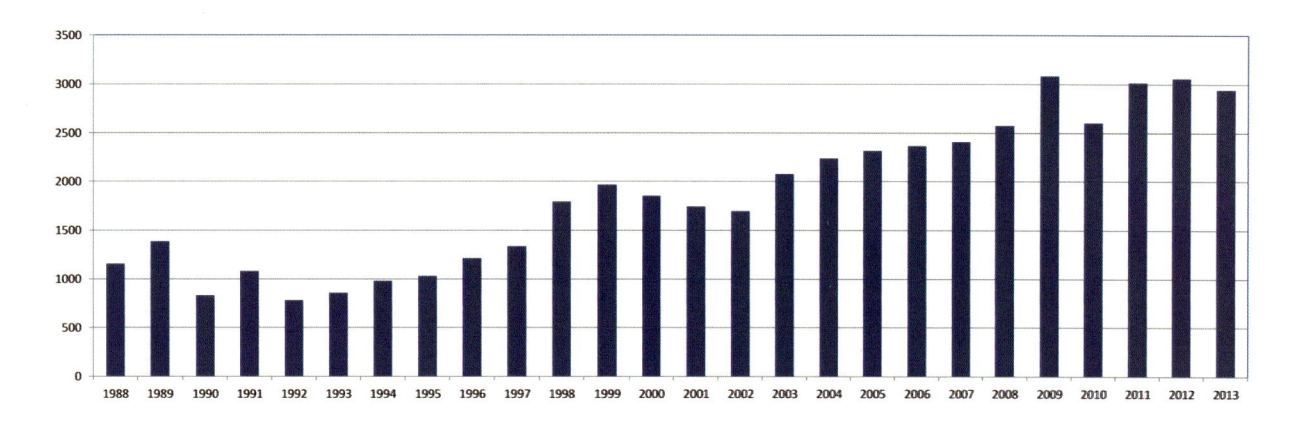

Figure 5.3: Actions recorded in the Internet Database of Archaeological Fieldwork.

the only exception being natural persons, and even then only in those instances when the activities that necessitate the excavations are not related to their enterprises. In most cases, the activities in question relate to the construction of family houses, garages or swimming pools. In these cases, the archaeological rescue work expenses should be covered by the organisation conducting the excavation. This exception, currently understood by the majority of archaeologists as a certain type of relief for less wealthy builders, represents a characteristic example of socialist law being adapted to the free market environment. By dividing the building owners (builders) into the two abovementioned groups, the Act's authors only wanted to differentiate items in the state budget that would be used for covering the costs of the fieldwork. In the first case, resources of state-owned firms would be used, in the other the expenses would be paid from the budgets of state-owned organisations such as museums and the Institutes of Archaeology.

The rapid increase in construction activities as well as the emergence of private, licensed organisations has required the establishment of a fund that can be used for covering the expenses of rescue excavations conducted at construction sites of non-profit-making natural persons. This fund was created by the Ministry of Culture of the Czech Republic and annually amounts to approximately €100,000370,000. If the total sum of expenses for conducting archaeological rescue work is taken into consideration, it seems clear that those paid by building owners definitely prevail. Annually, the builders (building owners) pay approximately €74 million for rescue archaeology.

Conclusions

If national heritage care laws from post-communist countries of Central and Eastern Europe are compared, the Czech law represents a quite unique feat. Even though it was created under the conditions of Real Socialism, with a centralised and state-controlled economy, it is still, after more than 25 years, valid and applied in a democratic state and free market. During this time, the overall conception of the law has not been significantly changed. As far as protection and care for the archaeological heritage is concerned, only amendments of a predominantly technical nature have been made to this law.

Adaptation of the law to new social as well as economic conditions has mostly taken place with the approval of all involved parties. A series of regulations have been adopted that are more or less generally respected. On the other hand, their real enforceability relies rather on moral appeal than on the letter of the law. Generally speaking, the current state of the archaeological heritage care system can be defined as extremely fragile and unsustainable in the long-term perspective.

Even though a whole series of attempts at a fundamental amendment of the existing law or preparation of a completely new Act have occurred since 1987, these efforts have not been, for various reasons, successful. For the time being, the last example represents a bill on national heritage protection that has been in preparation since 2012. Among the positive elements of this bill is an attempt to incorporate in the law various structures as well as approved mechanisms that are currently valid but without support in the existing law.

Besides the obvious, the abovementioned mechanisms include a concept of a central register of archaeological fieldwork and the principle of reporting as well as observing all actions threatening archaeological finds. Among the negative but logical consequences of this effort is a significant increase in bureaucratic duties. Even though a series of other problems can probably be described, approval of the current bill can be considered an absolute prerequisite as far as archaeological heritage protection is concerned. According to the plans of the Government of the Czech Republic, the new law could come into force in 2018.

References

Mařík, J. & Prášek, K. 2014: Management of archaeological excavations and control in the Czech and Slovak Republic, in V. M. Van der Haas & P. A. C. Schut (eds): *The Valletta Convention: Twenty Years After – Benefits, Problems, Challenges*, EAC Occasional Paper No. 9, Brussels, 113–18.

Niederle, L. 1919: Státní Archeologický ústav československý [State Archaeological Institute of Czechoslovakia]. *Památky archeologické* 31, 116–17.

Rataj, J., Šolle, M. & Vencl, S. 2003: Vzpomínky pracovníků Státního archeologického ústavu v Praze [Memories of workers of the State Archaeological Institute]. *Archeologické rozhledy* 55, 139–65.

Sklenář, K. 2011: Vývoj péče o archeologické památky v českých zemích do roku 1989 [Development of archaeological heritage care in Bohemia prior to 1989]. *Acta Musei Nationalis Pragae. Series A – Historia* 65, 3–106.

Vencl, S. 2002: Lothar Zotz: o něm i o nás [Lothar Zotz: about him and about us]. *Archeologické rozhledy* 54, 837–50.

Willems, W.J.H. 1997: Archaeological Heritage Management in the Netherlands: past, present and future, in W.J.H. Willems, H. Kars & D.P. Hallewas (eds): *Archaeological Heritage Management in the Netherlands*, Amersfoort/Assen, 3–34.

Legal Acts

State decree No. 274/1941: http://ftp.aspi.cz/opispdf/1941/091-1941.pdf (accessed 10.07. 2015).

Act No. 20/1987: http://www.mkcr.cz/assets/kulturni-dedictvi/pamatky/2013-Cultural-Landmark-Conservation-Act_2013.docx (English version, accessed 10.07. 2015).

http://www.mkcr.cz/assets/kulturni-dedictvi/pamatkovy-fond/legislativa/29--Uplne-zneni-s-judikaturou.doc (Czech version, accessed 10.07.2015).

Websites

Link 1: http://idav.cz (accessed 16.09.2015).

6 | Archaeological research in the Slovak Republic – positives and negatives

Matej Ruttkay, Peter Bednár, Ivan Cheben and *Branislav Kovár*

Abstract: In the Slovak Republic, archaeological research was only minimally regulated by law until 2002. However, the situation changed considerably after the introduction of Act No. 49 in 2002, amended later in 2010 and 2014. The Act brought some positive changes, but also many counter-productive results. In this paper, we try to evaluate its contribution to archaeological research in the Slovak Republic. We outline some problematic aspects of the Act, namely the introduction of archaeological licences, the opening of archaeology to private companies and the pressing issue of looting and metal-detecting at archaeological sites.

Keywords: law, archaeological research, private archaeological companies, looting of archaeological sites

In the Slovak Republic, until 2002 archaeological research was regulated by Act No. 27 of 1987, on the State Care of Monuments and Historic Sites. According to this law, the principal authority in the field of archaeological research was the Institute of Archaeology of the Slovak Academy of Sciences (IA SAS). Any other state institutions, such as museums, could carry out excavations only after approval by the IA SAS. At the same time, the IA SAS was also funding and executing most of the research. The law was codified according to the so-called Verursache principle ('polluter pays'), i.e. the investor is required to bear the costs of archaeological excavations. In practice, this was rarely applied because the IA SAS had sufficient resources for the research, and in the case of large building projects (e.g. the Gabčíkovo-Nagymaros Dam on the Danube) it was given enough extra funds. Another distinct advantage of this Act was that it gave a clear definition of the subject of protection: protection was applied not only to designated monuments, but also to any subject meeting the criteria of a cultural monument.

A significant change in the approach to archaeological research took place in 2002, when the new Act on the Protection of Monuments and Historic Sites (No. 49/2002) was passed by parliament. The very name suggests that priority was given to the protection of what was declared a national cultural monument. Protection of archaeological sites was vaguely defined. The above mentioned law was amended six times (most recently in 2014) in an attempt to correct this problematic situation.

Act No. 50/1976, with later amendments, specifies the procedure for the implementation of archaeological research induced by construction activities. Article 127 specifies that in the event of archaeological finds, the construction company is obliged to notify the relevant building authority, the Monuments Board of the Slovak Republic or the Institute of Archaeology. If during construction, a find of an extremely important cultural significance is made, after confirmation of its importance by the Ministry of Culture of the Slovak Republic, the building permission can be changed or revoked. The Ministry shall decide on the manner and reimbursement of costs incurred by the investor.

An archaeological site is defined as immovable property on topographically defined territory with excavated or un-excavated archaeological finds in their original archaeological context. Archaeological research may be conducted only by a legal entity, for example, by a company or state institution (never by a private person). Such a company must employ a person with special professional competence (a licence) to perform archaeological research.

Positives and negatives of key standards guiding the research and protection of archaeological sites. Do we protect archaeological sites adequately?

The priority of Act No. 49/2002 on the Protection of Monuments and Historic Sites was the introduction of control and improvement of research quality. The Act introduced new elements and institutions to regulate the research and protection of sites: licences for field archaeologists, an Archaeological Council as an advisory body to the Ministry of Culture, and the monitoring of excavation reports by the Monuments Board.

One of the key changes was that the Heritage Institute of the Slovak Republic, previously working as a methodological centre without decision-making powers, was replaced by the Monuments Board – a regular branch of specialised state administration, responsible for the protection of cultural monuments. This resulted in a narrowing of the competences of the Archaeological Institute, which remained administrator of the Central Register of Archaeological Sites in the Slovak Republic and also a kind of highest scientific authority, as the newly-constituted Monuments Board could issue decisions regarding archaeological sites only after consultations with the Archaeological Institute (Ruttkay, M. & Šmihula 2009, 365–76). The

Ministry of Culture of the Slovak Republic began issuing authorisations of special professional competence for archaeologists (licences) and permissions for legal entities to carry out excavations.

Originally, the only organisation authorised to carry out archaeological research by the law was the Archaeological Institute. In 2014, this right was granted to the Monuments Board, too. The Monuments Council and Archaeological Council were established as advisory bodies to the Ministry. Their decisions serve solely as recommendations to the Ministry and are not binding.

In Slovakia, organisations currently carrying out archaeological research can be divided into two main categories: governmental and non-governmental. Governmental organisations include the Archaeological Institute of the Slovak Academy of Sciences, the Monuments Board, the Slovak National Museum, the Mining Museum in Banská Štiavnica and others.

The second, much larger, category consists of non-governmental institutions, which can be divided into public universities, museums (belonging to the self-governing regions or municipalities) and private companies.

This division is particularly important in relation to finds obtained by archaeological excavations. These are all state property. Government research institutions automatically become the administrators of archaeological finds. Other organisations transfer the finds to the Monuments Board, which becomes their custodian no later than the date on which the field report is handed over. However, this process has not yet been fully mastered in practice.

The biggest and the best technically and personally equipped institution is the Archaeological Institute (founded in 1939), which in addition to rescue archaeological excavations also carries out systematic (non-rescue) excavations and research. Currently, the Institute employs 35 researchers with special professional competence to carry out archaeological excavations and 21 archaeologists without this authorisation. The Institute also employs a whole range of specialists from other disciplines anthropology, zoology, botany, geology, geophysics, conservation, numismatics, museology and the like.

The Monuments Board, using its network of regional branches, performs the main tasks associated with the protection of monuments and administrative work. To a lesser extent, it carries out scientific work. Currently, it employs 8 licensed archaeologists. The third crucial institution is one of the branches of the Slovak National Museum – the Archaeological Museum, which employs 4 licensed archaeologists.

Although the network of regional museums is relatively dense, only a few of them are authorised to carry out archaeological research since only a few archaeologists are licensed to carry out excavations.

Figures 6.1: Private archaeological companies also receive significant government contracts. Excavations at Bratislava Castle. (© Archaeological Institute in Nitra)

Most university departments are focused on implementation of systematic research, and only to a lesser extent are they involved in rescue excavations. Overall, the universities together employ only 7 professionals with competence to carry out archaeological excavations.

In the first years after the passing of Act No. 49/2002, although legislation now provided the option to grant licences to private institutions (including non-profit organisations or foundations), none were awarded. Finally, the Ministry began to give in to pressure and gradually started granting authorisations to private institutions, despite the negative recommendation of the Archaeological Council at the Ministry of Culture of the Slovak Republic. In Slovakia 15 private archaeological companies operate at the moment (Figure 6.1).

A new law should be introduced with the aim of improving the work of all these archaeological entities. At the same time it should also address the relationship between archaeologists and investors in the construction industry and the general public.

Archaeological licences

By law, anyone who wants to carry out archaeological research must pass an exam of special professional competence, thus becoming holder of a licence. For its acquisition it is required to prove practical experience and expertise in conducting archaeological excavations. Applicants must hold a postgraduate degree in archaeology. They must demonstrate knowledge of the laws and submit field reports from excavations under the supervision of licensed archaeologists.

Verification of practical and professional knowledge of applicants for licences is carried out by a committee for the verification of special professional competence set up by the Ministry of Culture of the Slovak Republic. Committee members are appointed and dismissed by the Minister, selecting them from the field of archaeology experts designated by the Archaeological Institute of the Slovak Academy of Sciences, universities, the Monuments Board, and the Slovak National Museum or other museums. The term of a commission member is three years.

So what are the experiences with licensing? It is certainly positive that, unlike other types of historical research, archaeological excavations can be conducted only by a legal entity. In this respect, the idea of licensing is not bad. Recently, however, there were instances when the ministry granted a licence in spite of a strong negative opinion of the Archaeological Committee at the Ministry of Culture, which assesses the candidate's professional ability. Another problem is that although formally led by a licensed archaeologist, in reality it is a large number of unlicensed archaeologists who are doing the real fieldwork, without the licensed archaeologist being present on site at all. An attempt to eliminate this deficiency was made with an amendment in 2014, which specified the maximum number of ongoing excavations per archaeologist at five all excavations being counted as ongoing until official submission of the field documentation to the Monuments Board. This, yet again, poses another problem, particularly as the size of the excavations is not considered. In practice it can lead to situations in which construction companies or private builders will have trouble finding contractors for small-scale excavations. So, paradoxically, simple inspection of a trench for the water pipe to a house counts the same as an extensive research project on a 30-km section of motorway with dozens of sites, while these are obviously qualitatively and quantitatively completely different cases.

The introduction of licences of special professional competence has some positives, especially in the sense that the directors of archaeological excavations should not be people without adequate field experience. Alas, we cannot talk about improving the quality of fieldwork. There would be a lot more sense in raising the quality of field-practice teaching at universities. On the negative side is also the fact that the Ministry of Culture inconsistently handles the recommendations of its expert committees.

Execution of archaeological excavations

According to the latest amendments to the law (2014), regional branches of the Monuments Board are responsible for supervising the execution of archaeological research. There are eight regional branches, and they are directly supervised by the Monuments Board. The good thing is that their decision should clearly determine the type and scope of research or its stages. However, a major problem is that the need for excavations is often negated on the grounds that no archaeological finds were recorded in neighbouring territories, which is especially worrying in the case of larger constructions (e.g. kilometres-long lines of water pipes, sewerage, etc.). Thus a large volume of data is lost. A central register of archaeological sites is administrated by the Archaeological Institute.

The costs of archaeological research are covered by the entities who initiat it rescue excavations are usually paid for by the investor or executor of a project. This applies to all types of constructions: from small family houses to large industrial parks, from infrastructural developments to highways. Systematic excavations (non-rescue) are paid for by whoever carries them out, i.e. scientific institutions, usually in the context of different research projects. The question remains as to what exactly falls under the costs of archaeological excavations? There is no doubt about fieldwork. But the question comes with the costs of conservation, restoration, storage of finds, and publication of results. Reimbursement of these costs generally depends on the success of negotiations with the investor.

As a negative of the current system we consider the fact that the selection of a research institution is completely in the hands of the investor or construction company. Unfortunately, practice shows frequent human error, as has happened in a number of cases where it was not the quality of research but other factors, such as costs and time, that were most important for the investor. Often it is even the time only, while the price is not decisive. In these instances it is the institution which is able to execute excavations planned for a few months in the course of a few days that gets the contract.

Archaeological excavations are carried out under the formal supervision of the relavent Regional Monuments Board, according to its decision, which determines the exact conditions. In fact, the competent authority often has neither the human nor the technical resources to cover all ongoing archaeological activities in its region.

Officials from the regional branches of the Monuments Board have the right to enter, at any time, an archaeological site, the premises of immovable cultural monuments, the properties at a historic site and any area where construction work or other economic activity is being prepared or carried out (the only exception is if it is an inhabited dwelling, in which case the consent of the resident is needed).

A permanent problem is the insufficient number of licensed archaeologists at the regional branches of the Monuments Board. Even with the best of intentions, these licensed archaeologists, who are additionally burdened with considerable bureaucracy, have no chance of overseeing all ongoing excavations and construction activities in their regions. There are certain suspicions that as a result of various pressures, excavations are not being carried out to the quality the situation required.

If the investor, despite a decision of the regional branch of the Monuments Board, fails to arrange archaeological excavations, he can be fined with considerable penalties, especially if construction activity resulted in the damage of a cultural monument. In practice, however, the human factor fails more often than not, and, for example, the maximum fine has never been issued, despite some well-known damage or destruction of archaeological sites. In the case of small private building projects, the fines can be so low that paying a fine can be a cheaper alternative compared to paying for regular archaeological excavations.

Field reports are reviewed by an experts' committee composed of employees from several institutions, which has an advisory role to the Monuments Board. Here, a certain disparity is noticeable, as they often pay more attention to the quality of formal documentation than to that of the fieldwork itself.

However, this is where probably the most significant improvement has occurred. Formalised requirements for field documentation have led to the improved quality of field reports and have significantly improved timely submitting of the finalised reports. Unfortunately, this crucial aspect varies from case to case, depending on the archaeological contractor.

A common problem in archaeological practice is the very definition of an archaeological find to be protected by law. An archaeological find was initially defined in Slovak law as 'a movable object which provides evidence of human life and related activities from the earliest times until modern times' (Act No. 49/2002 on the Protection of Monuments and Historic Sites, Article 2.5). In practice, however, cases occurred when, rather than archaeologists, investors called on explosives technicians to deal with discoveries of potentially dangerous military devices. Therefore, to avoid confusion, an amendment was made to the Act effective from 1 June 2009 – extending the definition of an archaeological find (Michalík 2009, 528). The amended Article 2.5 now reads: 'The term "archaeological find" shall mean any movable object that provides evidence of human life and activities from the earliest times until 1918 and which was or is situated in the earth, on the earth's surface or under water. Weapons munitions, ammunition, parts of uniforms, military equipment and other military material found in the earth, on the earth's surface or under water and dating from before 1946 shall also be considered archaeological finds.'

Another complication is the disclosure and presentation of results of archaeological works. With few exceptions, private companies are rarely involved in presenting the results of their work, and generally they are not interested in publishing. For these reasons, a lot of important information from field research is completely lost to both professionals and the general public.

Activity of amateur 'treasure hunters'

Concerning the abovementioned law, one of the biggest problems is the question of so-called amateur 'treasure hunters'. Using metal detectors they have inflicted considerable damage on several notable archaeological sites (Figure 6.2).

Slovak law prohibits the unauthorised excavation or study of cultural monuments, sites and zones under heritage protection, archaeological finds and archaeological sites, as well as the unauthorised collection, transfer and possession of movable finds, and the unauthorised search for finds using metal detectors (Act No. 49/2002 on the Protection of Monuments and Historic Sites, Art. 39.6).

The law is very strict against any activities of amateur collectors, which has unfortunately proved counter-productive. Moreover, the law has grouped commercially motivated 'treasure hunters' together with ordinary 'passer-by' enthusiasts, who had been conducting surface surveys of their local area without metal detectors and reporting their results to regional museums and other institutions. In consequence, the law has managed to cut off an important source of scientific information, but the protection of archaeological sites has not improved. Thanks to this equation of 'treasure hunters' with casual enthusiasts, professional archaeologists have lost an important group of regional collaborators (Figure 6.3).

The strict approach to treasure hunters seems to be futile. In the past, several amateur detectorists at least brought their finds to professional archaeologists for evaluation or for the sake of documentation. The archaeological site was damaged, but scientists were at least informed of the discovery of artefacts and they could proceed with excavations at the site. The threat of jail or heavy fines led looters to refuse providing finds for evaluation or publication. Realistically, we cannot protect the whole of Slovakia, and everyday

Figures 6.2: Thanks to cooperation with the owners of metal detectors, it was possible to obtain significant archaeological finds and record their location so subsequent excavations could be carried out. Gold plaque from Bojná, Early Middle Ages.
(© Archaeological Institute in Nitra)

Figures 6.3: Hoard from Pruzina, Early Middle Ages. (© Archaeological Institute in Nitra)

practice shows that the looting of archaeological sites is still prevalent, practically without any serious threat of sanctions. In the future, it would be appropriate to modify the law to limit the looting of archaeological sites, but at the same time not to lose essential scientific information.

In an uneasy position are also archaeologists themselves, who are often aware from different sources of the contents of some private treasure hunters' collections, but the information cannot be used, because in doing so they would likewise commit an offence or a crime under the terms of the law (Michalík 2012, 252).

Recent amendments are also greatly complicating the work of licensed archaeologists, who, in order to use a metal detector in exploration activities or to trace looted sites, need to obtain a whole range of permits. To most professionals, the whole process is so complicated and lengthy that it discourages exploration activities, and in particular it detracts from the verification of archaeological sites identified by use of various non-destructive methods. This situation is a great shame for archaeological research.

Legal Act No. 49 of 2002, with all later amendments, has brought many positives to the work of archaeologists in Slovakia. The main benefit is the improvement of field documentation and the timelier manner of its submission. On the other hand, there are still many negatives that will have to be modified in the near future.

References

Michalík, T. 2009: Právne aspekty ochrany archeologického kultúrneho dedičstva v Európe. *Archeologické rozhledy* LXI, 524–46.

Michalík, T. 2012: Súkromné zbierky archeologických nálezov. Je východiskom slovinský model? *Studia Historica Nitriensia* 16/12, 248–60.

Ruttkay, M. & Šmihula, V. 2009: Významné jubileum Archeologického ústavu Slovenskej akadémie vied, *Slovenská archeológia* LVII2, 365–76.

Legal Act No. 49/2002 on the protection of monuments and historic sites.

Legal Act No. 50/1976 on land-use planning and building order (the Building Act).

7 | French preventive archaeology: administrative organisation, role of the stakeholders and control procedures

Bernard Randoin

Abstract: In the 1990s, the 1941 law on French archaeology had to be changed to reflect the modern evolutions of both society and the discipline of archaeology. The new legislation has been extensively discussed in Parliament over a long period and is the result of various choices that were not made by archaeologists but by the representatives of French society.

This paper concentrates on the description of the system of preventive archaeology, which covers the administrative organisation and the different roles of the various actors in decision-making, fieldwork and quality control.

Keywords: organisation of archaeology, Heritage Code, quality control, operators

The present organisation of French preventive archaeology is prescribed by the law approved by Parliament in 2001 and modified in 2003. From the 1970s to the 1990s, the development of preventive archaeology emphasised the obsolescence of the legislative texts ruling French archaeology and dating back to 1941.

In 1998, under pressure from building contractors who demanded a clear legal framework, the French government submitted a draft law which was discussed for several months before it was put through to Parliament in May 1999. The examination and amendment process went on for another 20 months and was submitted to 6 plenary examinations and votes by the National Assembly or the Senate before its final adoption on 17 January 2001.

Two years after its adoption, the members of Parliament decided to revise the law mainly to give greater consideration to local authorities, many of which had archaeological services that were not included in the first legislation, and also to introduce commercial archaeology.

Once again a large debate went on for 16 months in Parliament, and the law which now rules French archaeology was adopted in August 2003 after 4 plenary examinations and votes by the National Assembly or the Senate.

A few adjustments have then been made to the legislation governing archaeology. All these legal provisions are now collected in the Heritage Code, under Book V (*Code du Patrimoine* 2003).

The originality of the French legislation on archaeology, if we may say so, lies in the way it has been built, with Parliament organising long and thorough discussions involving a very wide range of stakeholders, making strategic and political choices and approving them by formal votes.

Although the law does not explicitly refer to the Valletta Convention, Parliament has enshrined this convention in the French system, and it has been a constant point-of-reference during all debates.

Setting the scene

The Ministry of Culture and Communication (MCC) is responsible for the management of archaeological heritage in the French territories, but the system establishes a clear distinction between the entities responsible for decision-making and control and the operators who execute the fieldwork.

The responsibility of the MCC in taking decisions and controlling fieldwork is entrusted to the Regional Archaeological Services (*Service régional de l'archéologie* SRA) in the regional directorates of cultural affairs present in each of the 26 regions (22 regions in Europe and 4 regions overseas). These services are dependent on the regional prefect, who represents the government in the region. The French government is now reforming the administrative organisation of its territories. Some regions will be merged and on 1 January 2016 there will only be 17 regions (13 regions in Europe and 4 regions overseas). At the time this paper is due the final reorganisation is still under discussion. Therefore, this paper describes the current organisation in March 2015.

For scheduled research excavations the MCC services issue annual authorisations, control the fieldwork and assess the scientific results. For preventive archaeology they are responsible for the archaeological impact assessment of the proposed infrastructure or construction projects; they prescribe the measures to be taken to counteract the destruction of archaeological heritage; they control the fieldwork and assess the scientific results. The archaeological regional services employ trained archaeologists who also conduct research projects. Their scientific skills and expertise guarantee the relevance of their assessments.

The law also stipulates that SRAs carry out archaeological assessments of planning applications and infrastructure projects, which they receive automatically depending on their nature or size.

The assessment is made on the basis of information collected by the National Archaeological Inventory (Carte archéologique nationale) a national database and GIS recording all archaeological information, finds, excavations or observations made in the French territories. This inventory is maintained by the SRAs, which introduce the new discoveries as soon as they are available.

When, in the course of this assessment, they identify a potential impact on archaeological heritage, the relevant SRA prescribes an archaeological intervention which is imposed on the developer by means of an official act signed by the regional prefect.

Parliament decreed that the developers who had to make, and pay for, archaeological excavations, should be fully informed of their cost and duration. They also admitted that such an evaluation was impossible to make on a potential archaeological site without a minimal knowledge of its characteristics.

Therefore, the law makes a distinction, in the procedure of preventive archaeology, between two stages: diagnostics and excavation.

The diagnostic operation is generally the first archaeological intervention prescribed after the positive archaeological assessment of a planning project. It must determine the presence of archaeological remains, their extent, their date and nature and evaluate their state of preservation. These are the minimal data needed to assess whether an archaeological excavation is required and estimate its scientific objectives, the methodology to use, the necessity of special investigations or treatments (anthropology, specific conservation, etc.).

On the basis of the results of the diagnostic operation the SRA can decide whether an excavation is necessary and write the specifications of the excavation. The specifications are attached to the excavation prescription signed by the prefect of the given region. Again, according to the law, the construction works cannot start until the excavation is completed.

During this phase, the scientific relevance of the decision to prescribe an excavation and of its specifications is submitted to the evaluation of the Interregional Commission for Archaeological Research (Commission interrégionale de la recherche archéologique CIRA): an entity which will be presented further on when discussing the quality control system.

The members of Parliament decided that diagnostics can only be carried out by public entities: INRAP (Institut National de Recherches Archéologiques Préventives) and local authority archaeological services (which will be presented when discussing the topic of stakeholders) and financed by public money. They considered diagnostic operations as a way for society to determine if measures to offset the destruction of archaeological heritage had to be imposed on developers.

They also considered archaeological excavations as commercial services. Archaeological excavations are thus open to public entities as well as to commercial firms, provided they are accredited by the Ministry of Culture and by the Ministry of Research.

Each year since the implementation of the present legal provisions in February 2002 the state archaeological services have assessed between 30,000 and 35,000 development projects and prescribed between 2,000 and 2,500 diagnostics, which generated an average of 500 to 600 excavations.

The public financing of the diagnostics went through various discussions and several adjustments. The main objective was to minimise the financial impact of archaeology on each individual construction project or work. The presence of archaeological heritage and its importance, and therefore the cost of archaeology, is not evenly distributed within the French territories, and the fact that one construction would have to pay a lot for archaeology when some other projects would not pay much, if anything, was looked upon as unfair.

Under the present system tax is paid (above a certain size threshold and with very few exceptions) by every building or development project impacting the ground, whether or not it is going to generate archaeological interventions. In 2015, this tax amounted to €0.53 per square metre. It is recalculated every year on the basis of the national construction cost index. The tax finances the diagnostic operations made by accredited public entities and is also used to create a fund dedicated to help some developers pay for the excavations they are obliged to undertake.

The tax, implemented in 2004, is expected to produce an income of €118 million a year, but has not yet reached its full potential and so far has raised between €70 and 90 million a year.

Parliament also considered that, after the diagnostics, the cost of archaeological excavations could in some cases be too onerous for certain projects and could also potentially cause setbacks to some other government policies, such as the housing policy. This is why 30% of the income raised by this tax goes to a special fund: the National Fund for Preventive Archaeology (Fonds national d'archéologie préventive – FNAP).

This fund automatically finances the whole of the archaeological cost of excavations generated by private individuals who build their own house. It also automatically finances 50% of the cost of excavations necessitated by housing estates and 75% of the cost of excavations for social housing estates built under the dispositions of the national policy for housing.

Balancing the stakeholders

The state archaeological services are not allowed to undertake the required fieldwork since they are in charge of the prescription and control of the diagnostic operations and excavations. The fieldwork is carried out by a national entity created by law in 2002: the National Institute for Preventive Archaeological Research (*Institut National de Recherches Archéologiques Préventives* – INRAP, Link 1). Since 2003 other operators are also allowed to undertake preventive fieldwork provided they are accredited by the Ministry of Culture and by the Ministry of Research. This means that local authority archaeological services can undertake diagnostic operations and preventive excavations in their territory. Accredited private firms can only carry out excavations.

INRAP is a public body placed under the supervision of the Ministry of Culture and the Ministry of Research. It employs approximately 2,000 persons, 1,500 of which are trained archaeologists and professional excavators. Being the national public archaeological operator, it must be able to undertake excavations anywhere within the French territories, including overseas, and therefore has to employ specialists in all chronological periods from the Palaeolithic to the Second World War. If an advertisement for an excavation made by a developer under the public works contract regulations does not receive any answer, INRAP must undertake the excavation. If the operator selected by a developer goes bankrupt or fails, for any reason, to finish the excavation until the final report is completed, INRAP has to take over.

These public legal missions have been given to INRAP in order to ensure that no excavation will remain unfinished through lack of an accredited operator. The mission of INRAP also includes research, publication of the results and their dissemination amongst the public.

Accreditation is given to archaeological operators for five years on the basis of their application file, which includes a detailed presentation of the operator, its strategy, its technical, financial and human resources and also the curriculum vitae of its scientific staff. The applications are examined and discussed by the National Council for Archaeological Research (*Conseil national de la Recherche Archéologique* – CNRA) a national advisory commission which issues recommendations on the strength of which ministerial decisions are subsequently prepared.

The law implicitly recognises that the skills required may vary according to the chronological period in question. The accreditation can be given for one or more of the following chronological periods: Palaeolithic, Neolithic, Protohistory (Bronze Age and Iron Age), Antiquity, Medieval, Post-Medieval and Modern.

The accreditation allows an operator to tender for a public works contract when an excavation concerning one of the periods of their accreditation is required. The scientific and technical adequacy of the offer is checked, for each contract, by the state services which issue the excavation authorisation.

The local authority archaeological services can be accredited for both diagnostic operations (since they are considered as a public mission) and excavations. They are only allowed to undertake diagnostics in their own territory, but they can undertake excavations anywhere in the whole of the French territories. Most of them, however, mainly concentrate on carrying out excavations within their own territory.

In the 1980s a lot of local authorities created an archaeological service or employed an archaeologist in order to take into account their archaeological heritage, along with the state services. Most of the time they were undertaking research excavations on archaeological sites owned by the local authority, or working on the promotion of the archaeological heritage by organising guided tours, conferences, etc.

When the law created the possibility, some local authorities decided to ask for accreditation to become preventive archaeology operators. Their policy was initially to do the diagnostics and excavations on their own development projects, but they soon extended their activity to development projects which were considered as being strategic for the development of the territory (Figure 7.1).

Various levels of local authorities (towns, groups of towns or counties) now have their own accredited archaeological service. In the administrative organisation of the country, the smallest unit is the town (*Commune*). The next one up is the group of towns (*Communauté de communes* or *Groupement de communes*) an association between towns, usually a town and several villages or smaller towns in the immediate vicinity. Above towns or groups of towns is the county (*Département*). The region, which regroups several counties, would technically also be allowed to have an archaeological service but none of them has decided to do so.

The size and activity of the 67 accredited local authority archaeological services vary a lot. Some of them employ less than 10 persons, others more than 50. The variety is not necessarily related to the size of the territory: for example, one of the town services employs 52 persons when most only employ less than 10 persons. It depends very much on the policy of the local authorities and on the missions they assign to their services.

The number of local authority archaeological services accredited since 2005 has steadily grown from 31 to 67. For most of them the accreditation has been re-examined and maintained after the advice of the CNRA, which illustrates, on the one hand, that their activity was assessed positively and, on the other hand, that the local authority takes an interest in the activity of its service.

Although there are no national consolidated data, the local authority archaeological services are estimated to employ just over 700 persons. Accredited local authority services now cover 30% of the national territory, where 50% of the population lives.

Before 2001 only 2 private archaeological firms existed in France. They both specialised in a narrow chronological and geographical range. Between 2002 and 2003 they were not allowed to do any preventive archaeology, but they were reactivated in 2003. Most of

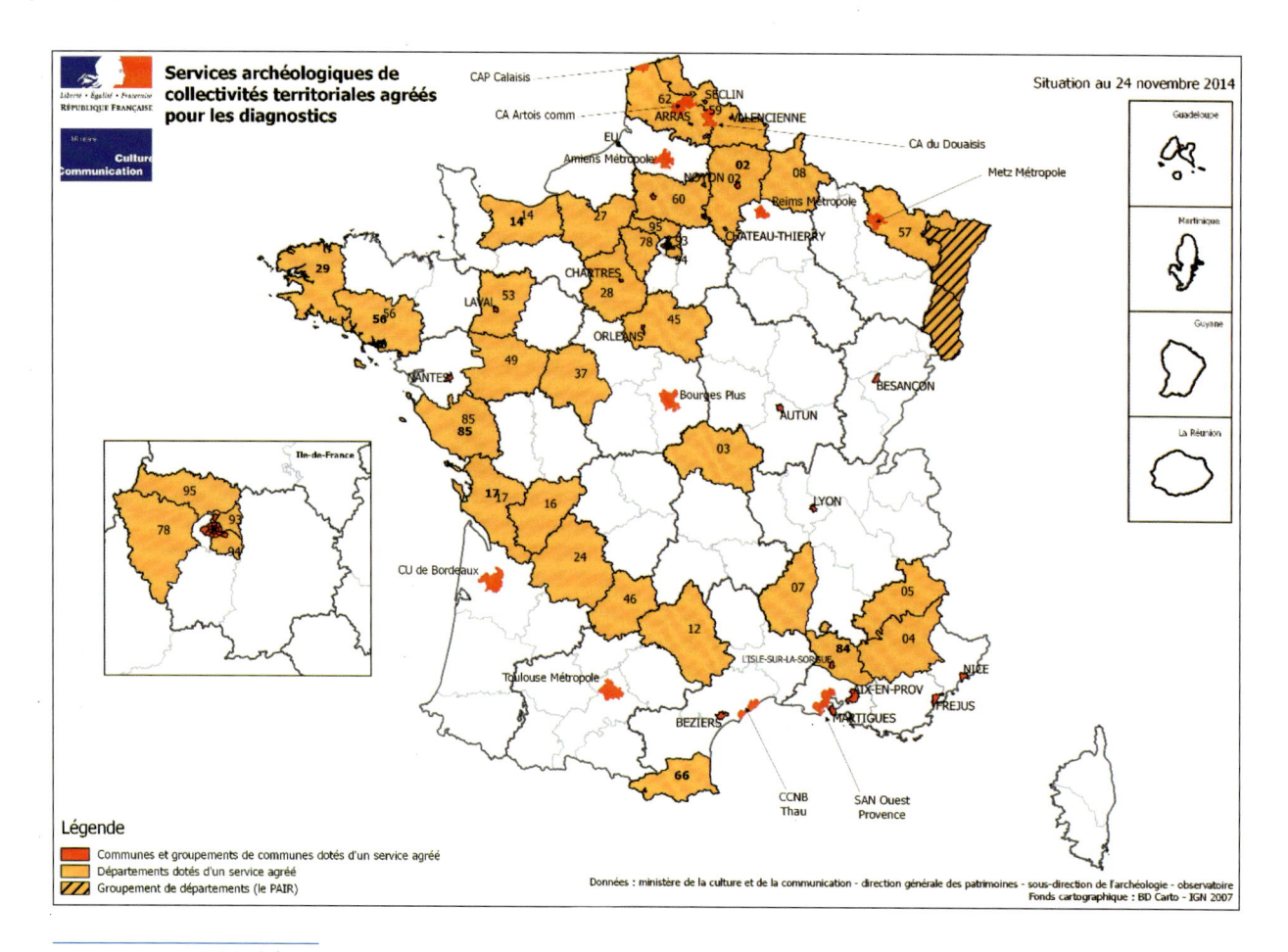

Figure 7.1: Geographical distribution of the territories covered by various local authority archaeological services (© MCC-SDA).

the other firms were created afterwards. In 2005, only 4 private firms were accredited. Now 19 private firms are accredited. They are only allowed to undertake excavations and obtain their excavation markets directly from private developers or through tenders organised by public sector developers under the public works contract regulations (Figure 7.2).

Most of the firms are accredited for chronological periods spanning from Protohistory to the Post-Medieval period, others just for the Medieval and Post-Medieval period. Some of them (6) are also accredited for the Neolithic period. Only 2 firms are accredited for the Palaeolithic, one of them also being accredited for the Neolithic and Protohistory (mainly Bronze Age).

The size of the private firms ranges from a few employees to over 150 employees. Although there are no reliable statistics for employment in this very fluctuant field, the private firms are estimated to employ over 500 people.

The majority of the private firms operate around their central office, they only try to gain contracts at a reasonable distance from their base to cut down their costs. Only 3 firms have obviously chosen to try and operate all over the country. They have created up to 10 regional offices; they also employ more people.

Over the last 6 years the private firms have gained around 30% of excavation contracts, which represents approximately 25% of the overall cost of excavations.

The 7 biggest firms share approximately 80% of all excavation markets gained by private firms.

The competition between INRAP, local authority services and private archaeological contractors has become tougher with the reduction of major works or building projects due to the economic crisis (Figure 7.3).

Assuring quality

All of the debates that took place in Parliament emphasised the crucial necessity of ensuring the quality of preventive archaeology operations that had to be considered as part of scientific research.

This gave rise to several legal provisions that were introduced in 2000 and have not been substantially modified since then. To ensure the quality of archaeological operations the members of Parliament agreed to oblige archaeological service providers to obtain an accreditation which guarantees that they are capable of undertaking excavation on sites of one or more chronological periods. This has been completed by the obligation imposed on the operators to write a scientific project for each excavation they tender for, this project being evaluated and authorised by the state archaeological services.

The legal provisions upheld the organisation which the Ministry of Culture had set up in 1994. The responsible body at national level is the National Council for Archaeological Research (*Conseil National de la*

The "operators" in preventive archaeology

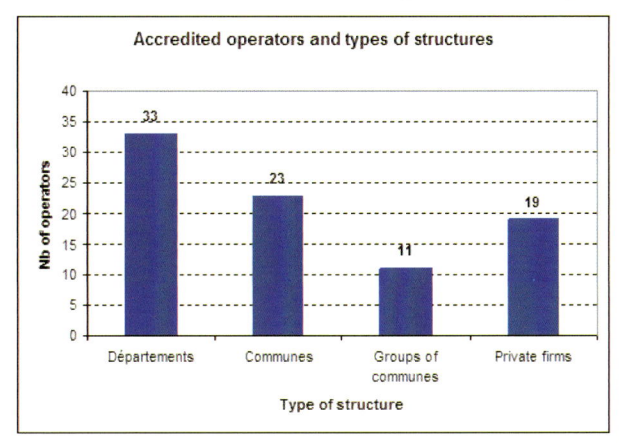

Accredited operators and types of structures
- Départements: 33
- Communes: 23
- Groups of communes: 11
- Private firms: 19

The fieldwork in preventive archaeology is done by INRAP and 86 entities
- 67 local authorities services
- 19 private firms

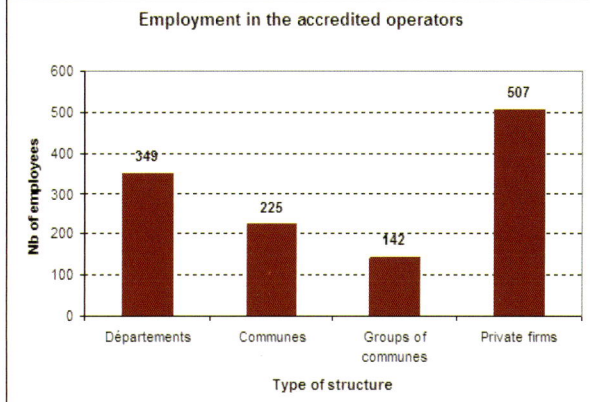

2000 persons are employed by INRAP
- 716 by the local authorities services
- 507 by the private firms

Employment in the accredited operators
- Départements: 349
- Communes: 225
- Groups of communes: 142
- Private firms: 507

Figure 7.2: Number and type of accredited archaeological services operating in the French territories, and number of personnel employed by the respective operators (© MCC-SDA).

Recherche Archéologique – CNRA), whilst at regional level there are 7 scientific commissions (*Commissions Interrégionales de la Recherche Archéologique* CIRA).

The CNRA is an advisory council for the Minister of Culture. It is responsible for writing and updating the national research agenda, and for evaluating the applications made by entities wishing to be accredited in preventive archaeology or to renew their accreditation. More generally, the council advises the minister on the national archaeology policy and on any subject the minister submits to the council.

The Ministry of Culture has, since 1994, wanted the council to not only reflect the position of the ministry but also to represent a synthesis of various points of view. It is composed of 32 persons amongst which are representatives of the Ministry of Culture (administration, archaeology, museums), the Ministry of Research, the National Centre for Scientific Research

Figure 7.3: Proportion of diagnostics and excavations carried out by different operators (© MCC-SDA).

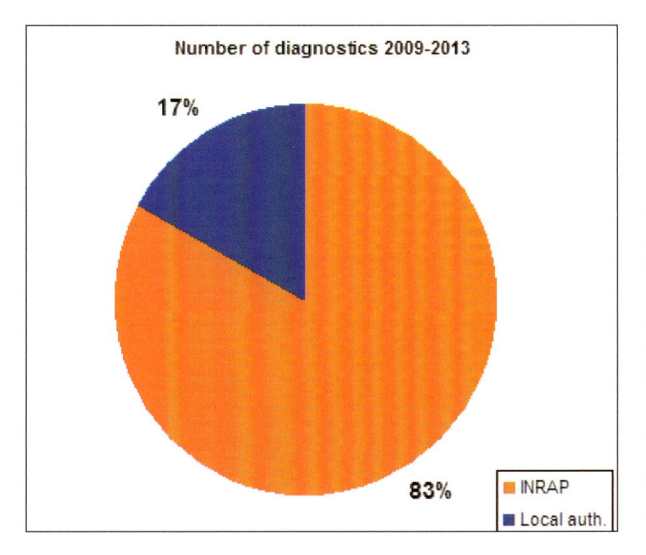

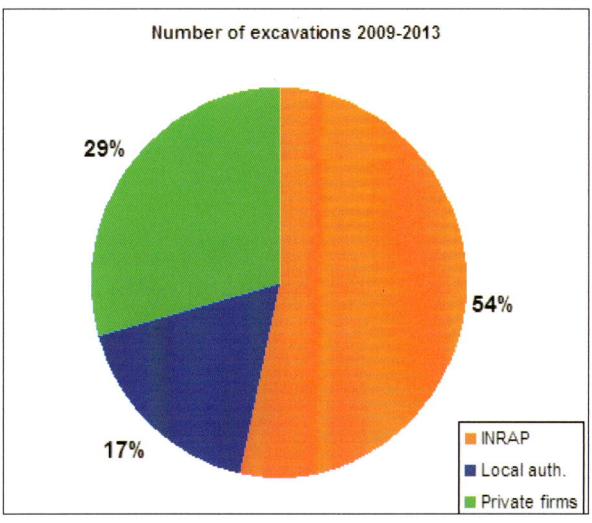

Number of diagnostics 2009-2013
- INRAP: 83%
- Local auth.: 17%

Number of excavations 2009-2013
- INRAP: 54%
- Local auth.: 17%
- Private firms: 29%

(*Centre national de la recherche scientifique* – CNRS), the Ministry of Education, universities, local authority services, INRAP and 2 elected members from each CIRA. The Council holds 6 plenary sessions a year.

Owing to this composition, the CNRA is considered by all stakeholders (archaeologists, developers, administrations, etc.) as a well-balanced body, which adds credence to its advisory role.

At regional level, quality control is carried out by the 7 Interregional Commissions for Archaeological Research (*Commissions Interrégionales de la Recherche Archéologique* CIRA). The French metropolitan territory is covered by 6 commissions which regroup from 2 to 5 administrative regions, one commission covers all the overseas territories (French Guiana in South America, Guadeloupe, Martinique, Saint-Barthélémy and Saint-Martin in the Caribbean, La Réunion and Mayotte in the Indian Ocean, and Saint-Pierre-et-Miquelon off the Canadian coast). This organisation may vary slightly when the new administrative divisions of the country come into effect after 1 January 2016.

They are advisory commissions which advise the SRAs and the regional prefects on the archaeological excavations they prescribe. They assess the relevance of the prescription of an excavation on the basis of the results of diagnostic operations and the adequacy of the objectives assigned to this excavation, as well as the suggested methodology. After the excavation is completed they also evaluate the results and the final report. The CIRAs also examine applications for research excavations and assess the final excavation reports. Their advice is required by law in some circumstances, such as, for example, the obligation of *in situ* conservation of remains found during a preventive excavation.

The composition of the CIRAs also preserves the plurality which prevails in the CNRA. Each metropolitan commission is composed of 8 experts: one from the CNRS, one from a university, one from the Ministry of Culture (SRA), one from a local authority service, one from INRAP and three other experts. The CIRA dedicated to overseas territories is smaller, with only 6 experts. Depending on the size of their territory, the commissions hold between 6 and 8 sessions of 2 or 3 days each year. One expert examines the applications and reports on them in detail; the report is then discussed by the commission before written advice is issued.

This organisation has proved to be efficient from the very beginning and has largely contributed to the acceptance of archaeology by society. One of the directors of a major development company has declared that he would never contest the prescription of an excavation that has been examined through this process and he never has.

The last element of the quality control system which operates on a daily basis is the scientific and technical control carried out by the SRAs. They inspect every diagnostic operation and excavation and have scientific discussions with the field archaeologists about the way the excavations are conducted and about the results. In case the works do not correspond to the expectations of modern archaeology they can make recommendations; they can even stop the excavation and compel the operator to change the site supervisor. Fortunately, this has only happened 5 times in the last 10 years.

No system is perfect and the French preventive archaeology system is criticised by some developers, and also by some archaeologists. The major subjects of contention are the cost of some excavations and the duration of the process. But since this system has been voted in by the national representatives after long debates and extensive consultations with stakeholders, a lot of principles have been accepted by society and have not been questioned in the last 10 years. A lot of adjustments still have to be made which have led the Minister of Culture to order two reports on preventive archaeology. The first report, dating from 2013, was the *Livre blanc de l'archéologie préventive*, written by a commission composed of 28 individuals, among them archaeologists, university professors and researchers, known for their interest in the organisation of French archaeology and their contrasting opinions (Link 2). This report makes a thorough evaluation of the state of French archaeology and some 50 suggestions to improve it.

In 2015, a member of Parliament issued another report on the same subject; the proposals of both reports have been evaluated by the government and some of them will be examined by Parliament in 2015 and 2016 to try and improve the legal provisions (Link 3).

References

Code du Patrimoine 2003: *Code du Patrimoine*, Livre V, archaeology is covered by articles L510-1 to L544-13 and R522-1 to R545-59. Consolidated version at www.legifrance.gouv.fr/affichCode.do?cidTexte=LEGITEXT000006074236 (accessed 10.10.2015).

Links

Link 1: http://www.inrap.fr (accessed 10.10.2015)

Link 2: http://www.ladocumentationfrancaise.fr/rapports-publics/134000214/index.shtml (accessed 10.10.2015)

Link 3: http://www.ladocumentationfrancaise.fr/rapports-publics/154000366/index.shtml (accessed 10.10.2015)

8 | A view from Turkey on the Valletta and Faro Conventions: effectiveness, problems and the state of affairs

Mehmet Özdoğan and *Zeynep Eres*

Abstract: Turkey is proud to own a rich and varied archaeological heritage. Even though each year there are numerous large-scale scientific excavations, the number of rescue excavations is minimal compared to the pace construction activities have taken. According to Turkish legislation, the state assumes the legal authority and responsibility for all archaeological heritage; however, for a site to be under protection, it has to be registered. Site registration is an extremely bureaucratic procedure, the total number of registered archaeological sites all over Turkey in 2015 being only 12,757. The problems that are encountered in Turkey for the preservation of archaeological heritage are far greater in scale and more complex than in most other European countries: there is no near-to-complete inventory of sites, the governing system has not adjusted to running salvage operations efficiently, and the sites are of prodigious dimensions.

Keywords: Turkey, archaeological heritage, rescue excavations, site registration, public awareness

Introduction: the cultural setting

Turkey comprises two peninsulas: Thrace and Anatolia, the former being an extension of Europe towards Asia and the latter of Asia towards Europe, bridging between the Near East and south-eastern Europe. The Sea of Marmara, with the narrow straits of the Dardanelles and the Bosphorus, not only marks the dividing line between Asia and Europe but also represents the main maritime route connecting the Pontic steppes of Eurasia with the Aegean and the Mediterranean. It is mainly due to its critical location that the archaeological heritage of Turkey is indispensable in understanding cultural interaction among distant geographic regions. Moreover, some of the most consequential developments that laid the foundations of our present-day civilisation, such as the emergence of farming village communities, took place in Turkey. Accordingly, the knowledge embedded in the archaeological sites of Turkey is of critical importance not only at a local level but also in understanding the progressive stages of civilisation.

A significant bias has been to view the Anatolian peninsula as a uniform geographic entity acting either as a bridge between continents, transmitting ideas, technologies and people, or becoming a cultural frontier between the Balkans and the Near East (Özdoğan 2007). Turkey comprises a number of distinct geographical zones covering vast areas, each having its particular cultural identity, developing together with the major cultural formation zones surrounding the peninsula: the Caucasus and Iran in the east; Syro-Mesopotamia, the Levant, and the Circum-Mediterranean in the south; the Aegean in the west and south-eastern Europe with the Pontic hinterland in the north. Due to this multifarious cultural mosaic, the type of archaeological remains varies from region to region. For example, sites in the south-eastern parts consist of huge settlement mounds up to 70 metres high that accumulated from the remains of mud-brick architecture; in the northern parts, due to the extensive use of wood as a building material, mound formations give place to flat settlements, as in most of Europe. Likewise, along the Aegean-Mediterranean littoral, the ruins of the Hellenistic and Roman periods cover vast areas, hindering the visibility of earlier occupations. In defining archaeological heritage, the extreme diversity in the type of sites led to certain biases in setting criteria, either looking for ruins with monumental architectural remains or for mound sites (in the Near Eastern sense) overlooking flat or inconspicuous settlements. Before going into the details of present-day problems in preventive archaeology, we find it necessary to present a conspectus of the historic development of archaeological heritage management in Turkey, as it differs considerably from that in Europe.

A background to the historic development of archaeology in Turkey

In contrast to most countries of Western Europe, archaeology had developed in the Ottoman Empire by the second half of the 19th century as a component of the package of westernisation; thus it was a top-to-bottom development that continued until the early republican period as an elite pursuit (Eldem 2004; Özdoğan 1998). At the time when Ottoman elites began developing an interest in archaeology, Europe was living through a period of Graecism with the excitement due to the acquisition of the Elgin Marbles; thus to the Ottomans, the concept of archaeology was directly related to the remains of the Hellenistic-Roman period and, to a lesser degree, to the monumental remains of Mesopotamian cultures. Cultural remains that related to their national heritage those of the Turkic or Islamic periods were not considered within the framework of archaeology but were left to the domain of art history and ethnography (Eldem 2011; Özdoğan 2004). Even the Ottoman Imperial Museum was built as a copy of a Hellenistic temple. With the foundation of a national state the Republic of Turkey in 1923, the concept of archaeological heritage was expanded to cover the entire cultural sequence of Anatolia from the earliest period onwards (Çığ 1993; Özdoğan 1998). Accordingly, it was not focused on ethnic or religious heritage but fostered a concern for all cultures that had lived in Turkey in the past. Even though the newly founded Republic was a national state, in strong contrast to the multi-ethnic identity of the Ottomans, the Republic developed an ideology based on Anatolism. Contrary to the other national states in the Balkans, permits were issued to cover all time periods from early prehistory to late Byzantine (Özdoğan 2004). Archaeological excavations, both by Turks and foreigners, were encouraged during the early years of the Republic, and a number of museums were established throughout the country; however, there was an apparent lack of trained personnel (Tanyeri-Erdemir 2006). To overcome this deficiency a number of students were sent to Europe to study various fields of cultural heritage, from prehistory to classics to linguistics; likewise, eminent German scholars escaping the Nazi regime were invited to Turkey and given positions in the country's universities as the chairs of the newly established archaeology and philology departments. So archaeology in Turkey developed under the strong influence of German tradition.

The background of the bureaucratic setup

As briefly noted above, archaeology began as a top-to-bottom endeavour organised by the state. Prior to the establishment of an archaeology service, there were already a number of foreign excavations in almost all parts of the Ottoman Empire (Kuban 2013; Martin 2013; Özdoğan 2013a). Here it is worth noting that up to the end of the 19th century, most of the Near East, the Aegean and Southern Balkans the prime target areas of western archaeologists were still within the domain of the Ottomans. In the beginning, the Ottomans had no interest in archaeological remains; however, with time, the removal of antiquities from the empire to European museums became a matter of disquiet. The main concern of the Ottoman elite was to prevent the export of antiquities and to bring them together in the newly established Imperial Museum. Thus, the first Ottoman legislation on antiquities had its focus on controlling foreign missions and preventing unauthorised excavations. The Ottoman Antiquities Law of 1869 prohibited all antiquities from leaving the country; this met with considerable resistance from foreign missions, one of the most publicised cases being the smuggling of the famous treasure of Troy by Schliemann (Easton 1994). Still later, when it became clear that some of the western archaeologists were also working together with the foreign services of their countries, the Ottomans became more cautious and suspicious of all foreigners travelling around the empire looking for antiquities (Trümpler 2010). The feeling of distrust towards foreign archaeologists persisted until quite recently, though there have always been foreign expeditions in Turkey. The Turkish Republic inherited the centralised Ottoman system. The Ottoman Antiquities Law of 1906 remained in use up to 1973, the central state being the sole decision-making authority, regulating and maintaining strict control of all archaeological research projects (Eres & Yalman 2013).

According to Turkish legislation the State assumes the legal authority and responsibility for all archaeological heritage. Even though the Ministry of Culture and Tourism (previously the Ministry of Culture) has branches, councils and museums in every province, all excavation and survey permits are issued by the General Directorate of Monuments and Museums of the Ministry of Culture and Tourism in Ankara. Moreover, all non-rescue excavation permits, both for Turkish and foreign research teams, have to be approved by the Council of Ministers. Likewise, neither local museums nor local offices of the Ministry have the authorisation to issue survey or rescue permits, even for sites that are under immediate threat of destruction. Needless to say, initiating any field project requires considerable bureaucracy; in the case of rescue operations, this either brings investments to a standstill for an unknown period of time or results in the destruction of sites.

The problem of site registration

According to Turkish law the state is responsible for all sites that are registered, thus all kinds of development projects that would have an impact on a registered site have to go through the Ministry of Culture and Tourism. It is up to the Ministry to decide whether the site should be protected thus necessitating revision or even cancellation of the investment project or whether a rescue excavation should take place. The final decision, either completely or partially destroying the site or changing the project, is taken after assessing the results of the rescue excavation. Turkey ratified the Valletta Convention in 1999 and the provisions of the convention are, more or less, met; however, the problem is the limited range of registered sites. Registration of a site is an extremely time-consuming, bureaucratic procedure that has to be carried out by the local councils. Up to 1973 there was only listing of historic and/or archaeological sites, but no official site registration. In the earlier years, the listing of sites was regarded to some extent as the compilation of an inventory of cultural assets. The earliest attempt at inventorying cultural property took place in 1917 and encompassed solely the urban remains in Istanbul. The domain for inventorying was expanded in the 1970s to cover all of Turkey; however, the prime concern of the inventory was still to list monumental historic remains within urban areas. Inventorying archaeological sites or vernacular architecture was either completely overlooked or extremely rare.

The Ottoman Antiquities Law of 1906 remained in force until 1973. When designing the new law in 1973, current international conventions and charters, such as the Venice Charter, and various European legislations were taken as models. While the earlier law was solely concerned with historic monuments, now, for the first time, the concept of site preservation was introduced to the Turkish legal system, defining archaeological, urban and natural sites. The new law also demanded the registration of all sorts of cultural properties, particularly archaeological ones. To this purpose, a central office was established to carry out the registration of sites and monuments and to take decisions on all sorts of intervention, restoration or renovation. However, there was neither the necessary capacity nor funding to cover all of Turkey. Most of the work was concentrated in the main urban centres and focused on the registration of historic buildings. Later, in 1983, the antiquities law was considerably revised, making local councils responsible for the registration of cultural properties within their domain. Unfortunately, they were understaffed and were also under the burden of urban problems. So registration of archaeological sites was carried out on an *ad hoc* basis, primarily depending on the interests of the council members; likewise, archaeological heritage that lay buried under present-day urban coverage remained either unnoticed or totally overlooked.

The number of archaeological sites registered up to the year 2000 was less than 7,000, while the number of published archaeological sites was over 100,000. The Turkish Academy of Sciences (TÜBA) regarded cultural inventory as a major problem in 1999. The Academy initiated a multifarious undertaking for the registration of all sorts of cultural heritage archaeological, architectural and natural designing a major database in accordance with the criteria set by expert academics (Başgelen 2002). The project was tested during 2000–2002 in two small districts using intensive surface surveys; the number of inventoried sites exceeded the total number of registered sites in the whole of Turkey; however, none were registered later. The project was interrupted in 2004 by the central government on the basis that the Academy was not authorised to carry out such a task. However, with this undertaking of the Academy the problem of site inventorying became a part of the agenda in Turkey, and various governmental and non-governmental bodies began inventorying historic remains within their region. It is to be regretted that each of these organisations worked independently, devising their own sets of criteria with no possibility of combining their work into a national database. At present, both the disparate efforts to develop regional inventories and central government's official registration programme are ongoing, though the latter proceeds at a slow pace. By 2015, the total number of registered archaeological sites had risen to 12,757, which is still minimal compared to what is known to exist. In consequence, there is no possibility to learn either from the Ministry of Culture and Tourism or from academic institutions how many mounds or castles, for example, exist in Turkey.

The problem of preventive archaeology

Turkey is experiencing a rapid pace of development. Urban, touristic and industrial centres are expanding, and new highways, pipelines and energy lines are being constructed in almost every part of the country. Almost all of the mountains are being quarried for building materials, and dams are being built in every possible location. In keeping with the Turkish Law of Antiquities, in accordance with the Valletta Convention, prior to commencement of any construction the developing agency has to apply for clearance from the organs of the General Directorate for Monuments and Museums. However, the decision is based on whether or not there is an already registered site within the construction area. Thus, all others are vulnerable to destruction. On some occasions, when an unregistered site is somehow noticed and the information is passed on to the media, a last minute rescue operation takes place, usually in spite of protests by the investors. Nevertheless, during the last few decades rescue excavations have taken place in various parts of the country; however, mostly on an *ad hoc* basis (Özdoğan 2010; 2013b). The fact that the total number of rescue excavations which took place in the whole of Turkey in 2014 was only 203 clearly presents how drastic the picture is.

In 1993, to overcome the problem of saving unregistered sites, regulations were amended, requiring Archaeological and Environmental Impact Assessment Analysis to be carried out in all major development areas. Even though it became mandatory in 1993, until 2002 it was hardly ever implemented, and even later it was only carried out extremely inefficiently, mostly by private companies paid by the developers. The impact analysis has been implemented efficiently only in projects financed by international bodies such as the World Bank or in projects tendered for by EU companies: the Baku-Ceyhan pipeline, the Ilısu Dam Reservoir and the Yenikapı salvage projects being some of the successful cases that will be further exemplified below. In 2014 the legislation of Archaeological and Environmental Impact Assessment Analysis was modified providing central government the right to exempt development projects of critical importance, such as the new highway system around Istanbul.

Major archaeological rescue projects

The first major organised rescue project took place within the reservoir area of the Keban Dam on the Upper Euphrates. The project was initiated in 1967 by a joint undertaking of the Middle East Technical University and Istanbul University. Work began by surveying two of the major alluvial plains, covering roughly 65% of the reservoir area (Erder 1967). Fifty-two archaeological sites, mostly settlement mounds, two medieval mosques and a Roman bridge were recorded during the survey; later the number of endangered archaeological sites mounted to 63. The collaborating institutions decided to continue the project by making an international call for cooperation; from 1968 to 1976, with the participation of two British, one German, one American and several Turkish teams, it became possible to carry out not only rescue excavations but also documentation of regional vernacular architecture. The two mosques (Figure 8.1) and the Roman bridge were dismantled and moved to new locations. Within the framework of the project, one of the settlement mounds recorded during the survey was fully excavated, 10 were substantially excavated (Figures 8.2–8.3), and 8 were tested by soundings. The results of rescue operations were published in yearly reports. The success of the Keban project led to a similar initiative at the Karakaya and Atatürk Dams along the Middle

Figure 8.1: One of the medieval mosques at Pertek within the reservoir area of the Keban Dam. The mosque was moved to a new location in 1968.[1]

Figure 8.2: Tepecik was recorded as a medium-sized mound during the survey of the Keban Dam reservoir in 1967. After 8 years of work, it was understood to be the combination of three distinct mound formations: the earliest, dating from the Pottery Neolithic to the Uruk period, being deeply buried under alluvial deposits. A large part, about a third of the Early Bronze Age settlement, could be excavated, but the early Neolithic horizon was encountered only in a 4x4 m sondage 10 metres below the surface of the plain.

---

[1] All photos are from the author's archive.

Figure 8.3: Last days of excavation at Korucutepe when inundated by the rising waters of the Keban Dam; the site was the capital city of Isuwa Kingdom a tributary of the Hittite Empire.

Euphrates. Of the total number of 580 sites recorded, 38 of them could be excavated, but only two extensively. Along with the settlement mounds, there were three castles, a caravanserai, 8 Roman aqueducts, an Assyrian rock inscription and numerous historic mansions, none of which could be removed from the dam reservoir area. Here, it is worth noting that archaeological sites within the reservoir area of the Atatürk Dam were considerably bigger than those recorded along the Upper Euphrates. Among them, Samsat the site of ancient Samosata was the most impressive (Figure 8.4). The central mound, measuring 600 metres in diameter and 52 metres in height, had occupation layers dating from the Neolithic to the late medieval period, and was encircled by a fortified lower town of about 5 kilometres in diameter. In spite of the efforts of the archaeological team, the main excavation could only reach Iron Age levels, and in limited soundings to the level of the Uruk horizon. Samsat, like other big mounds such as Lidar, Kurbanhöyük and Gritille, is now submerged with all the unrecovered archaeological information. None of these sites, including Samsat, had not been officially registered (Özdoğan 2000). If they had already been registered, then the rescue excavations could have started before the construction of the dam, and even if it would not have been possible to excavate all of the archaeological deposits, at least excavation could have reached the Uruk horizon, where written documents were to be expected. In 1984 the joint salvage project came to an end and a new one was not undertaken prior to the recent construction of Turkey's next major dam the Birecik Dam on the Euphrates. The reservoir area of the Birecik Dam had already been surveyed, with 32 sites having been recorded, but no organised rescue operations were initiated. The famous historic sites of Zeugma and Apamea ad Euphrates were among those to be inundated by the Birecik Dam. Their presence had been known since the early 19th century, and limited excavations had taken place in during the late 1980s and 1990s, exposing numerous examples of mosaic floors of the Roman period; however, neither of these sites was registered. A few months before the completion of the Birecik Dam one of the mosaics of Zeugma came to the attention of the press, triggering unexpected excitement both in Turkey and abroad. An immediate rescue operation was organised, in which several foreign teams were invited to participate and, working up until the last moment, a significant number of Roman villas were excavated, salvaging the mosaic floors now exhibited in the newly constructed museum at Gaziantep. After the turmoil caused by the loss of Zeugma, rescue operations were reorganised within the reservoir areas of two other dams: Karkamış on the Euphrates and Ilısu on the Tigris, again under the patronage of the Middle East Technical University, making it possible to conduct large-scale rescue excavations, firstly at Karkamış and then at Ilısu (Tuna & Öztürk 2001). (Figure 8.5).

Prior to the initiative taken by the Middle East Technical University in 1999 there had been almost no archaeological excavations on the upper reaches of the Tigris the region that represents Upper Mesopotamia. In spite of the paucity of known archaeological sites, the

Figure 8.4: Samsat: the central mound with still-standing medieval castle on top. The level reached by excavation is visible.

presence of the impressive medieval town of Hasankeyf located on the Tigris, with its castle, palaces, Artukuid Bridge, mausoleums and other remains, had long been known (Figure 8.6). Hasankeyf had always been considered one of the most impressive and picturesque archaeological sites in Turkey; however, at the time when the construction of the dam began in the late 1990s it was not registered and, moreover, there were no registered sites within the reservoir area. In planning the dam, the State Hydraulic Works (DSI) had consulted the General Directorate of Monuments and Museums as early as 1990 and received clearance to proceed, being informed that there were no historic relics within the reservoir area of the Ilısu Dam. However, even the preliminary survey carried out under the auspices of the Middle East Technical University revealed the presence of hundreds of sites and monuments within the planned reservoir. In the beginning, even though the General Directorate of Monuments and Museums was inclined to issue rescue excavation permits, DSI was reluctant to provide funding. Later, with the pressure exerted by the press and the recommendations of the World Bank, DSI took the decision to subsidise all rescue operations. At present there are numerous ongoing rescue operations making ground-breaking discoveries that have almost forced the entire history of Near Eastern civilisation to be rewritten. The 11th-millennium BC finds from Körtiktepe stand out as the most exciting, complimenting the recent discoveries made at Göbeklitepe. Even though rescue excavations

Figure 8.5: Mezraa Teleilat near the Euphrates: an extensive Neolithic settlement with a Neo-Babylonian palace on top.

Figure 8.6: Hasankeyf, a medieval town near the Tigris, which will be inundated by the Ilısu Dam.

of the settlement mounds have now reached a level that might be considered satisfactory, what should be done with the monumental architectural remains of Hasankeyf remains an unresolved problem. A similar case, though on a smaller scale, had been encountered at Allinoi within the reservoir of the Yortanlı Dam. The site had been extensively excavated revealing hitherto unknown monumental remains of the Roman period; there had been considerable public concern to save the site by changing the location of the dam; however, the final decision was to cover the remains with sand and to inundate them.

Although largely efficient salvage operations took place prior to the construction of dams on the Euphrates and on the Tigris, hundreds of other dams have either been built or are still under construction (Figure 8.7) at sites where little archaeological assessment has been carried out. Very few of these sites have been surveyed and even fewer have been examined in rescue excavations (Özdoğan 2000), (Figure 8.8). On the other hand, the most efficient preventive archaeology, in full compliance with international regulations, was carried out on the Baku-Ceyhan pipeline project. The full extent of the pipeline was surveyed, and on some occasions the course of the line was altered. A team of archaeologists were assigned to conduct watching briefs during the construction, and rescue excavations were carried out where it was not possible to change the course of the pipeline. However, it is worth noting that the foreign contractor company responsible for the pipeline's construction had incorporated provisions in the tender process for effective implementation of preventive archaeology throughout the project.

Another successful case of preventive archaeology has been the Marmaray-Yenikapı project a major undertaking for the construction of a railway tunnel below the Bosphorus, connecting the historic centre of Istanbul with the urban areas on the Asian side. At the planning stage, the ancient Byzantine harbour, which had been completely filled in during the 13th century by the alluvial deposition of the Bayramdere-Lykos River, was chosen for the location of the tunnel. Here again, the international construction company heading the project had stipulated in the tender process that utmost care should be given to the preservation of all sorts of cultural remains. Some Byzantine structures were encountered even in the early stages of the tunnel's construction; as some of them were considered to belong to the fortifications along the Sea of Marmara, the extent of the construction area was slightly modified to protect the remains (Rose & Aydıngün 2007). In the later stages, when the construction reached the ancient surface of the harbour's sea bed, 38 well-preserved remains of Roman and Byzantine ships, most of them with their entire cargo, were recovered. In spite of the pressure from political circles, utmost care and time was taken over the full documentation and removal of the shipwrecks. Still later on, below the level of the ancient harbour and of the present sea level, at minus 6 to 9 metres to the surprise of all the remains of an early 6th-millennium Neolithic village were encountered (Kızıltan & Polat 2013). Even though the Neolithic of Istanbul is rather well documented, Yenikapı was an exceptional case revealing a rich assemblage of well-preserved organic materials: not only wooden artefacts but also various trees, plants and cereals. The recovery of well-preserved skeletal remains lying on wooden constructions stirred great public excitement, as

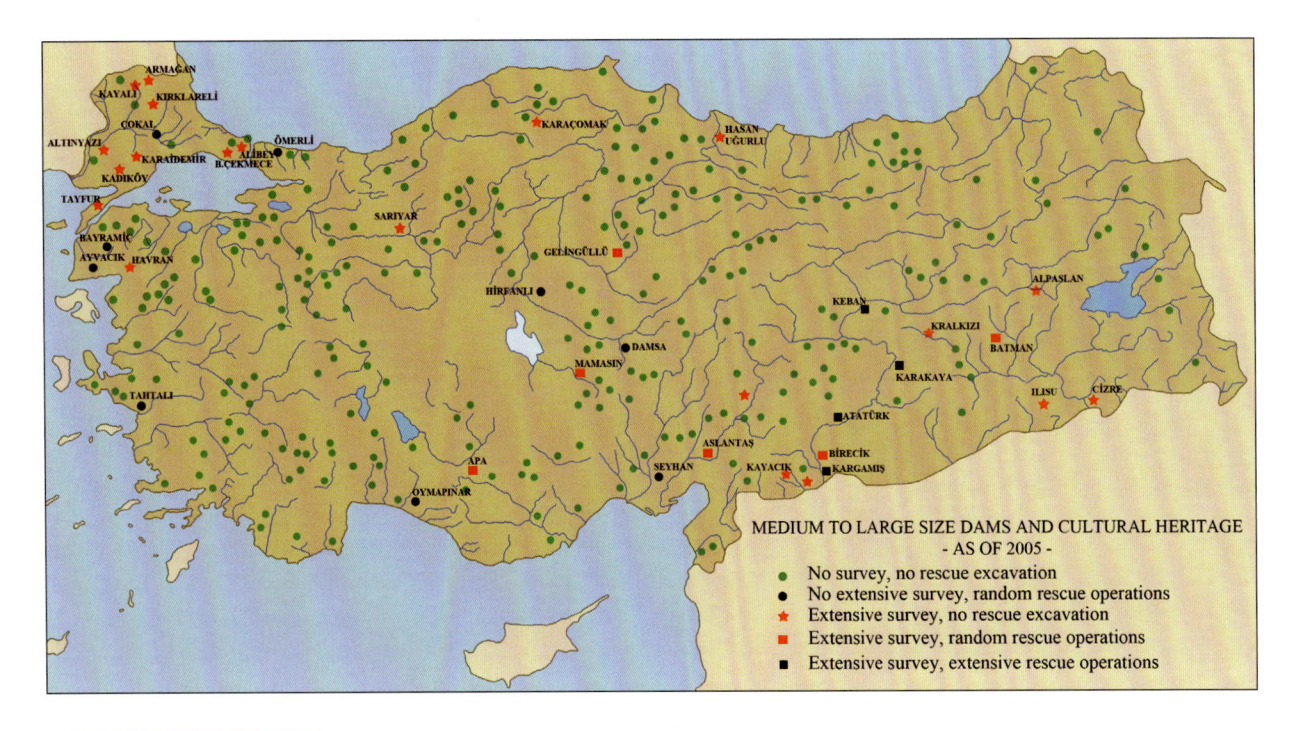

Figure 8.7: Map of major dams sorted according to archaeological projects as of 2005; small dams are excluded.

they represented the earliest inhabitants of Istanbul. Moreover, the recording of hundreds of footprints belonging to the inhabitants of the Neolithic village was immediately publicised and rather quickly attained a symbolic value. With the synergy generated by the Yenikapı discoveries it became possible to carry out rescue excavations throughout the entire Marmaray project. The most significant was the work conducted for the new station at Sirkeci, where, for the first time, a full cultural sequence was recorded in Istanbul, measuring about 20 metres thick. On the Asian side, at the location of the Üsküdar station, a late Byzantine church with a graveyard was fully excavated, as was a Neolithic settlement with over 67 burials at the Pendik station. However, although success has been achieved with the abovementioned rescue operations, it is not

Figure 8.8: Kumkale: a small Crusader castle, fully excavated, but completely left under the reservoir of the Aslantaş Dam on the Ceyhan.

possible to say the same for other construction sites around Turkey.

Problems due to the scale of sites

The dimensions of archaeological sites pose a major problem in preventive archaeology and conducting rescue excavations. In Turkey, as is the case in most Near Eastern countries, settlements have developed to form substantial mounds which can extend for several kilometres and stand up to 72 metres high, as in the case of Sultantepe. Even some of the single-period settlement sites, such as the Pre-Pottery Neolithic Göbeklitepe, can comprise over 20 metres of building remains. Due to the extensive presence of settlement mounds of considerable heights, inconspicuous archaeological sites that do not stand out as topographical features are either overlooked or ignored. Settlement mounds are not the only archaeological remains of considerable dimensions; in almost all of the country there are historic ruins consisting of monumental stone architecture covering several square kilometres. Likewise, Iron Age tumuli, either singular or several of them together, can be tens of metres high. It should be considered that at some of these large-scale sites (either settlement mounds or ruins with stone architecture) scientific excavations have been going on for decades. Excavations at sites such as Boğazköy-Hattusha, Ephesos, Pergamon, and Miletus have celebrated their hundredth anniversaries, whilst what has been exposed represents only a relatively small part of each site.

It is worth noting that most important archaeological settlement sites developed on alluvial plains along perennial rivers. Evidently, at present these are the most vulnerable areas, if not because of the location of dams, then because they face mechanical threats – irrigated agriculture. Accordingly, in Turkey the scope of problems encountered in preventive archaeology is far greater and more multifaceted than in most European countries. The construction of a major dam, including its planning stage, usually takes less than 10 years not enough time for an adequate rescue excavation, even at a settlement mound of modest dimensions. In most other parts of the world small excavations, soundings and/or core drillings can be considered sufficient to understand cultural sequences and the cultural modalities of a settlement. However, huge settlement sites in Turkey are sure to include archaeological materials such as written documents, cuneiform archives, sculptures and various artworks, the loss of which would be inestimable to the cognizance of the mainstream of civilisation. In many parts of the world destruction of an undocumented site is the loss of local cultural heritage; however, in the case of monumental key sites, the obliteration of archaeological deposits has consequences on a global scale.

The controversy between scientific and rescue excavations

In accordance with the Turkish law on antiquities, only academic institutions, and, in some cases, museums are eligible to receive excavation permits, even if it is a rescue operation. As Ankara is reluctant in allocating major rescue operations to museums, as in the case of the Keban Project, salvage excavations are conducted by academics, of course only during the time of the summer holidays. On the other hand, in major projects that turned out to be prestigious for Turkey, such as those concerning Zeugma, Marmaray-Yenikapı or the Ilısu Dam, local museums were authorised to keep on excavating throughout the year. The scope of this work necessitated the employment of archaeologists beyond the capacity of local museums. Development contractors were asked to employ archaeologists, however, only as workers with no vocational rights, as contract archaeology, (or private archaeology) is not recognised in the Turkish legal system. Thus, the abovementioned projects could be managed in spite of the drawbacks of the mandatory legislation. In this respect there is still considerable resistance, both from academics and the government to introduce the concept of private archaeology.

Conceptualising the Faro Convention in Turkey

Publicising the Faro Convention has not been a part of the agenda in Turkey; however, in spite of the general reluctance in keeping up with current developments in the management of cultural properties, there is a growing tendency in Turkey to raise public awareness of heritage issues (Ahunbay et al 2006; Eres 2010; Özdoğan & Eres 2012). Until a few decades ago there was no interest in Turkey either to manage archaeological remains as open-air museums or to develop public awareness; the main concern was to attract tourists. Archaeological remains were something of an abstraction to local communities. Major archaeological sites with monumental architecture located in the vicinity of tourist itineraries were made accessible to visitors. It could be argued that attracting tourists might have been the initial motive; however, during the last decades, there has been an unexpected quest by non-governmental and administrative bodies to learn more about local histories, supporting site management projects and even, in some cases, archaeological excavations.

In this respect the Karatepe Aslantaş Project of Halet Çambel, initiated as early as the 1950s, remains a unique example of designing a site as a diverse open-air museum and also of developing public awareness (Özdoğan 2014). The site was on a remote mountaintop with no road connection, inhabited by seminomadic groups. Çambel undertook the arduous task of educating and training the local community, restoring architectural remains, managing the entire region as Turkey's first national park in combination with an archaeological open-air museum, as well as displaying the tangible and intangible assets of local communities (Çambel 1993). Later, Peter Neve, working at the Hitite capital Boğazköy-Hattusha, took Çambel's work as a model and implemented an impressive conservation scheme by training the local workforce (Neve 1998; Seeher & Schachner 2014). Taking Boğazköy as a model, another cultural heritage management programme was initiated at the Pre-Pottery Neolithic site of Çayönü in 1989, preserving the Neolithic remains by burying the original architectural features and making one-to-one copies over the fill, thus protecting the archaeological

Figure 8.9: Aşağı Pınar open-air museum with visitors from local villages.

Figure 8.10: Aşağı Pınar in Eastern Thrace: part of the open-air museum.

fabric while enabling visitors to view the prehistoric settlement in its original environmental setting, only at a slightly elevated level (Özdoğan 1999; 2006). In this respect, the ongoing programme in developing local awareness run by the Çatalhöyük team under the direction of Ian Hodder needs to be mentioned, as its modalities are totally different from all others. The main focus of the project is to appreciate and to understand the perception of local communities concerning the archaeological remains within their region, however, without imposing novel concepts or trying to educate them (Atalay et al. 2010). Çatalhöyük stands as an exceptional case of running a heritage programme which limits the level of intervention to preservation rather than restoration.

Another major undertaking is the multifaceted project at two neighbouring sites in Eastern Thrace: the Neolithic settlement of Aşağı Pınar and the Early Bronze Age site of Kanlıgeçit (Eres 2014; 2013; Özdoğan 2006). The project design covers preservation, restoration, replicating architectural remains as previously done at Çayönü, training, stimulating experimental archaeology, and recreating the environment of the Neolithic era (Figures 8.9–8.10). The main difficulty encountered in running the programme was the obsession of the local communities that there was nothing of archaeological importance in Eastern Thrace, as there were neither ruins nor mounds. Textbooks had always placed emphasis on the archaeological heritage of Anatolia without even mentioning Thrace. It took almost 10 years to convince the local community that even though there are no monumental buildings, the knowledge of the past imbedded in their region is of utmost importance. Firstly, as it is 8,000 years old and signalled the spread of early farming communities from Anatolia to Europe. The success of the Aşağı Pınar and Kanlıgeçit projects has stimulated similar undertakings at a number of other early prehistoric sites including Aktopraklık, Aşıklı Höyük, and Yeşilova (Karul et al. 2010; Özbaşaran et al. 2010; Derin 2010).

Concluding remarks

In spite of the accepted importance of archaeological remains and Turkey being proud of having some of the most renowned sites in the world, the dilemma in site preservation and development is acute. As has been noted in some detail above, there is no near-to-complete inventory of sites in Turkey. There is a tendency to consider what is already exposed and open to visitors as being more than sufficient. Thus, any new archaeological site within a construction area is considered a burden to be overlooked unless there is insistent pressure at local or international level, the most effective being foreign developers that have to abide by the conventions in their own countries. Besides the problems due to the conceptual approach of the state agencies, evidently there are other problems that hamper efficient implementation of preventive archaeology: the lack of an archaeological inventory, the scale of archaeological sites, deficiencies in the legal system, and bureaucracy are among them.

Addressing certain concerns, such as ameliorating legislative and bureaucratic procedures, could be possible; however, how the management of settlement mounds tens of metres high should be tackled is difficult to resolve.

References

Ahunbay, Z. & İzmirligil, Ü. (eds) 2006: *Management and Preservation of Archaeological Sites*. ICOMOS Türkiye, YEM Yayınları, İstanbul.

Atalay, S., Çamurcuoğlu, D., Hodder, I., Moser, S., Orbaşlı, A. & Pye, E. 2010: Protecting and Exhibiting Çatalhöyük, *TÜBA-KED* 8, 155–66.

Başgelen, N. (ed.) 2002: *Birecik-Suruç Türkiye Kültür Envanteri Pilot Bölge Çalışmaları 1/1* [Turkish Academy of Sciences, Cultural Sector Initiative, Cultural Inventory of Turkey Birecik – Suruç Pilot Project], TÜBA TÜKSEK Yayınları, İstanbul.

Çambel, H. 1993: Das Freilichtmuseum von Karatepe-Aslantaş, *Istanbuler Mitteilungen* 43, 495–509.

Çığ, M. İ. 1993: Mustafa Kemal Atatürk und die Archäologie in der Türkei, *Istanbuler Mitteilungen* 43, 517–26.

Derin, Z. 2010: İzmir Yeşilova Höyüğü'nde Yeni Bir Eğitim Yöntemi: Zaman Yolculuğu / New Educational Methods at İzmir-Yeşilova: Time Travel (with English summary). *TÜBA-KED* 8, 263–374.

Easton, D. F. 1994: Priam's Gold: the Full Story. *Anatolian Studies* XLIV, 221–43.

Eldem, E. 2004: An Ottoman Archaeologist Caught Between Two Worlds: Osman Hamdi Bey (1842–1910), in D. Shankland (ed.): *Archaeology, Anthropology and Heritage in the Balkans and Anatolia: The Life and Times of F. W. Hasluck, 1878–1920*, ISIS Press, İstanbul, 121–49.

Eldem, E. 2011: From Blissful Indifference to Anguished Concern: Ottoman Perceptions of Antiquities, 1799–1869, in Z. Bahrani, Z. Çelik & E. Eldem (eds): *Scramble for the Past – A Story of Archaeology in the Ottoman Empire, 1753–1914*, SALT, Garanti Kültür A.Ş, İstanbul, 281–329.

Erder, C. (ed.) 1967: *Doomed by the Dam*. Middle East Technical University, Ankara.

Eres, Z. (ed.) 2010: Özel Konu: Türkiye'de Tarihöncesi Kazı Alanlarında Koruma ve Sergileme Çalışmaları / Special Section: Protection and Public Display of Excavated Prehistoric Sites in Turkey (with English summary). *TÜBA-KED* 8, 101–300.

Eres, Z. 2013: Managing the Cultural Heritage of the Istranca Region, in H. Angelova & M. Özdoğan (eds.): *Where are the Sites? Research, Protection and Management of Cultural Heritage. International Conference Centre for Underwater Archaeology, 5 8 December 2013, Ahtopol*, 155–196, http://issuu.com/vasilantonov/docs/conference_proceedings (accessed 11.10.2015).

Eres, Z., 2014: Conservation and Site Management Studies in the 20th Year of the Kırklareli Aşağı Pınar and Kanlıgeçit Excavations, in M. Bachmann, Ç. Maner, S. Tezer & D. Göçmen (eds): *Heritage in Context*, Miras 2, Deutsches Archäologisches Institut Istanbul, Ege Yayınları, İstanbul, 89–130.

Eres, Z. & Yalman, N. 2013: National Concerns in the Preservation of the Archaeological Heritage within the Process of Globalization: A View from Turkey, in P. F. Biehl & C. Prescott (eds): *Heritage in the Context of Globalization Europe and the Americas*, Springer, 33–41.

Karul, N., Avcı, M.B., Deveci, A. & Karkıner, N. 2010: Bursa Aktopraklık Höyük'te Kültür Sektörünün Oluşturulması: Çok Yönlü Bir Proje / Developing Cultural Sector at the Bursa Aktopraklık Mound: A Multifaceted Project (with English summary). *TÜBA-KED* 8, 241–62.

Kızıltan, Z. & Polat, M. A. 2013: The Neolithic at Yenikapı Marmaray-Metro Project Rescue Excavations, in M. Özdoğan, N. Başgelen & P. Kuniholm (eds): *The Neolithic in Turkey. New Excavations & New Research, Vol. 5: Northwestern Turkey and Istanbul*, Archaeology and Art Publications, İstanbul, 113–65.

Kuban, Z. 2013: Zu Diensten: Carl Humann. Der Archäologische Entrepreneur. *Der Anschnitt. Anatolian Metal* VI, 71–80.

Martin, L. 2013: Max von Oppenheim Diplomat, Orientalist und Ausgräber? *Der Anschnitt. Anatolian Metal* VI, 57–70.

Neve, P. 1998: Restaurierungen in Boğazköy-Hattuşa, in G. Arsebük, G. Mellink & W. Schirmer (eds): *Light on the Top of the Black Hill – Studies Presented to Halet Çambel*, Ege Yayınları, İstanbul, 515–30.

Özbaşaran, M., Duru, G., Teksoz, D. & Omacan, S. 2010: Yaşayan Geçmiş: Aşıklı Höyük / The Living Past: Aşıklı Höyük (with English summary). *TÜBA-KED* 8, 215–28.

Özdoğan, M. 1998: Ideology and Archaeology in Turkey, in L. Meskell (ed.): *Archaeology Under Fire. Nationalism, Politics and Heritage in the Eastern Mediterranean and Middle East*, Routledge, London, 111–23.

Özdoğan, M. 1999: Preservation and Conservation of Prehistoric Sites. Two Experimental Cases: Çayönü and Kırklareli-Aşağıpınar, in M. Korzay, N.K. Burçoğlu, Ş. Yarcan & D. Ünalan (eds): *International Conference on Multicultural Attractions and Tourism* I, Boğaziçi Üniversitesi Yayınları, İstanbul, 179–95.

Özdoğan, M. 2000: Cultural Heritage and Dam Projects in Turkey: An Overview, in S. A. Brandt & F. Hassan (eds): World Commission on Dams – Dams and Cultural Heritage Management Final Report August 2000, 58–61, http://www.eca-watch.org/problems/asia_pacific/china/culture_dams_forWCD2000.pdf (accessed 11.10.2015).

Özdoğan, M. 2004: Heritage and Nationalism in the Balkans and Anatolia or Changing Patterns: What has Happened Since Hasluck?, in D. Shankland (ed.): *Archaeology, Anthropology and Heritage in the Balkans and Anatolia: The Life and Times of F. W. Hasluck, 1878–1920*, ISİS Press, İstanbul, 389–405.

Özdoğan, M. 2006: Organizing Prehistoric Sites as Open Air Museums. Two Experimental Cases: Çayönü and Kırklareli Aşağı Pınar, in Z. Ahunbay & Ü. İzmirligil (eds): *Management and Preservation of Archaeological Sites (4th Bilateral Meeting of ICOMOS Turkey ICOMOS Greece)*, Side Foundation for Education Culture and Art, İstanbul, 50–7.

Özdoğan, M. 2007: Amidst Mesopotamia-Centric and Euro-Centric Approaches: The Changing Role of the Anatolian Peninsula between the East and the West. *Anatolian Studies* 57, 17–24.

Özdoğan, M. 2010: Some Considerations on Current Trends in Cultural Heritage: A View from Turkey, in A. Akdemir & O. Koç (eds): *2010 World Universities Congress. Proceedings I* (20/24 October 2010, Çanakkale, Turkey), Pozitif Matbaacılık, Ankara, 1361–7.

Özdoğan, M. 2013a: To Contemplate the Changing Role of Foreign Academicians in Turkish Archaeology. A Simple Narrative from Scientific Concerns to Political Scuffles. *Der Anschnitt. Anatolian Metal* VI, 35–40.

Özdoğan, M. 2013b: Dilemma in the Archaeology of Large Scale Development Projects: A View from Turkey, *Papers from the Institute of Archaeology (PIA)* 23/1, 1–8.

Özdoğan, M. 2014: In Memoriam Halet Çambel (19162014), *Paléorient*, 40/1, 9–11.

Özdoğan, M. & Eres, Z. 2012: Protection and Presentation of Prehistoric Sites. A Historic Survey from Turkey, *Origini* XXIV/V, 471–87.

Rose, M. & Aydıngün, Ş. 2007: Under Istanbul. One of the Largest Urban Excavations in History Exposes a Massive Byzantine Port, *Archaeology* 60/4, 34–40.

Seeher, J. & Schachner A. 2014: Boğazköy/Hattusa – Fifty Years of Restoration and Site Management, in M. Bachmann, Ç. Maner, S. Tezer & D. Göçmen (eds): *Heritage in Context*, Miras 2: Deutsches Archäologisches Institut Istanbul, Ege Yayınları, İstanbul, 131–58.

Tanyeri-Erdemir, T. 2006: Archaeology as a Source of National Pride in the Early Years of the Turkish Republic, *Journal of Field Archaeology* 31/4, 381–93.

Trümpler, C. (ed.) 2010: *Das Grosse Spiel. Archäologie und Politik zur Zeit des Kolonialismus (1860–1940)*. Ruhr Museum, Essen und DuMont Buchverlag, Köln.

Tuna, N. & Öztürk, J. (eds) 2001: *Salvage Project of the Archaeological Heritage of the Ilısu and Carchemish Dam Reservoirs, Activities in 1999*. TAÇDAM Middle East Technical University, Ankara.

9 | Everything you always wanted to know about commercial archaeology in the Netherlands

Marten Verbruggen

Abstract: From 1995 onwards, commercial archaeology was informally introduced in the Netherlands. The introduction went hand in hand with the implementation of a number of principles of the Valletta Convention, such as the 'polluter pays' principle and that of a direct interaction between archaeology and spatial planning. The new system, only incorporated in law in 2007, was a reaction to the failing system of archaeological heritage management in the previous period.
In 2011 the Dutch implementation model of Valletta was evaluated positively by an independent research bureau: the policy of preservation *in situ* has proved fruitful, and the publication rate of what is excavated is very high. However, recently there is concern over declining prices and their effect on research quality.

Keywords: commercial archaeology, Netherlands, Valletta Convention

Introduction

In 2007 the Netherlands formally incorporated a number of the principles of the Valletta Convention into Dutch law. These principles were: better protection of the archaeological heritage; direct interaction between archaeology and spatial planning; and the 'polluter pays' principle. Moreover, the Netherlands chose to introduce market principles for archaeological services, which marked the start of commercial archaeology in the Netherlands.

In recent years, the Dutch implementation model of Valletta has been discussed by several authors: Kristiansen (2009), Van Den Dries (2011), and Willems (2014). These discussions focus mainly on the advantages and disadvantages of the model, viewed through the eyes of the archaeologist working at a university or with the government, where the main focus is on producing knowledge. In this article the author sheds light on the workings of the Dutch model from the perspective of the commercial archaeologist, where archaeology, developer, society and the political-economic context play an equal role. The author witnessed the changes which have taken place in archaeological heritage management in the last 20 years from nearby. Firstly as a teacher at Leiden University, secondly as a member of staff of the Dutch State Service for Archaeology (ROB), and presently as the managing director of RAAP Archeologisch Adviesbureau, the 30-year-old and biggest commercial unit of the Netherlands.

Since a commercial archaeologist is used to producing bite-sized chunks within a limited amount of time, the author has chosen to shape this article like a self-interview.

When and why was commercial archaeology introduced in the Netherlands?

Formally, commercial archaeology was introduced with the acceptance of the Archaeological Heritage Management Act in 2007. A number of companies, however, could already excavate from 1995 onwards using permits issued by universities or ROB. Rick van der Ploeg, Secretary of State for Culture, worded the introduction in 1999 at the Inaugural Meeting of the Europae Archaeologiae Consilium as follows (Van der Ploeg 2000):

'I have decided that the existing potential market for archaeological services shall be opened up, but when I talk about "cultural entrepreneurship", I don't intend this to be a simple matter of privatisation and introducing economic competition. What I want to achieve is not that irreplaceable cultural heritage is dealt with as cheaply and rapidly as possible. I do want, however, to stimulate the archaeological community to work cost-effectively, to think about the quality of their work and how to improve it in such a way that the end-result is an improvement in the way the heritage is being dealt with and at the same time no undue burden is placed on those that have to pay for it all. Competition stimulates creativity and generates new ideas, it provides a pressure cooker that is not functioning when all is state-controlled.'

The amendment of the Heritage Act in 2007, therefore, was based on a practice that had already existed for several years. The reason why the government allowed, and even encouraged, this situation can be found in the economic and political developments of the time. Ironically, the introduction of market principles is a direct result of a treaty signed in 1992. No, not the one of Valletta, but the Treaty of Maastricht, which was signed only a few weeks later. For it was in Maastricht that the Economic and Monetary Union (EMU) would be formed, together with provisions for the creation of a common currency. For that reason tight budget discipline was agreed upon, including the maximum allowed budget deficit (the famous 3%).

Around 1990, the Netherlands suffered high unemployment, did not meet the EMU-qualifications by far, and therefore had to get its finances in order. One of the methods used to achieve this was by introducing market elements into fields that before had been a government responsibility – a process that started in 1994 and is continuing even today. One could therefore argue that the archaeological contents of the Dutch implementation model are determined by the Valletta Convention, but the styling is strongly influenced by the Treaty of Maastricht.

In the same speech, Van der Ploeg announced a number of other changes in the system of archaeological heritage management which also sprang from the political-economic situation of the day: decentralisation and deregulation. The responsibility for archaeological heritage moved from national government to local government, and a direct interaction between archaeological heritage management and the process of spatial planning had to be established.

The introduction of market principles in Dutch archaeology cannot be separated from the failing policies of the national government in the previous period. In this period of reconstruction after the Second World War, the building of houses, infrastructure and the agricultural land consolidations destroyed the archaeological heritage on a large scale. Between 1950 and 1990, 30% of the archaeological information in the soil was lost (Groenewoudt et al. 1994), and of the archaeology which was excavated, 50% was neither analysed nor published (Hessing & Mietes 2003).

These were frantic times for rescue archaeology: there was a total lack of financial means and capacity, and there was no political and social interest in solving this problem. Van Dockum and Willems (1997) described this period by using a Dutch saying which roughly translates as 'mopping a floor with the tap still running', i.e. a waste of time and effort, and instead of investing in more mops it was decided to try to get some hands on the tap.

How would you characterise the Dutch implementation model of Valletta?

Kristiansen (2009) calls the Dutch model a 'capitalist model' on the assumption that archaeology as a whole is left to the free market. In reality, market elements were only introduced in operational archaeological research. The Netherlands deliberately chose to lay the care for heritage with municipalities, thereby making it a public responsibility. Both policy and all choices about the nature and scale of the archaeological research funded by developers, therefore, are the task of the local authorities.

However, for reasons of capacity and efficiency an open market has been created for conducting research, which is supervised by the Ministry of Education, Culture and Science (OCW). For this purpose, quality standards have been established throughout the field: the Dutch Archaeological Quality Standard, to which all contractors have to comply. These standards determine *how* the research has to be conducted, and, more specifically, the process and the archaeologists who are allowed to conduct the separate steps in the process. Special specifications apply to research publications. These contain the requirements for the manner in which the prime research documentation has to be supplied, which chapters the report has to contain, and the requirement that all specialist sub-reports have to be included in full. Each report has to submitted digitally by uploading it to ARCHIS (the Dutch archaeological database) and the National Library of the Netherlands in The Hague. Also, a paper version has to be supplied to the archaeological depot, together with the finds. In this manner, *all* publications are accessible to the scientist and the public. The nature and scale of the research, and thereby to a large extent the costs, are determined by the municipality on whose soil the excavation takes place. Figure 9.1 shows the different roles in this process and how they interlink.

The process commences when a developer applies for planning permission with the local authority. The local authority can impose archaeological research on the developer, including the manner in which the research must be conducted, and which scientific questions have to be answered in the report.

These requirements can be found in the project requirements. These conditions are stipulated in the project requirements specified by the local authority. The developer then puts the research out to tender and choses a contractor. Obviously, the developer wants a low price, but also wants research work that will be approved, since after its completion the developer must submit the archaeological report to the local authority, which will subsequently verify it against the project requirements. After approval, the developer has fulfilled his application requirements. The contractor hands in the final report to the Minister for Culture (ARCHIS), in accordance with the Heritage Act, and deposits the archaeological material at a designated archaeological depot. A striking yet important detail is that there is no formal relation between contractor

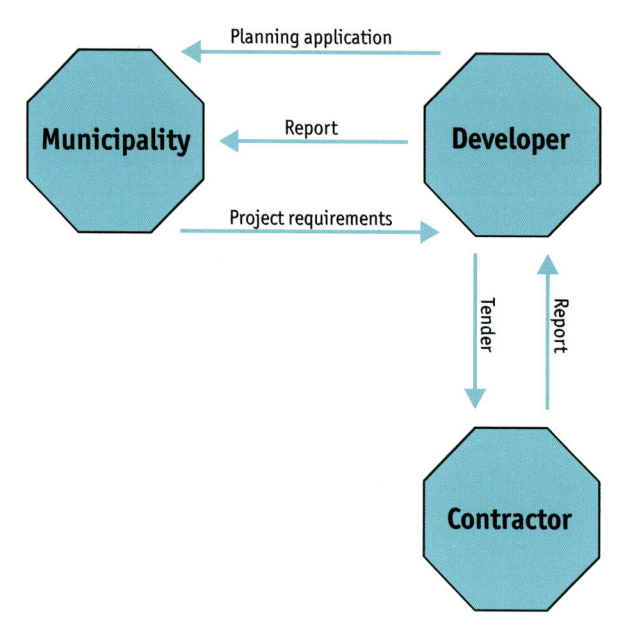

Figure 9.1: Roles in Dutch archaeological heritage management, © RAAP Archeologisch Adviesbureau.

and municipality. The archaeological process runs via the developer, unless the developer authorises the contractor to maintain direct contact with the local authority.

To summarise, I would call the Dutch model neither socialist nor capitalist, but a hybrid, where the authority lies with the municipality and the execution of the archaeological research is to a large extent in private hands. In Van der Ploeg's vision, the choice for private contracting is solely rooted in economic motives. It is an effective and efficient means of fulfilling the government's responsibility concerning archaeology, without taking up production itself.

What did the introduction of market elements stir up?

By opening up the market, a quickly growing sector has evolved in which private companies, and also municipal archaeology departments and a number of universities, operate. Unfortunately, there are no reliable annual figures of the total turnover of the commercial market and the number of staff, so we have to rely on sources that deal with different aspects of the market.

Van Den Dries et al. (2010) provide an elaborate analysis of the turnover of the Dutch commercial sector in 2008, which amounted to €50 million and approximately 600 employees. A second source (Van Londen et al. 2014) estimates the number of staff in the commercial market at approximately 500. The decline may be explained by the effects of the economic crisis, which resulted in a wave of redundancies in the commercial sector. This has led to a major increase in the number of self-employed archaeologists, as a result of which the number of companies has risen to well over 100. Besides these companies, there is also a group of municipal archaeologists with a combined turnover of approximately €24 million and 247 jobs. Finally, 4 universities hold an excavation permit, but their share in the commercial market is nil. These universities, however, have set up commercial units, which are regarded to belong to the commercial sector.

The commercial market has led to an innovation spree in Dutch archaeology. Sander van der Leeuw (2005) describes this innovation not so much as the development of entirely new methods, but rather as the introduction of already existing methods in archaeology. This means not just the adaptation of existing techniques, but also the integration thereof in the archaeological routine in a commercial fashion. In this manner, surveying techniques, coring, resistivity and magnetometer surveying and automated phosphate analysis entered archaeology early on, quickly followed by robotic total stations and GPS, field computers with databases and digital terrain models (Figure 9.2).

In 2014 a Dutch exchange standard for the documentation of archaeological excavations was

Figure 9.2: A commercial excavation with the use of a robotic total station and field computer (2008, Maastricht-Aachen Airport), © RAAP Archeologisch Adviesbureau.

introduced, through which companies, archaeological depots and universities could share and re-use each other's data more easily. The most important change caused by the introduction of commercial archaeology, however, was the change from commitment to obligation when contracting archaeological research. Where previously a sum of money had been handed over to finance an excavation, now the archaeological process was cut up into segments and the funding was made conditional on the progress of each segment. The final 10–20% was only paid after completion of the publication. A non-commercial archaeologist can hardly envisage the disciplinary effects this has. No publication means no invoice, no invoice means no income, and no income means no salary! The result of this change was that for the bulk of the archaeological companies the publication rate reached nearly 100%. Just compare that to the earlier situation of mere commitment, where money had run out before one could even start producing a publication of the results.

Obviously, introducing market principles has also lead to strong competition. In the early years, up to about 2009, there were clear signs of an emerging market in which most companies could make a profit. Mind you, not a big profit, for not a single company struck it rich in archaeology. In 2010 the effects of the economic downturn hit Dutch archaeology with full force. Four companies went bankrupt and the equity capital of the 13 largest companies evaporated into thin air. Based on public accounts of the Dutch Chamber of Commerce, at the end of 2013 the average equity of these companies amounted to a mere €215,000 per company. It is feared that in the next two years several companies will call it a day.

The effect of this competition is that the price of archaeological research, especially fieldwork, has dropped a few dozen percentage points since 2010. Whether the drop in price has had an effect on the quality of the research has not (yet) been examined. In my personal opinion this is only true in a limited amount of cases. After all, every report is verified by the local authorities against the project requirements and the Dutch Archaeological Quality Standard. Moreover, I highly respect the ethics of the archaeologists, whether employed by a private company or by a municipal department. After all, their names will be forever tied to the publication that will be read by many of their peers. They will do anything within their power to produce a proper product.

Is the Dutch implementation model doing its job?

That very much depends on who you ask. Scientists are, of course, disappointed in the scientific quality of the developer-funded reports. The farming community and the building sector regularly grumble about the high costs of excavations and think that the joys and burdens are unequally, and thus unfairly, shared. After all, they have to pay for something everyone benefits from. In my personal opinion, however, this is all part of the deal. No one wants to individually pay for a public good, that is why it is regulated by law. A law that has to unite a number of conflicting interests: better heritage protection, a financial stimulus (for the 'happy' few) for *in situ* preservation, sound scientific research, and, if feasible, all this at the lowest possible cost. How could it possibly please everyone?

Be that as it may, figures show that the Dutch model works. In 2011 the independent research bureau RIGO extensively evaluated the workings of the Archaeological Heritage Management Act of 2007 (Keers et al. 2011). In his letter to the Dutch parliament, the minister of Education, Culture and Science (Zijlstra 2012) endorsed the conclusion of the researchers that better protection of the archaeological heritage has proved possible due to the present legislation. In spatial planning, archaeology is taken into account more often, and more archaeological finds are preserved *in situ* due to the introduction of the developer-funded principle. Subsequently, he handed out a few gifts, among which was the funding of a number of retrospective studies whereby developer-funded studies are reprocessed into a synthesis. So far so good then; one might even call it a success story.

Last question: you feel you have an extensive knowledge of commercial archaeology in the Netherlands. Is there something you do not know?

Unfortunately, yes. Van der Ploeg facilitated a market for archaeological services on the assumption that it would enhance innovation and efficiency. His economic outlook on the matter was a major shock for archaeologists who at the time were working for his own department, the ROB, which openly opposed the marked principle (ROB 1995, 39). The opponents were of the opinion that archaeology could not be left to the market, because the customer is not really interested in the product, and the market was created by law. In my opinion, Van der Ploeg proved to be absolutely right. The market is innovative, works efficiently and the customers get what they pay for. Moreover, far fewer archaeological sites are unknowingly lost, and the publication rate is very high. To bring back to memory the simile of the open tap and the mop: the tap has been closed half way and a wet-and-dry vacuum cleaner has been bought. In this manner, the market – as an instrument – is of important social importance. It works!

And yet there is concern in the commercial sector, most importantly, and for many years, over declining prices. I do not know whether this is caused by the economic crisis, and therefore is temporary in nature, or whether it is inherent to the archaeological market and is presently masked by the economic crisis, in which case there is a weak spot in the archaeological market.

I will elaborate on what I mean. Firstly, the market sector is extremely fragmented. There are approximately 35 licensed companies with between 5 and 100 members of staff, which all offer the same products (Links 13). Secondly, the level of quality is set because it is outlined in the project requirements and the Quality Standard. Price has, therefore, become the only distinctive bench-mark for the developer. It will be obvious that there is a major demand for the lowest price. Presently, it seems that most companies choose to uphold a high level of quality, despite the low prices, by eating into their equity. When this situation prevails for too long, however, quality will fall to the minimum level formulated in the project requirements, and which the authorities are prepared to enforce. In the long run, the quality could thereby wither away, since not all quality aspects can be put down on paper beforehand.

Optimists estimate that the decline in prices is the sole result of the economic crisis. If that is the case, a second restructuring in the commercial sector will have to bring supply and demand back into balance in order for the profits of the companies to return to an acceptable level. Pessimists, on the other hand, recognise the effects of the crisis, yet believe there is a flaw in the make-up of the archaeological market. When the pessimists are proven right and it is indeed difficult for the developer to measure the quality of the archaeological research the authorities will have to intervene in the system by enforcing the quality aspect more stringently. Fortunately, that very well fits a hybrid system in which the authorities remain responsible for the system of archaeological heritage management. In that case, however, it is advisable to place the monitoring of the quality level in private hands.

References

Groenewoudt, B., Hallewas, D.P., & Zoetbrood, P.A.M. 1994: *Interne Rapporten no. 13*, ROB, Amersfoort.

Hessing, W.A.M., & Mietes, E.K. (ed) 2003: *Project Odyssee: een zoektocht naar de achterstand in uitwerking van archeologisch onderzoek in Nederland*.

Keers, G., Van der Reijden, H., Van Rossum, H., 2011: Ruimte voor archeologie. Themaveldrapportages evaluatie Wamz. RIGO-rapport P18090. RIGO Research en advies BV, Amsterdam.

Kristiansen, K. 2009: Contract archaeology in Europe: an experiment in diversity, *World Archaeology* 41(4), 641–48.

ROB, 1995: Het Verleden Zeker.

Van Den Dries, M. 2011: The good, the bad and the ugly? Evaluating three models of implementing the Valetta Convention, *World Archaeology* 43(4), 594–604.

Van Den Dries, M.H., Waugh, K.E., & Bakker, C. 2010: A crisis with many faces. The impact of the economic recession on Dutch archaeology, in N. Schlanger & K. Aitchison (eds), *Archaeology and the economic crisis*, 55–68.

Van der Leeuw, S.E., 2005: Why Dutch archaeology isn't what it was forty years ago, in M.H. Van den Dries & W.J.H. Willems (eds), *Innovatie in de Nederlandse archeologie*, 7–22.

Van der Ploeg, F. 2000: The management of the archaeological heritage, in J.W.H. Willems (ed.), *Challenges for European Archaeology*, 45–51, Zoetermeer.

Van Dockum, S. & Willems, W.J.H. 1997: *Laag voor laag: de kracht van complementair bestuur in de archeologische monumentenzorg*, Den Haag.

Van Londen, H., Vossen, N., Schlaman, M. & Scharringhausen, K. 2014: *Discovering the archaeologists of the Netherlands 2012–14*, University of Amsterdam.

Willems, J.H. 2014: Malta and its consequences: a mixed blessing, EAC Occasional Paper No. 9.

Zijlstra, H., 2012: Kamerbrief 07-02-2012/OCW

Websites
Link 1: www.voia.nl/ (accessed 09.09.2015).

Link 2: www.opgravingsbedrijven.nl/ (accessed 09.09.2015).

Link 3: www.SIKB.nl (accessed 09.09.2015).

Session 2

Balancing stakeholders

The 'Time Stairs' in the underground car park at Rotterdam Markthal: archaeological finds displayed at the levels at which they were excavated (from the Dutch case study, see Wesselingh).
Photo: Bas Czerwinski

10 | Scotland and a 'national conversation'

Rebecca H Jones

Abstract: In 2014, Scotland had a national conversation about its place in the United Kingdom, with a referendum on independence that saw a majority in favour of staying in the political union. 2015 marked an important year for archaeology in Scotland, with a year-long celebration of archaeology, the European Association of Archaeologists' annual meeting and the launch of Scotland's first Archaeology Strategy.

Keywords: Scotland, archaeology, strategy, policy, engagement

Introduction

The reality of archaeology in Scotland in recent years has seen the same trends as elsewhere in the UK and Europe: a reduction in the amount of developer-led archaeology undertaken due to the economic downturn and a loss of skills as archaeologists turned to alternative employment. Whilst things have been improving in recent years, there is recognition of a potentially looming 'skills gap' as archaeologists retire in the coming few years with a reduction of numbers to replace them. Yet archaeology remains extremely popular, buoyed by numerous media productions, with a high level of community interest, which is continuing to be developed through a range of projects, both locally and nationally. 2015 is being seen as a pivotal year in Scotland: Scotland's Archaeology Strategy was launched by Fiona Hyslop MSP, Cabinet Secretary for Culture, Europe and External Affairs, at the European Association of Archaeologists' annual meeting in Glasgow in September, all during a year-long celebration of archaeology (*Dig It! 2015*), led by the Society of Antiquaries of Scotland and Archaeology Scotland, with over 100 partners the length and breadth of Scotland (Figure 10.1). In 2017, there will be a government focus year on History, Heritage and Archaeology, providing further opportunities to celebrate archaeology, hopefully building on successes in 2015.

Policy and strategy context

The publication of Scotland's Archaeology Strategy fits into a wider strategic landscape. In 2014, Scotland's first Historic Environment Strategy – *Our Place in Time* – was published. This high-level framework, developed collaboratively across the historic environment sector, sets out a ten-year vision intended to ensure that the cultural, social, environmental and economic value of Scotland's historic environment continues to make a strong contribution to the wellbeing of the nation and its people. A number of committees and working groups have been set up to look at the delivery of the strategy, and the overarching high-level Historic Environment Forum, comprising senior stakeholders, is chaired by the Cabinet Secretary for Culture, Europe and External Affairs, demonstrating the importance that the Scottish government places on the historic environment.

The Historic Environment Strategy itself arose from a desire by the sector to have a clear strategy that all parties could get behind in the challenging situation in which we find ourselves, with the major impact not only of the financial downturn but also of climate change. These discussions themselves arose out of an evidence-based review of the policy context for the historic environment.

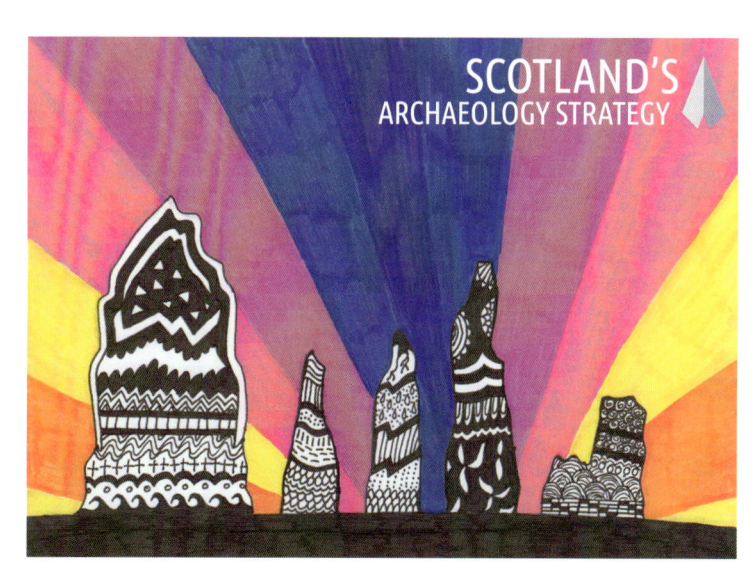

Figure 10.1: Postcard produced to promote Scotland's Archaeology Strategy – a winning drawing by Darcey Axon of the Callanish Stones from the 'Dig Art! 2015' competition run as part of *Dig It! 2015*. This competition inspired people of all ages to create artistic responses to Scotland's past.

Policy for archaeology in Scotland is set out in the Scottish Historic Environment Policy (2011) and the Scottish Planning Policy (2014). Both set out the policy framework that informs the work of a wide range of public sector organisations. Archaeologists based in Historic Scotland, a government agency, deal with the designation of nationally important sites (Scheduled Monuments) and consent for work undertaken on the list of over 8,000 scheduled sites. Other designations include Listed Buildings, Historic Marine Protected areas, and the Inventories of Gardens and Designed Landscapes and Historic Battlefields. All involve close liaison with other interested parties and owners. Heritage Management also includes the co-ordination of Scotland's World Heritage Sites.

In local government, archaeologists advise planners on archaeological works that need to be undertaken in response to development, as well as providing local expertise in a wide range of areas, from community archaeology to rural and forestry concerns. They maintain Historic Environment Records, and a Strategy for Scotland's Historic Environment Data (SHED) was developed across the sector and launched in 2014. The key aim for this strategy is to work in partnership in order to protect, promote and enhance Scotland's historic environment through coordinated activity to improve the data that underpins decision-making and research, and the associated systems and processes.

Separate to these developments, Museums Galleries Scotland, the national development body for the museum sector in Scotland, created a National Strategy for Scotland's Museums and Galleries – *Going Further* – in 2012.

Recognising this strategy landscape, the Archaeology Strategy, developed in response to a review of the Historic Scotland Archaeology function in 2012, attempts both to support delivery of the Historic Environment Strategy, and also articulate with the Museums Galleries Scotland Strategy, recognising that the products of archaeological activity almost always end up in museums, and that this is providing huge challenges to the museum sector in a time of diminishing resources.

The creation of Historic Environment Scotland

In 2012, the Cabinet Secretary, Fiona Hyslop MSP, announced that Historic Scotland, a government agency, and the Royal Commission on the Ancient and Historical Monuments of Scotland (RCAHMS), an independent non-departmental organisation financed by the Scottish Parliament, responsible for recording, interpreting and collecting information about the historic environment, would merge to form a new body: Historic Environment Scotland. Created by the Historic Environment Scotland Act 2014, the new body is a non-departmental public body with a Board established early in 2015 and with staff and functions all transferred on 1 October 2015. Historic Environment Scotland will lead on *Our Place in Time*, and continue a broad range of functions, including the statutory role as regulator and advisor to Scottish Ministers, as well as maintaining a portfolio of 345 Properties in Care on behalf of Scottish Ministers. The creation of Historic Environment Scotland is being seen as 'the most significant reorganisation of Scottish archaeology in over 100 years' (Driscoll 2015).

Scotland in 2015

Scotland, as a nation, was global news in 2014. The referendum on independence from the United Kingdom, held in September, facilitated a national (and international) debate on what kind of Scotland people wanted to see, and the turnout (just under 85% of the voting population) was the highest recorded for an election or referendum in the United Kingdom for generations. That the majority (55%) voted to stay in the United Kingdom is well known, but the legacy of the campaign has been the continuation of an energised debate. An increased turn-out for the 2015 United Kingdom Parliament elections saw a massive swing to the Scottish National Party, which now holds 56 of the 59 Scottish seats in Westminster (London).

In the archaeology sector, the arrival in Glasgow of over 2,000 archaeologists and heritage managers from over 80 countries for the European Association of Archaeologists' annual meeting in September was one of the most significant events to happen in Scottish archaeology, giving us a platform to discuss and present our archaeology on a European (indeed, global) stage, situating our archaeological activities within the wider world and bringing benefits and ideas back to our day-to-day archaeological activities in Scotland.

Balancing stakeholders

At the *Europae Archaeologiae Consilium* meeting in Lisbon in March 2015, there was a discussion on how to balance the expectations of stakeholders, examining whether the 'delivery model for preventative archaeology is still a scientific endeavour, or just another pre-construction service'. Yet archaeology is 'the study of the human past through its material remains', and that study remains a scientific endeavour. Furthermore, our stakeholders are not only the developers and funders of the archaeological activity, but the public for whose benefit we seek to gather this knowledge, and whose taxes pay for some archaeologists and heritage managers. Moreover, the level of engagement in archaeology in Scotland is such that developer-led projects increasingly have communication and public engagement built into project designs.

But is 'preventive archaeology' just about planning and development? As noted earlier, climate change is having an impact on archaeology (Figure 10.2). We hold two positions within the climate change debate:

- firstly, our sites can be adversely affected by climate change itself, such as coastal erosion and increased precipitation and sometimes the mitigation measures introduced to conserve buildings and monuments;
- secondly, archaeology has much to tell us about past climate change impacts on the landscape and environment of Scotland.

Figure 10.2: Excavation of Salt Pans at Brora, Sutherland, by the SCAPE Trust and local community volunteers. This stretch of beach has been the subject of rapid coastal erosion. Crown Copyright HES, photograph taken in August 2011.

An increased number of storms is creating challenges in managing Scotland's fragile coastal heritage. The SCAPE Trust (Scotland's Coastal Archaeology and the Problem of Erosion), based at St Andrew's University with grant funding from Historic Scotland, is particularly focused on remains threatened by coastal erosion. Their award-winning Scotland's Coastal Heritage at Risk project encouraged volunteer citizen archaeologists to monitor, record and submit information about their local coastal heritage. Fieldwork projects and excavations are carried out with local volunteers and archaeology groups, resulting not only in an increased number of people with archaeological skills, but also a motivated, mobilised army of volunteers, keen to report sites to the Trust and to their local authority archaeological colleagues.

A conversation about archaeology

There is no doubt that climate change and responses to climate change are among the biggest challenges facing archaeology in Scotland. But of equal concern are the financial challenges which have already resulted in a reduction in the number of archaeologists employed in the public sector: in historic environment services and museums, locally and nationally.

The launch of Scotland's Archaeology Strategy, in September 2015, was intended to be part of an open conversation about archaeology's contribution to society in Scotland and the importance of situating our heritage in a global context. Delivery will focus around five strategic aims:

1. Delivering Archaeology to broaden and deepen the impact and public benefit of archaeology within and beyond Scotland
2. Enhancing Understanding: to increase knowledge, understanding and interpretation of the past
3. Caring and Protecting: to ensure that the material evidence of the human past is valued and cared for by society and managed sustainably for present and future generations
4. Encouraging Greater Engagement: to enable and encourage engagement with our past through creative and collaborative working, active involvement, learning for all ages and enhanced archaeological presentation
5. Innovation and Skills: to ensure that people have the opportunity to acquire and use the archaeological skills that they need or desire, and that those skills provide the underpinning for innovation in the understanding, interrogation, learning and funding of archaeology.

We want to be ambitious, but in the course of a national conversation about archaeology, we recognise that some difficult questions will need to be asked and answered. This is a conversation that will continue for years to come.

References

Driscoll, S. 2015: Welcome from the Organisers, in *EAA Glasgow Abstracts*, 67, http://eaaglasgow2015.com/wp-content/uploads/2015/07/EAA-Glasgow-Abstract-Book.pdf (accessed 01.09.2015).

Historic Scotland 2011: *Scottish Historic Environment Policy, December 2011*, Edinburgh.

Scottish Government 2014: *Our Place in Time. The Historic Environment Strategy for Scotland*, Edinburgh.

Scottish Government 2014: *Scottish Planning Policy*, Edinburgh.

Scottish Strategic Archaeology Committee 2015: *Scotland's Archaeology Strategy*, http://archaeologystrategy.scot/files/2015/08/ScotlandsArchaeologyStrategy.pdf (accessed 01.09.2015).

Sites and Monuments Records Forum Scotland 2014: *Scotland's Historic Environment Data Strategy*, http://smrforum-scotland.org.uk/wp-content/uploads/2014/03/SHED-Strategy-Final-April-2014.pdf (accessed 01.09.2015).

11 | The General Directorate of Cultural Heritage's competencies in the context of safeguarding and promoting the Portuguese archaeological heritage

Maria Catarina Coelho

Abstract: The General Directorate of Cultural Heritage (DGPC), established in 2012, is responsible for protecting the archaeological heritage of mainland Portugal. Its tasks include research, management, protection, preservation and dissemination of information about the country's archaeological resources. The DGPC strategy for the management and safeguarding of national archaeological heritage favours contact and dialogue between the various actors in society committed to the protection and promotion of archaeological heritage, as a result of which partnership agreements have been established with higher education institutions, as well as with local and regional institutions, as these play a vital role in improving awareness among the communities where they are based.

Keywords: Portugal, archaeological heritage, management, heritage dissemination, local communities

The General Directorate of Cultural Heritage is nationally responsible for safeguarding the archaeological heritage of the Portuguese mainland, ensuring its study, management, protection, preservation and dissemination (Link 1).

The ratification by the Portuguese state of the Valletta Convention in December 1997 initiated the process of adopting national legislation incorporating the precepts laid out therein, which led to the creation of the Portuguese Institute of Archaeology in the same year.

The Regulation of Archaeological Works published in 1999 (Decree 1999) was the first decree to incorporate the principles of the Valletta Convention regarding the concept of archaeological heritage, the different types of archaeological interventions and the definition of the requirements for its implementation.

In the last two decades, guided by the spirit of the Valletta Convention and following its ratification by the Portuguese state, there have been major changes in the national archaeological scene. An autonomous body was established to deal with archaeology management and archaeological heritage protection, accompanied by the creation of laws and regulations (Law 2001; Decree 1999; Decree 2000a; Decree 2014) that boosted growth in archaeological activity, the number of archaeologists and the development of private archaeological companies.

The creation of specific legislation has enabled the mandatory execution of archaeological fieldwork at the developer's expense prior to building and construction works, infrastructure implementation, large public or private projects, as well as rural land-use projects or other minor private works that take place within areas of archaeological sensitivity and thus are perceived to have a damaging impact on the surrounding heritage. This in turn has permitted the development of a policy of prevention and protection through the identification and recording of archaeological sites and remains (Bugalhão 2009).

The creation of specific legislation allowing for the designation of heritage status has also enabled the implementation of more effective protection for archaeological sites and their surroundings (Decree 2009a; Decree 2009b).

The protection of archaeological sites is also achieved by the use of buffer zones within land-use planning. These zones may include *non aedificandi* areas (where no building development is permitted). It is required that interventions at heritage sites are carried out by interdisciplinary teams in order to safeguard the diversity of heritage both in urban areas and in the countryside.

In 2000 a multidisciplinary research centre was created which is dedicated to the study of the past (Archaeosciences Laboratory, Link 1). Activities undertaken by this laboratory include various complementary disciplines. Its principal objective is to improve our understanding of the way of life of our ancestors – their economy, social organisation, culture and biology, as well as their relationship and interaction with the environment.

The adoption of the Valletta Convention recognised the importance of preventive archaeological activity and emergency interventions, as distinct from scientific and planned operations. Since then, the widespread application of the principles of preventive archaeology has led to an extraordinary increase in contract archaeology and the emergence of companies dedicated to carrying out archaeological work.

Order Archaeological Work Authorization

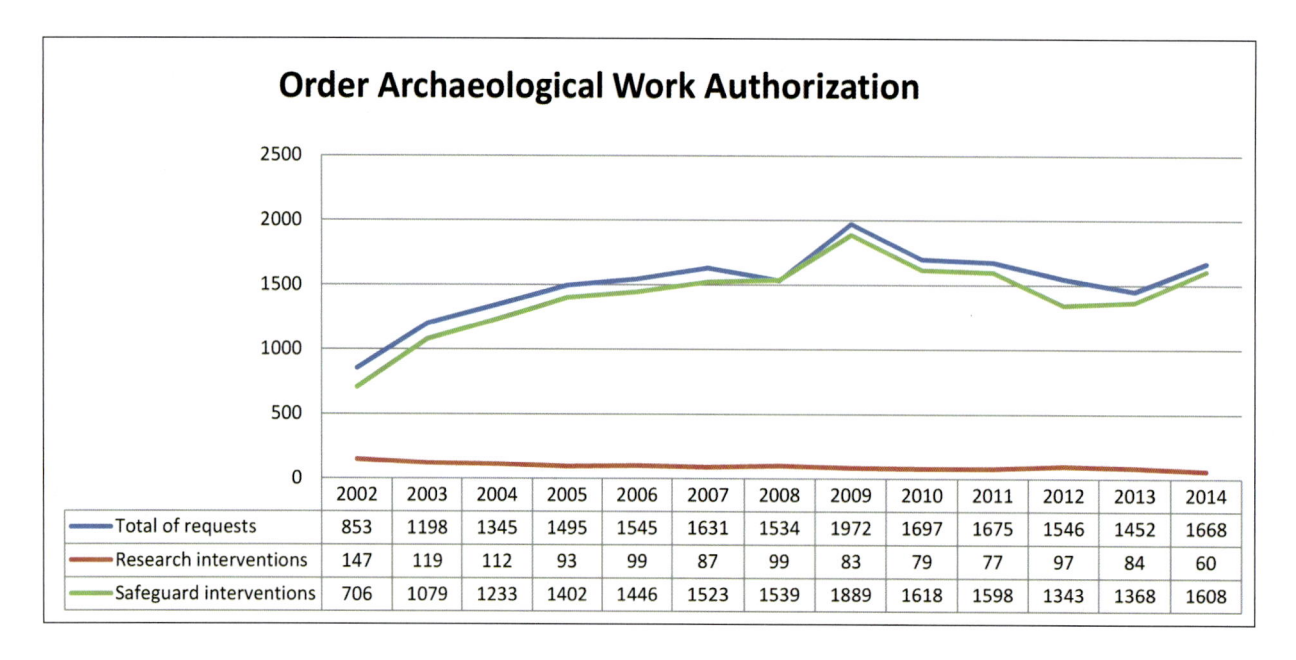

	2002	2003	2004	2005	2006	2007	2008	2009	2010	2011	2012	2013	2014
Total of requests	853	1198	1345	1495	1545	1631	1534	1972	1697	1675	1546	1452	1668
Research interventions	147	119	112	93	99	87	99	83	79	77	97	84	60
Safeguard interventions	706	1079	1233	1402	1446	1523	1539	1889	1618	1598	1343	1368	1608

Figure 11.1: Statistics for archaeological work permit applications over the past 12 years.

In 2014 alone, 1,697 applications for archaeological work permits were submitted to the DGPC, of which 99.78% concerned preventive interventions. These figures reflect less, in my view, the importance of consolidation of preventive archaeology in Portugal, and more the disinvestment recorded in archaeological excavations under the Multi-Annual Scientific Research Projects (Figure 11.1).

Evaluating the situation after almost 20 years of archaeological heritage management along the lines of Valletta highlighted the need to implement more advanced procedures to manage relevant data through proper use of digital media and the safeguarding and development of the Archive of Portuguese Archaeology, which integrates the documentary resources of different public institutions involved in the management of the archaeological heritage.

The decline in the number of archaeological interventions of a preventive nature stems from the decrease in public and private works recorded since 2011 in the context of the economic and financial crisis that occurred in Portugal.

The inclusion of archaeological heritage as one of the factors to take into consideration in an environmental impact assessment (EIA) has enabled well-thought-out and balanced decisions to be made regarding the viability of development projects. Carrying out the assessment requires the gathering of information, identification and prediction of impacts on the heritage and the formulation of measures to avoid, minimise or offset any potential negative effect of the proposed development's implementation.

The transposition into Portuguese law of the European directive on the assessment of the effects of certain public and private projects on the environment

Figure 11.2: Statistics for archaeological work permit applications relating to Environmental/Heritage Impact Assessments over the 12 years.

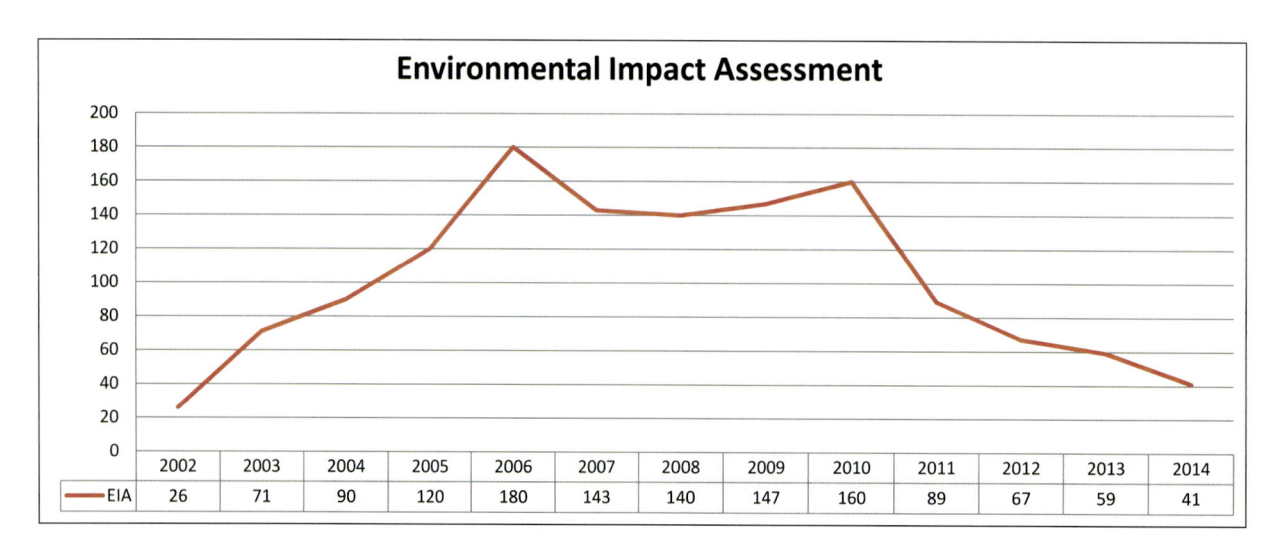

Environmental Impact Assessment

	2002	2003	2004	2005	2006	2007	2008	2009	2010	2011	2012	2013	2014
EIA	26	71	90	120	180	143	140	147	160	89	67	59	41

(EEC 1985; Decree 2000b), which approved the legal framework for EIAs of public and private projects likely to have significant effects on the environment, was a key preventive tool for sustainable development policy. It has become a fundamental instrument in preventing the adoption of environmental policy that could have a major negative impact on heritage asset protection.

Since its introduction, progressive and meaningful participation of the public bodies responsible for the management of cultural and archaeological heritage has been observed in the context of EIA procedures (Figure 11.2).

Today, archaeology is understood as a territorial resource, and archaeological – activity as a means of territorial management, achieved through an ongoing relationship between scientific activity and social participation. The aim of spatial planning is to promote efficient use of space and the responsible management of existing resources based on an interdisciplinary programme of study and planning, in which archaeology plays a key role through the integration and evaluation of heritage resources.

Indeed, in the current national urban and rural planning policy, implemented through a territorial management system built around territorial management instruments (TMI), the archaeological heritage is identified as a territorial resource that is relevant to the memory and identity of communities. The TMIs establish the measures necessary to protect and enhance this heritage, ensuring their integrity and determining the use of surrounding areas (Figure 11.3).

The new Regulation of Archaeological Works published in late 2014 imposed the adoption of new and efficient normative order compliance procedures and technical principles to be followed in carrying out archaeological work (Decree 2014). It also clarified policy regarding requirements to disseminate the results of archaeological work, produce scientific publications and engage in awareness-raising activities and heritage education.

Article 7 – **Application instructions**
1 – The application for authorisation to conduct archaeological work is accompanied by the following information and documents:
vii) Plan for public disclosure of archaeological work among the community;

Article 15 – **Content of reports**
1 – The final report contains the following elements:
p) Description of disclosure and publication actions, if any, to raise awareness and heritage education. (Decree 2014).

The DGPC strategy for the management and safeguarding of national archaeological heritage favours contact and dialogue between the various actors in society committed to the protection and promotion of archaeological heritage, as a result of which partnership agreements have been established with higher education institutions, as well as local and regional institutions, as these play a vital role in improving awareness among the communities where they are based.

The creation of a Portuguese national archaeological database – *Endovélico* – happened at a time when new heritage policies were being formulated in Portugal during the late 1990s (Neto et al., 2007) (Figure 11.4). These events were triggered by the contagious European spirit of heritage awareness and also by the impact that the Côa Valley Rock Art site had in Europe and around the world, bringing Portuguese archaeology to the forefront of newspapers worldwide (AA.VV, 2002; Bugalhão & Lucena, 2006). This database was especially developed to collate information on nationwide archaeological records, and it is regularly optimised and daily updated. *Endovélico* is accessible to all archaeologists, both from the public sector and private enterprises, but it is also available, with some restrictions, to the general public.

More recently, the *Portal do Arqueólogo* was developed as a digital platform to speed up archaeological

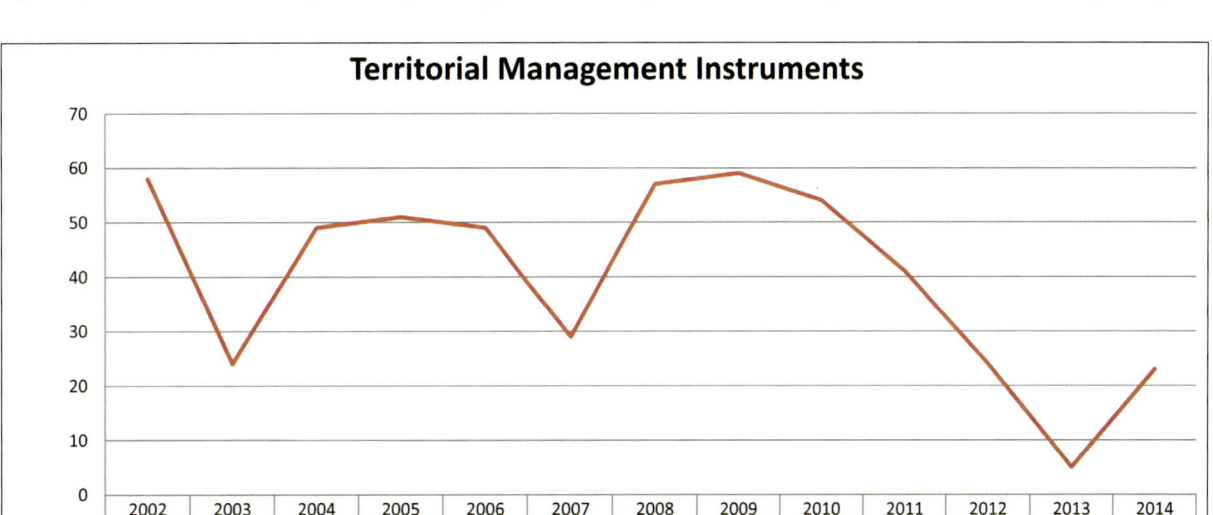

Figure 11.3: Statistics for archaeological work permit applications relating to Territorial Management Instruments over the past 12 years.

Territorial Management Instruments

	2002	2003	2004	2005	2006	2007	2008	2009	2010	2011	2012	2013	2014
TMI	58	24	49	51	49	29	57	59	54	41	24	5	23

Figure 11.4: There are 31,928 entries for archaeological sites in the Endovelico database (©DGPC).

effectively democratised procedures relating to the licensing of archaeological activity (Link 3).

As mentioned above, the new Regulation of Archaeological Works, recently approved in November 2014 (Decree 2014), promotes and encourages the dissemination of results of archaeological excavations of a preventive nature, in addition to their scientific publication, seeking to ensure that the archaeologists involved communicate these results to the public (Sousa 2013).

However, as early as 1998, to address the dissemination of scientific data, the *Revista Portuguesa de Arqueologia* (*Portuguese Journal of Archaeology*, Link 4) was created with the purpose of publishing partial or brief final reports of archaeological work results. A monographic series was also launched under the title *Trabalhos de Arqueologia* (*Archaeological Works*, Link 5), targeting the publication of monographs, including final reports of research projects, university theses and congress papers.

These publications enjoy a circulation that goes far beyond the borders of the country, since the papers are published and shipped to national and foreign libraries through an effective system of publications exchange, as well as being published online, thus making them accessible to millions of potential readers.

For several years now, the DGPC has annually devised a range of cultural heritage dissemination activities, some of them with a particular focus on archaeology. Examples include events held as part of the European Heritage Days, and the *Encontros com o Património* (Meetings with the Heritage) – a radio programme broadcast every Saturday morning for the past 7 years on one of the leading radio stations in Portugal (Figures 11.5–11.6).

licensing procedures, and to provide researchers and professionals with access to spatial data and brief descriptions of archaeological sites and works. Applications for archaeological work authorisations can also be submitted through this platform, which has

Figure 11.5: Promotional image of the radio programme Encontros com o Património (©DGPC).

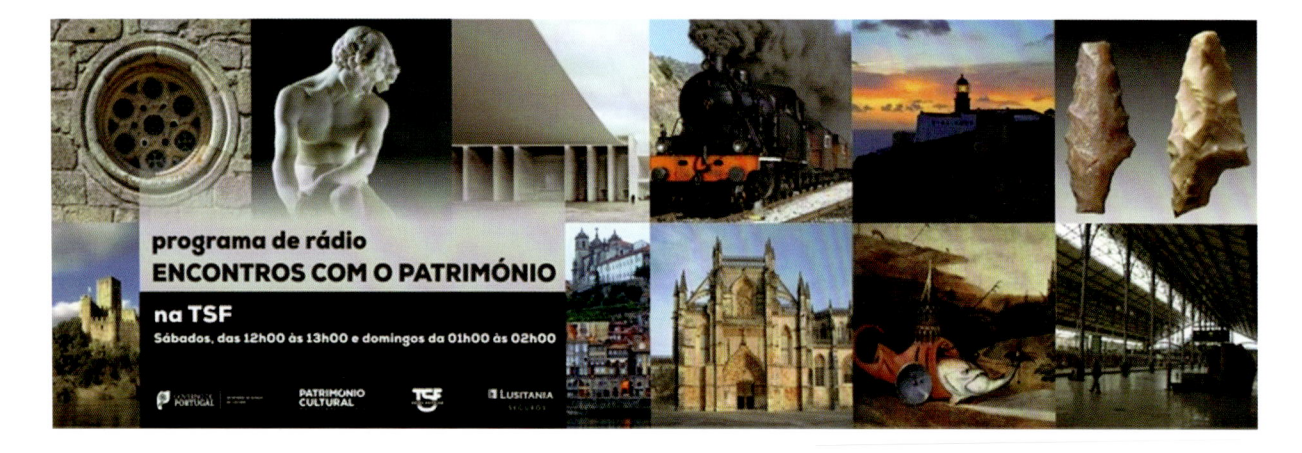

Figure 11.6: Poster of the European Heritage Days 2014 (© DGPC).

References

AA.VV 2002: Endovélico – Sistema de Gestão e Informação Arqueológica, *Revista Portuguesa de Arqueologia*, 5:1, Instituto Português de Arqueologia, Lisbon, 277–83.

Bugalhão, J. 2009: A Arqueologia Portuguesa nas últimas décadas em Portugal, *Arqueologia e História*, Associação dos Arqueólogos Portugueses, 60–61, Lisbon, 19–43.

Bugalhão, J. & Lucena, A. 2006: As Novas Tecnologias como Instrumento de Gestão e de Divulgação do Património: o exemplo do Endovélico – Sistema de Gestão e Informação Arqueológica in: *Actas dos Encontros Culturais do Baixo Tâmega*, Câmara Municipal de Baião. Baião:, 175–92.

Decree 1999: [The Regulation of Archaeological Work] Decreto-Lei nº 270/1999, 15 July, Diário da República, I série-A, No. 163, http://www.aparqueologos.org/images/PDF/Documentos/DL270_99.pdf (accessed 18.12.2015).

Decree 2000a: Decreto-Lei nº 287/2000, 10 November, Diário da República, I série-A, No. 260, https://dre.pt/application/dir/pdf1s/2000/11/260A00/63196319.pdf (accessed 21.12.2015).

Decree 2000b: Decreto-Lei nº69/2000, 3 May, Diário da República, I série-A, No. 102, https://dre.pt/application/dir/pdf1sdip/2000/05/102A00/17841801.pdf (accessed 18.12.2015).

Decree 2009a: Decreto-lei nº 140/2009, 15 June, Diário da República, I série, No. 113, https://dre.pt/application/dir/pdfgratis/2009/06/11300.pdf (accessed 18.12.2015).

Decree 2009b: Decreto-lei nº 309/2009, 23 October, Diário da República, I série, No. 206, http://www.patrimoniocultural.pt/media/uploads/legislacao/DL309_2009.pdf (accessed 18.12.2015).

Decree 2014: [The Regulation of Archaeological Work] Decreto-Lei nº 164/2014, 4 November, Diário da República, I série, No. 213, https://dre.pt/application/conteudo/58728911 (accessed 18.12.2015).

EEC 1985: Council Directive of 27 June 1985 on the assessment of the effects of certain public and private projects on the environment, 85/337/EEC, Official Journal of the European Communities, No. L 175/40, 5 July, http://eur-lex.europa.eu/legal-content/EN/TXT/PDF/?uri=CELEX:31985L0337&from=EN (accessed 18.12.2015).

Law 2001: [Heritage Law], Lei nº 107/2001, 8 September, Diário da República, I série-A, No. 209, https://dre.pt/application/dir/pdf1s/2001/09/209A00/58085829.pdf (accessed 18.12.2015).

Neto, F., Caldeira, N., Gomes, A.S., & Bragança, F. 2007: Sistemas de Informação e Gestão Arqueológica: Endovélico e SIG in: *Conhecer o Património de Vila Franca de Xira. Perspectivas de Gestão de Bens Culturais*, Câmara Municipal de Vila Franca de Xira, 117–24.

Sousa, A. C. 2013: A revisão do Regulamento de Trabalhos Arqueológicos e os contextos sociais da arqueologia portuguesa no século 21: uma breve reflexão. Revista Património, Direção-Geral do Património Cultural, vol. 1, Lisboa, 36–42.

Websites

Link 1: http://www.patrimoniocultural.pt/pt/patrimonio/patrimonio-imovel/patrimonio-arqueologico/ (accessed 20.11.2015).

Link 2: http://www.patrimoniocultural.pt/en/patrimonio/patrimonio-imovel/patrimonio-arqueologico/laboratorio-de-arqueociencias-larc (accessed 18.12.2015).

Link 3: http://arqueologia.patrimoniocultural.pt/index.php?sid=sitios (accessed 20.11.2015).

Link 4: http://www.patrimoniocultural.pt/pt/shop/search/?name=revista+portuguesa+de+arqueologia (accessed 18.12.2015).

Link 5: http://www.patrimoniocultural.pt/pt/shop/catalog/trabalhos-de-arqueologia/ (accessed 18.12.2015).

12 | Working for commercial clients: the practice of development-led archaeology in the UK

Dominic Perring

Abstract: This paper describes current practice in development-led archaeology in the UK. Key issues are explored with a focus on the strengths and weaknesses of market-based provision. It addresses concerns over the way in which the growth of archaeology as a business has not been accompanied by an equivalent growth in the public benefits of our activities. This is seen, in part, to derive from the way in which conservation policies have been applied, exacerbating a division between the archaeology of cultural resource management and a differently theorised academic sector.

Keywords: development-led archaeology, cultural resource management, competitive tendering, research design, regulation

Introduction

To the uninitiated there seems something unnatural in the idea of archaeology being practised on behalf of commercial clients. Whose values should prevail as money comes to the fore? What happens when our study of the material evidence of the past becomes a product that can be purchased and packaged to meet the needs of clients? This is, however, the situation in which most English archaeologists work, where the burden of supporting archaeological research shifted from the public sector and into the hands of property developers some 25 years ago. The consequences of this changed relationship are much debated, although they are perhaps not quite as profound as might be imagined.

The purpose of this paper is to review some of the strengths and weaknesses of development-led archaeology in the UK (particularly informed by practice in England). I take this to include both rescue (or 'salvage') excavations, and works undertaken to identify and protect cultural resources (including 'preventive archaeology'). An important characteristic of most development-led archaeology is that it is undertaken under contract to clients who pay towards the cost of the work undertaken. Most of this work is undertaken in advance of construction and engineering projects, where those who pay for the archaeology do so in order to manage and mitigate the impact of change to the historic environment rather than in the pursuit of academic understanding or for wider public benefit (which tends to be best served by the use values we can find for archaeological remains, where we convert discovery into understanding through original research and accessible forms of presentation). This narrow focus on management objective applies whether the clients are government agencies investing in works undertaken in the public interest or speculative property developers seeking profit. Archaeologists have long wrestled with the contradictions inherent in working on such projects, since our reasons for wanting to do the work do not always coincide with those of the commissioning bodies. This confusion of purpose lies at the heart of the issue raised by the organisers of the Lisbon conference where this paper was presented: *'whether the delivery model for preventive archaeology is still a scientific endeavour or whether it is just another pre-construction service'*.

I have had frequent cause to consider this problem as the director of a university-based archaeological unit that does most of its work for commercial clients. We brand ourselves as Archaeology South-East (ASE), but financially and organisationally we are an integral part of the UCL Institute of Archaeology. Approximately 120 professional archaeologists work for ASE, where we have to win sufficient funding to fully cover the costs of our operations and overheads. As a not-for-profit body, working within an academic institution, the success of ASE is additionally measured by its contribution to teaching, research and social impact (HEFCE 2011). The need to reconcile our dissonant academic and commercial objectives begs wider questions about the purpose of archaeological study, and highlights current challenges facing the development of our profession.

My understanding of where we may be succeeding and failing is also coloured by the experiences that carried me towards my current management position. I started working in urban rescue archaeology in the early 1970s and have been lucky enough to work on projects throughout Europe and the Middle East, spending many years working outside the UK (notably in Milan and Beirut), migrating between government, academic and private-sector employment. Despite the shared archaeological problems found in these different workplaces, the social rules that guide civic and professional behaviour are very differently understood and applied (different perspectives on these problems can be found in Pitta et al. 1999; Giddens 1984). This is particularly important in development-led archaeology, where academic and professional judgements can influence the commercial viability of vast construction projects. This can make archaeologists actors of consequence within differently constructed institutional frameworks, where we face a host of challenges in balancing the conflicting demands placed upon us (e.g. Perring 2010). These conflicts can only be addressed within the context of locally understood rules guided, of course, by our commitment to ethical professional practice. This means that my analysis of the strengths and weaknesses of British archaeology does not convert into a proscriptive view of what might be good and bad for archaeological practice elsewhere. Some of the ways in which we do archaeology in the UK – that work here because of established traditions, codes of conduct and regulatory arrangements would do harm elsewhere.

This point merits early emphasis because the political and funding arrangements that support development-led archaeology have encouraged highly polarised views over issues of principle. Prominent amongst these, because of its political resonance, is the ongoing debate between those who see a role for market-based competition in deciding on who should undertake archaeological investigations (competitive tendering) and those who believe that publicly directed archaeology can better meet public need (Vander Linden & Webley 2013; Demoule 2010; Zorzan 2010; Kristiansen 2009; Everill 2007). There are echoes here of the wider debate between proponents and opponents of neoliberal economic practices. When seen as an argument between private capital and social value it is entirely legitimate to question whether the market can be anything other than a corrupting force designed to reduce costs, devalue research and diminish the role of public institutions. On the other hand, the benefits of state-centralisation are equally open to question if we turn our concerns to power-dynamics and the social contract between state and citizen. State-supported archaeology has tended to privilege expert values over those of other communities of interest (Smith 2004). Does the involvement of the state, and the institutionalisation of what has been termed an authorised heritage discourse, necessarily impose top-down, bureaucratic procedures and approaches that alienate and disempower local communities and other stakeholders?

Martin Carver has explored how these opposed visions have affected archaeological field-practice, showing how approaches adopted in the UK and USA are differently conceived to models found elsewhere in Europe (Carver 2011, 66). He describes differences between unregulated, regulated and deregulated practice and sees the UK systems as essentially unregulated, by which he means no longer directly managed by the state. In practice the detachment of professional archaeological practice from state funding has only been accomplished by promoting a wider range of checks-and-balances, such that archaeology in the UK is more closely regulated now than ever before. Carver's main point stands, however, in that the process of delegating archaeological work to the private sector has exacerbated an intellectual divide between a development-led archaeology that undertakes its research to inform resource management decisions, and a post-processual academic discipline that locates archaeological study within the wider contemporary debate over material culture and social theory.

There is a consequent divergence of views between those who believe that the first purpose of our endeavours is to conserve the past for future generations to enjoy (e.g. Society for American Archaeology 1996; Hamilakis 2007, 26–7) and those who argue that the pursuit of knowledge through archaeological investigations can sometimes offer greater public benefit (Lipe 1996; Willems 2012). These goals may seem complementary, but involve such fundamentally different approaches to theory, method and outcome that we are close to disciplinary fracture. Cultural resource managers and their academic colleagues conceive, consume and produce archaeology so differently that it is increasingly difficult to see common ground in our parallel engagement with the material past (Bradley 2006; Carver 2011, 67).

Development-led archaeology in the UK

I will return to these themes later, but now turn my attention to the way in which development-led archaeology in the UK works. My comments are based largely on the situation in England, and it is important to note that each of the nations of the UK boasts a different history of political and organisational arrangements resulting in significant differences in practice (Fitzpatrick 2012). I will start with a brisk overview of the scale and structure of our industry, drawing attention to the way in which current arrangements were framed by changes that took place in the period 1970–90, before offering my views on present strengths and weaknesses in the way in which we do archaeology.

The commercial sector appears to have survived the recession in good shape and has recently returned to growth, although the benefits of this remain unevenly distributed. Approximately 5,000 developer-funded investigations take place each year, such that 90% of all archaeological investigations are undertaken by commercial organisations (Fulford 2010, 33; Aitchison 2012). This platform of funded work means that annual income is creeping back towards the pre-recessionary peak of £150 million, and the industry is cautiously optimistic about prospects for future business growth (Aitchison 2010, 26; Aitchison 2014). It is estimated that commercially-funded organisations currently employ nearly 3,000 people, representing some 60% of all those working in British archaeology. Development-led archaeology is where most archaeological research takes place, where new data are obtained, and where most archaeological careers are forged.

Our principal clients are found in the construction industry, chiefly in house-building and related areas of property development. These clients seek archaeological consultants and contractors in order to navigate and satisfy the requirements of a highly regulated planning system. The approaches adopted in England largely pre-dated the adoption of the Valletta Convention (Council of Europe 1992), where the publication of *Planning Policy Guidance* in 1990 represented an important turning point (DoE 1990). Prior to this date the main burden for supporting rescue archaeology had fallen on the public sector.

Rescue archaeology in the UK was largely an invention of the post-war period and saw rapid growth in the 1970s as public funding supported the creation of a series of local and regional teams of professional archaeologists, generally known as field units (Rahtz 1974; Jones 1984). This built on arrangements that had relied on volunteers coordinated by local museums and university departments (often working in association through excavation committees) that obtained charitable status as archaeological trusts. The result was that a variety of local solutions emerged, encouraged and grant-aided by the ministries and departments of central government but not centrally managed (Thomas 2007; Schofield et al. 2011). In many parts of the UK, local government became directly involved, and many districts (local) and counties (regional) appointed archaeologists. These posts were often located within local museums services, but local-authority archaeologists soon forged links with planning departments, firstly to anticipate where rescue excavations might be necessary and subsequently to reduce the scope for conflict over heritage management issues.

The successful introduction of archaeological advice into the planning system, which drew on the language and practice of cultural resource management and environmental impact assessment developed in the USA in the wake of the National Historic Preservation Act of 1966 (King 2004, 23), encouraged a separation between the decision-making that obliged developers to support archaeological works and the undertaking of those works. A division of responsibilities between 'curator' and 'contractor' was seen to reduce the scope for conflicts of interest. This process enabled the move towards developer funding, which was selectively used to supplement public funding from the late 1970s and became the main funding vehicle for archaeology in the 1990s. The rapid growth of the professional sector in this period encouraged the establishment of a professional body, now the Chartered Institute for Field Archaeologists (CIFA), with a remit to promote and raise professional standards.

It is impossible to overstate the importance that the period 1970–90 had for the shaping of British archaeology (Everill 2007 119–21; Everill 2009). Ideas, organisations and methods developed during this period established the platform that has been re-engineered to meet present requirements. This is clearly illustrated by the organisational histories of the archaeological companies that do most of the work. The latest *Yearbook* of the CIFA lists some 70 Registered Organisations (CIFA 2015). These are the bodies that have been subject to peer review and inspection, and successfully demonstrated their adherence to professional standards. Altogether these companies employ over 2,500 archaeologists. Forty-six of these organisations, employing 2,200 staff, undertake developer-funded archaeological excavations. Along with individual CIFA membership held by the managers of companies that are not individually registered, over 80% of development-led archaeological work falls within the orbit of CIFA regulation. This list of Registered Organisations working in the UK is dominated by business practices that were established in the 1970s as publicly-funded regional field units and which remain not-for-profit organisations structured to meet research and charitable objectives. Two-thirds of the professional staff working in developer-funded archaeology are employed by such bodies, which include eight out of the ten largest organisations that offer commercial services. Most of these bodies have specialist knowledge of regions within which they have worked continuously for over 40 years. A smaller group of more recently established companies do not trace their origins to the public sector provision of the 1970s, but have also inherited the research culture and commitment to local community engagement that characterises the longer-established archaeological companies.

The most important area of change has been in the growth of consultancies that provide planning advice to commercial clients and then manage the commissioning process. Here the inherited traditions of public-sector rescue archaeology are less in evidence, although the need to build value for high-profile clients can encourage innovative research. The Frameworks Archaeology work on behalf of the British Airports Authority is a widely cited example of a consultancy involvement leading to enhanced research ambition (Andrews et al. 2000). Most work is more mundane, and the needs of cost and risk management can discourage experimental approaches. Some consultants find it necessary to argue that research is an academic pursuit that should not burden their development-sector clients, whose responsibilities are best met by mitigating the impact of development through a mix of avoidance (preservation *in situ*) and routine recording (preservation by record).

In summary, the UK has developed a strong locally-based planning-driven platform for development-led archaeology, where professional archaeologists work effectively with construction industry clients. This has resulted in the development of a successful business sector, underpinned by widespread consensus over the conservation and commercial objectives of archaeological practice. Successful outcomes involve different actors (particularly planners, developers and archaeologists) coming together to negotiate regulatory hurdles in order to mitigate the adverse impacts of construction proposals. This is informed by a shared understanding of what constitutes good practice.

Regulation in practice

In most development-led projects two planning documents are used to structure and give intellectual coherence to the exercise. The first of these is the 'written scheme of investigation' (WSI), in which the archaeologists seeking to undertake the works convert their interpretation of the purpose of the archaeological study into a research design and methodology. Ideally, this document will incorporate a variety of views, drawing on regional or national research agendas as well as the specific needs of the client and their consultants. What is critical, however, is that this document meets the approval of the local planning authority acting on the basis of advice received from an appropriate 'curatorial' archaeologist (sometimes involving further consultations with other interested parties). Negotiations over the different drafts of a WSI create the circumstances in which research questions can be refined, methods agreed on, and costs established. The WSI can also be used as the basis for taking quotations from different archaeological contractors, and will allow the curatorial archaeologist to monitor the conduct of any agreed works. Normally, the document will identify the parameters of the exercise, so that unexpected results can be seen as falling outside of the original project design and consequently serve as a trigger for contingency, re-measurement or variation. Most larger-scale projects will go through the usual stages of assessment (desk-based impact assessment and background research), evaluation (test-pitting and deposit modelling), and excavation (fieldwork). Each of these exercises is guided by a separate plan of action, as identified within a WSI drawn up for that particular stage of works, and the reports on these different stages of work contribute equally to the formation of research questions and methods. This staged approach encourages an iterative approach to research design, informs conservation strategies with sound information on the significance and vulnerability of archaeological remains, and generates the information that ensures that estimates and quotations are based on an adequate knowledge of the scale of problem to be addressed.

The second critical document is the preparation of a post-excavation assessment (PXA) report. Based on English Heritage guidance, this management document is prepared from an initial quantified assessment of the potential of the finds recovered to address the research questions found within the original project design or WSI (English Heritage 1991; English Heritage 2006). Here new tasks and costs can be identified, and agreement reached on a forward programme of work that will best realise the research potential of the finds uncovered. Ideally, the restrictions placed on a development will only be discharged by agreement over the programme and funding of the works described in the PXA. It is not uncommon for an initial commission to include a contingency sum for post-excavation and publication work that is only released once a PXA has met with approval from the local planning authority acting on the professional advice of the relevant curatorial archaeologist.

The quality of the archaeological work undertaken on development-led investigations is therefore the product of four inter-related aspects of the way in which the archaeological work is organised. In the first place, archaeological teams wish to achieve a high standard in order to meet their own expectations of themselves: almost without exception we are in archaeology to do good archaeology, which we define as having research and public outcomes that demonstrate value to our clients and peers (formal Quality Assurance procedures are also widely used). Secondly, we are bound by adherence to a series of professional codes of practice that are agreed-on and externally monitored, mainly through the CIFA but also through a variety of related professional associations. Thirdly, we work within a highly regulated business, where local-authority 'curators' demand adherence to pre-determined project-specific research methodologies and will inspect work at various stages of implementation. Finally, we are employed by clients and their consultants, who may not always understand the detail of what we do, but will expect us to satisfactorily explain and justify our work to a wide range of stakeholders (often including an excited local press).

What are the benefits of development-led archaeology?

Commercial funding has allowed for the continued growth of archaeology as a professional activity at times when public spending has been in retreat. More archaeologists are employed and more archaeology is being done than would otherwise have been possible. It is impossible to imagine a situation in which a politically constrained tax-based funding system would have matched the pace of growth permitted by this private investment.

The fact that more of us have jobs in archaeology than would otherwise have been the case does not necessarily mean that we are doing better archaeology. There are, however, several practical benefits to current arrangements. Many of these derive from the closer relationship between developer and archaeologist forged by the commissioning process. Initially this ensures that when more construction work is taking place the funds for archaeological attendances increase commensurately. The very purpose of a market-based system is to marry supply to demand, and, since commercial clients are prepared to invest in archaeological research if it will speed the progress of a construction project, this gives archaeologists a cash-flow that is equal to the needs of the situation. The funding model that applied before 1990 was less flexible, and archaeological sites were destroyed without record because of difficulties in mobilising competent field teams. The laws of supply-and-demand are not entirely benign, however. The cyclical nature of the construction industry leaves us alternating between periods of skills-shortages and periods of staff redundancies, in ways that can be damaging to both individual careers and the wider professional structure (as illustrated by the situation in Poland reported on by Marciniak & Pawleta 2010, 92).

Clearer benefits emerge from the sense of shared responsibility that comes from being part of a project development team. As an employed contractor our construction-industry clients see the value of involving us in critical decisions. We participate in a dialogue over how best to integrate our works with those of other contractors, which reduces the risks of important archaeology being accidentally destroyed because of programming errors. Whilst I have met few developers who deliberately set out to damage an archaeological site, I have witnessed sites damaged because a main contractor had a deadline to meet and penalty clauses to avoid. This is an avoidable, and an increasingly rare event.

Collectively, and in response to commercial pressures, we have become more efficient at making sure that work is taken to conclusion. A crucial role is played here by project management practices imported to archaeology from our construction industry partners. There was something of a 'Ponzi scheme' to the way in which development-led archaeology functioned before 1990. It was commonly the case that the budgets available were not sufficient to see work through to completion, so that staff would nominally be redeployed onto the next funded project whilst actually continuing analytical studies on the results of earlier excavations. As long as the flow of new work continued to grow it was possible to direct a proportion of incoming funds to unpublished backlog projects, but only at the consequence of leaving a growing funding-gap as new fieldwork projects matured into analytical programmes of post-excavation study for which there was no budget. These deficiencies in project funding are responsible for a substantial body of unpublished fieldwork from the 1980s, and highlight the fact that the publicly-funded rescue archaeology model developed during the 1970s would not have been sustainable even without the public-sector funding cuts of the 1980s.

Now, however, the client's need to submit a report to secure the discharge of a planning condition converts an academic responsibility into a contractual one, requiring archaeologists to marshal and disseminate results to secure payment. Since this burden of work falls on the organisation rather than the individual, and we have a professional reputation to protect, most archaeological contractors are prepared to invest properly in establishing a management-driven dialogue over the progress of analytical study. Different specialist teams combine to keep work on track, where ideas are shared and no one is irreplaceable. This does not need to diminish the authorial voice of the director of the archaeological project, although there is a risk that it will, but it reduces the scope for projects to become orphaned by the disengagement of a principal investigator. The dialogue between an archaeological contractor and the consultants and curators involved provides a further testing-ground for the development of academic and professional arguments. Open access to both data and ideas is a necessary part of commercial archaeology especially when several different archaeological teams might find themselves employed on different stages of the same project.

From a research point of view we have also benefitted from being drawn into a wider range of landscapes than would otherwise be the case. This works in two ways. Development-led archaeology takes us into environments that are not normally accessible for research projects, as is particularly the case in the urban 'brown-field' sites that are only available for archaeological investigations in the brief interval between demolition and construction. My research interest in the archaeology of continuously inhabited cities depends entirely on the opportunities brought about by development-led archaeology (Perring 2015). It is also the case that the need to respond to the problems raised by our commercial clients takes us into landscapes that we might otherwise neglect. An illustration in point is the way in which the archaeology of London's hinterland, in particular the wetland landscapes of the Thames estuary, received scant attention before commercial funding helped change the research agenda by requiring us to assess development impacts in areas that had otherwise escaped academic interest.

The increased importance placed on the mediating role played by local-authority employed archaeologists is a direct product of concerns that a deregulated environment would diminish the quality of archaeological research. This has formalised the status of archaeology in local consultation processes, adding new mechanisms for stakeholder engagement and encouraging the identification of community benefit (Southport Group 2011; Perring 2014). It has also encouraged the elaboration of a broad platform of research agendas (Olivier 1996) and professional standards that are explicitly aimed at improving the quality of archaeological work undertaken (e.g. Baker & Worley 2014). In some areas, such as cemetery excavations and environmental research, the advances in the quality of our work have been significant. Whilst it is not difficult to identify ways in which similar improvements could have been achieved without commercial funding, the introduction of such funding has been a spur.

In sum, we now have a developed and well-funded professional sector, where organisational cultures and regulatory regimes combine to place considerable emphasis on public benefit rather than economic gain as the main driver of business activity. The greatest gains have been made in areas such as funding, professionalism, improved project and organisational management, stakeholder engagement and dialogue, and more open approaches to data sharing and dissemination. Commercial provision has, in some instances, improved transparency and accountability and added to the range of stakeholders consulted in the process of planning and implementing archaeological investigations.

The perils of market provision

There remains, however, a widespread perception that competitive tendering drives down the quality of archaeological work (vocally expressed by Chadwick 2000; Demoule 2010). There are undoubtedly cases where archaeologists have underestimated costs, offering unrealistically low quotations that result in funding agreements that fall short of need. As a consequence, work has been rushed to completion at the expense of sampling or post-excavation analysis. These are problems of poor planning, usually derived from inadequate assessment of potential, and can happen where budgets are fixed regardless of the means of procurement. I am not aware of any archaeological organisation that has deliberately set out to underestimate costs in a tender exercise, which would be a commercially suicidal strategy. It is, however, easier to misjudge the needs of a complex archaeological landscape when ignorant of its full potential, and a poorly informed tender is at risk of accidentally becoming the most cost competitive.

Fortunately, cost is not routinely the most important consideration, even if it is invariably a material one. Where a client is in receipt of several quotations, it is a prudent habit for them to challenge the viability of any proposal that appears substantially cheaper than the rest. Most clients with experience of dealing with archaeologically sensitive sites prefer to work with contractors who understand their needs and can be relied on to complete works and deliver reports that are fit for purpose to an agreed timetable. Risk management and speed of mobilisation and delivery come closer to the top of their list of requirements. For this reason many contracts are awarded on a single-tender basis. Many private-sector clients are not obliged to seek competitive quotations, and are prepared to pay a modest premium to retain the services of a trusted contractor. In such cases there is an understanding that the cost proposal will be competitive and transparent rather than cheap. In more complex projects it is also routine to be put through a complex pre-qualification exercise to test competence and professionalism across a range of scored criteria. In the advanced stages of procurement of works for large rail projects, such as Crossrail and HS2, we have had to provide detailed responses on questions concerning our quality assurance procedures, staff experience, sustainable procurement, organisational and financial capacity. In a recent exercise for the Thames Tideway Tunnel we failed to pass the qualification exercise because of a comparative lack of experience in dealing with the particular archaeological problems of the Thames foreshore. There are always risks attached to a fixed-price quotation, and alternative models are coming to the fore on larger projects, as in the application of measured contract practices to archaeological excavation (Heaton 2014). On the largest jobs, where both risks and rewards are greatest, it is common to find joint-venture approaches where several archaeological teams pool resources and share risks.

The suspicion that commercial consideration will result in cost-saving practices that compromise the quality and integrity of the archaeological work places us under constant scrutiny. Most archaeological companies are fearful of the reputational damage that they would suffer if cost considerations resulted in underperformance. This scrutiny, and our accountability to external stakeholders, exceeds the expectations placed on archaeological units before the advent of commercial competition (when budgets were often inadequate for the work that was undertaken). Within Archaeology South-East we work on about 450 commissions each year. We do not expect, however, to complete each and every project within budget. Some studies cost more to complete than expected, others cost less. When we are faced with significant archaeological discoveries in a situation where we have exhausted a project budget and contingencies, we complete the work by a managed overspend: drawing on surplus generated in operations elsewhere. This is common practice, and is facilitated by the funding flexibility found in the commercial sector.

We still face the challenge that our understanding of archaeological value, which may encourage us to invest in additional analytical study above and beyond the expectations of an original commission, is not matched by an equivalent awareness amongst our clients. The customers we serve may have only a partial understanding of what makes for good archaeology and do not judge quality using our criteria. We are selling what are known as credence goods, where the customer is obliged to take the detail of what we offer on trust (Dulleck & Kerschbamer 2006). This can give rise to situations where archaeological contractors are tempted into offering simpler and more inexpensive solutions to an archaeological problem, since these are more likely to secure a commission than a more sophisticated form of intervention. Examples can include reducing the range of post-excavation analytical studies. Here market forces can work against our academic interests.

We consequently rely on the public sector to regulate practice and to act against those who fail to meet quality thresholds. This is a fraught working relationship, and different local authorities have different expectations and guidelines. There are times when local government monitoring can seem unnecessarily intrusive and bureaucratic, but it provides the constraint mechanism that reminds our clients of the need to employ qualified archaeologists able to satisfy the regulatory requirements. It is therefore a matter of concern that pressures on local government spending have reduced investment in specialist heritage advice within local planning authorities at a time when workloads are rising (Aitchison 2014; CBA 2015). One of the consequences of this is that it can take local authorities longer to reach decisions and approve submitted proposals, inspect work in progress, and agree to the discharge of planning conditions. The solution presently emerging is for local authorities to charge developers directly for such services. This promises to generate funds to support the continuation of current procedures, but will tilt the balance of power towards developers and their consultants. A greater degree of accountability might, however, encourage greater consistency in the way in which archaeological work is regulated. In driving standards upwards it would help to see a clearer emphasis on research quality as a necessary proof of competence.

Most local authority archaeologists strive to ensure that community benefits are realised, archives and finds are suitably cared for, and results are published in academically sound peer-reviewed journals and monographs. These are all areas where problems occur, in part because the protracted time-scale of archaeological study makes it difficult to track progress and assure delivery. Although more archaeological work is being conducted than ever before, we struggle to make best use of the data that is being recovered. Primary archives are often retained by archaeological contractors because local and county museums lack the resources to curate the volumes of material being recovered (Edwards 2012). These resources are therefore inaccessible to the local public and their future is uncertain. It is also the case that too many excavations are left unpublished. In some cases, this is because works were undertaken for management and not research purposes, where, since the archaeological remains at issue were left *in situ*, there is no public requirement for additional reporting. In others, the programmes of post-excavation analysis have been delayed by the staged nature of the fieldwork. Other delays, however, are the product of commercial and management failures.

In a recent survey of a sample of commercial projects, Fulford (2010) found that only 6% of investigations carried out between 1990 and 1994 had reached final publication 12 years later (by 2006). The projects reviewed, however, included desk-based assessments and evaluations and watching briefs where no archaeological discoveries were made and where academic publication was never appropriate. A sub-sample of Roman period projects from one region established that a third of the archaeological excavations had resulted in formal publication. Some of the unpublished backlog is still receiving attention and will eventually be published. Improvements in project planning mean that more recent excavations have a significantly better chance of reaching academic publication. More needs to be done, but the present situation is considerably better than in the pre-1990 era of rescue archaeology and continues to improve (Thomas 1991; Fitzpatrick 2012, 153). Unfortunately, however, reports can be frustratingly slow to emerge. Reports prepared for planning purposes the 'grey literature' of assessments produced in evaluation and post-excavation phases can offer quicker and more comprehensive access to the salient results of recent fieldwork. This information can, however, be difficult to obtain and is poorly structured for most forms of research. The wider use of online technology is beginning to improve the situation, and many larger archaeological contractors now publish libraries of their reports through their websites (Hardman 2009). There are exciting moves afoot to provide better and earlier access to the digital archives on which research is based.

Despite problems with the publication of the data from individual excavations, this is not the main area of failure. The results of most archaeological excavations will eventually be made public. These are almost always, however, site-specific works of dense description. Development clients are responsible for 'preservation by record', which can be achieved by narrowly framed accounts that do little to advance wider understanding. There is a dangerous perception that problem-led research, and the task of wider synthesis, is the terrain of universities and beyond the competence and remit of development-led archaeology (Dries 2015). The pace of commercial fieldwork has, in any case, outstripped the capacity of individual researchers to come to terms with its results (Bradley 2006). We lack resources for research and synthesis, leaving us with a surprisingly modest return on the investment of hundreds of millions of pounds in the study of Britain's archaeological past.

An uncertain industry

These problems contribute to a wider sense of dissatisfaction amongst those who work in development-led archaeology (Chadwick 2000; Carver 2011, 75). This finds expression in complaints about low pay and the lack of career opportunities and job security:

archaeologists working in the UK are not as well-paid as their contemporaries in comparable professional employment (Aitchison & Rocks-Macqueen 2013; Everill 2007, 119). This is not a new problem, since low pay was a feature of the vocational and voluntary rescue archaeology of the 1960s and 1970s, but it is now felt more keenly. The advent of commercial archaeology has drawn attention to issues of professional status and pay.

Many field archaeologists move from project to project, working to a tight routine and with limited opportunity to follow-up on research opportunities. It is possible to develop a career based largely on 'negative' evaluations: trenching exercises where we discover and measure the absence of archaeological finds. Matters are not helped by a fragmented and discontinuous approach to the conduct of fieldwork. The staged approach to most development projects, potentially involving different archaeological contractors at different stages, puts distance between preliminary assessment, the framing of research objectives, and subsequent investigation and analysis. Whilst effective management can sustain research coherence, this risks diminishing the intellectual role of the staff involved. With so many parties involved in defining the programme of work, how can individual research interests be developed and pursued?

Commercial practice has also reinforced an earlier tendency within rescue archaeology towards the technicalisation of the work, where it can be argued that the primary duty of the excavator is to produce records, data and reports that serve the needs of others (Barker 1977). Despite increases in funding, we have seen surprisingly little methodological advance over the last 20 years. Excavations are conducted to high standards, but using ideas and techniques that were pioneered in the 1970s. Opportunities to innovate can be stifled by the standardisation of practice encouraged by the regulatory regime. This can crowd out opportunities for creativity, as in placing obstacles to the development of new digital technologies because of conservative expectations enshrined in locally imposed standards. There are even some areas of work where our research methodologies are less rigorous than was the case before 1990 because exceptional sampling approaches are not easily incorporated into standardised briefs and specifications, as in the study of topsoil archaeology (Evans 2012, 29).

We are hampered by the short-term goals of contract archaeology, where we are measured by our efficiency in getting holes dug, not against what we learn from digging holes. This encourages a casual approach to skills, where it is easy for employers to become more concerned about speed and effectiveness of delivery than about the quality of the work undertaken. An emphasis on excavation as a form of decontamination, involving low-grade archaeological data, distances professional archaeology from the academic sector. This results in discrepant approaches to fieldwork training, such that the skills learnt at university can be ill-suited for the different demands of contract archaeology (Sinclair 2010; Aitchison 2006). This in turn reduces opportunities for career progression, making it difficult for archaeologists to navigate between academic and contract employment. This all adds to a process of alienation, whereby field-archaeologists have a limited personal involvement in the way in which their work contributes to the final product (Zorzan 2010, 45, 221).

Escaping the constraints of preservation *in situ*

A key problem is that the demands of the marketplace tend not to coincide with professional and personal goals. A confusion of purpose clouds the procurement and delivery of contract archaeological services. Is archaeology about giving our commercial clients 'holes in the ground'? Is it about protecting community and public interest in local cultural resources? Is it about knowledge gain?

This confusion is not a consequence of the means of procurement, such as competitive tendering, but of the primacy given to conservation goals. Lipe (1996) and Willems (2012) have argued persuasively that preservation *in situ* has become an obstacle to problem-oriented archaeological research. Conservation policies were a necessary corrective to the unsustainable profligacy of rescue archaeology, and gave rise to the emphasis on sustainable management that has underpinned planning policy for the last quarter century. But in order to press the 'polluter pays' argument, it became dangerous to argue the value of destructive investigation. The expectation that preservation *in situ*, was nearly always to be preferred has resulted in investigation becoming a 'second best' option for 'second best' sites. Our avoidance of more interesting parts of the archaeological landscape, against unproven future benefit, has left us with an indigestion of data from a fragmented sample of landscapes of lower potential. As a consequence, much archaeological work has become a planning-driven routine of uncertain public benefit: a box-ticking exercise of fleeting attention that offers poor rewards to all involved. This is why it can sometimes seem as if we are offering a pre-construction service rather than participating in scientific endeavour.

This perception is an unfair one, since academic research underpins even the most technical exercises involved in locating and assessing the significance of heritage resources. We must, however, be more ambitious than this. Without using our holes in the ground to advance understanding there is no point in digging them. If we do not invest more in getting this right, and in convincing our clients of the central importance of research quality in our work, then we will eventually struggle to retain status as a material consideration within the planning process. Right now the UK is facing a housing shortage that is becoming an increasingly important issue in domestic politics, and this is adding to pressures to cut back on the planning restrictions that might delay new construction. Those of us working in archaeology and heritage conservation benefit from considerable levels of public support, but we take this for granted at our risk.

Our research and commercial interests depend on finding clear public benefit in our archaeological work. This is not just to be found in the evidential value of the resource, but in the use values that can be advanced (Southport 2011, 57). The last few years have seen energies directed into using our research to excite and engage with a wide range of audiences, through public outreach, access and communication (Sayer 2014), now reinforced by the goals of the Faro Convention (Council of Europe 2005). This is a welcome development that has been widely embraced by the leading companies involved in development-led archaeology. Archaeological practice has much to offer in promoting the goals of social cohesion and environmental wellbeing, adding values to our understanding and use of space and place. There is also scope to embrace new sponsors and partners, especially in community engagement projects, at a time when the political necessity of showing quality outcomes as the means of justifying planning-led constraint is increasingly evident.

I appreciate that some of this may sound like wishful thinking, and that our clients may not always be prepared to support these goals. A more flexible and ambitious regulatory regime, where evidential values do not outweigh use values, can and does help us towards these goals. All archaeological contractors can point towards success stories, where sensible planning decisions and supportive clients have allowed us to do exactly this sort of work.

We also need to see development-led archaeology re-integrated with university-based academic research. The main reason that the UCL Institute of Archaeology continues to be home to Archaeology South-East is to facilitate research collaboration and to draw on development-led practice in improving teaching and training. The economic recession denied us the opportunity to invest properly in this relationship, but a return to growth provides a trigger for doing more to make this arrangement succeed. Elsewhere, our commercial competitors are equally keen to develop partnerships with academic institutions, but with limited success. The potential for closer collaboration between university teaching departments and commercial archaeological concerns has become difficult to realise because of conflicting ideas of what constitutes value. This is a lost opportunity where both sides of the divide have been at fault, although Richard Bradley and Martin Carver deserve enormous credit for detailing creative ways in which we can overcome these problems (Bradley 2006; Carver 2011, 142–4). A greater understanding of how and why our aims have diverged is essential, and would do much to dispel an ill-informed demonisation of commercial practice as unreflexive and under-theorised, on the one hand, and of academic remoteness from the realities of British field archaeology, on the other.

Research remains at the heart of all we do. Archaeologists working in commercial employment no different to our public sector, museum and academic colleagues – aim to do intellectually rewarding work that meets the highest standards: an archaeology that has the promise of both adding to knowledge and improving the quality of life. We seek to engage with projects that challenge us, since this is where we will learn the most and can excite others in our work. In order to achieve this we need to escape the preciousness that we take to a resource that is less vulnerable than is sometimes assumed. Development-led archaeology needs to be given more scope to take risks in the pursuit of research goals, allowing us to promote a different dialogue with our many stakeholders, including both commercial clients and the different local communities whose landscapes we are privileged to study.

Acknowledgements

This paper owes an enormous debt to two key audiences. I am grateful to all who participated in the conference session of the Europae Archaeologiae Consilium at Lisbon, where the lively discussion drew my attention to the many ways in which the nature of commercial archaeology in the UK can be misunderstood. I am even more indebted to my colleagues at Archaeology South-East for making development-led archaeology both fun and rewarding, with my very special thanks to Isa Benedetti-Whitton, Kieran Heard, Matt Pope, Michael Shapland and Dan Swift for their comments on an earlier draft.

References

Aitchison, K. 2006: What is the Value of an Archaeology Degree?, with responses from J. Grant, D. Perring, T. Schadla-Hall & S. Shennan, *Papers from the Institute of Archaeology* 17, 16–27.

Aitchison, K. 2010: United Kingdom archaeology in economic crisis, in N. Schlanger & K. Aitchison (eds) *Archaeology and the global economic crisis. Multiple impacts, possible solutions*. ACE / Culture Lab Editions, Tervuren, 25–30.

Aitchison, K. 2012: *Breaking New Ground: How Professional Archaeology Works*. Landward Research Ltd, Sheffield.

Aitchison, K. 2014: *Heritage Market Survey*. Landward Research Ltd.

Aitchison, K. & Rocks-Macqueen, D. 2013: *Archaeology Labour Market Intelligence: Profiling the Profession 2012–13*. Landward Research Ltd.

Andrews, G., Barrett, J.C. & Lewis J.S.C. 2000: Interpretation not record: the practice of archaeology, *Antiquity* 74, 525–30.

Baker, P. & Worley, F. 2014: *Animal bones and archaeology: Guidelines for best practice*. English Heritage, Swindon.

Barker, P. 1977: *Techniques of Archaeological Excavation*, London.

Bradley, R. 2006: Bridging the Two Cultures – Commercial Archaeology and the study of Prehistoric Britain, *Antiquaries Journal* 86, 1–13.

Carver, M.O.H. 2011: *Making Archaeology Happen, Design versus Dogma*, Left Coast Press, Walnut Creek CA.

Chadwick, A. 2000: Taking English archaeology into the next millennium a personal review of the state of the art, *Assemblage* 5, www.assemblage.group.shef.ac.uk/5/chad.html (accessed 13.11.2015).

CIFA 2015: *Chartered Institute for Archaeologists Yearbook and Directory 2015*, Reading.

Council for British Archaeology 2015: *Archaeology Matters: Local Heritage Engagement Network Toolkit 3: Local authority historic environment services: what they do and why are they important?* CBA, York.

Council of Europe 1992: *European Convention on the Protection of the Archaeological Heritage (revised)*. European Treaty Series 143. Council of Europe, Strasbourg.

Council of Europe, 2005: *The Framework Convention on the Value of Cultural Heritage for Society*, European Treaty Series 199. Council of Europe, Strasbourg.

Demoule, J-P. 2010: The crisis – economic, ideological, and archaeological, in N. Schlanger & K. Aitchison (eds) *Archaeology and the global economic crisis. Multiple impacts, possible solutions*, ACE / Culture Lab Editions, Tervuren, 13–7.

DoE 1990: *Planning Policy Guidance Note 16: archaeology and planning*, HMSO, London.

Dries, T. 2015: Dare to choose: make research the product, in A.C. Schut, D. Scharff & L.C. de Wit (eds) *Setting the Agenda: Giving New Meaning to the European Archaeological Heritage*, EAC Occasional Papers No. 10, 69–74.

Dulleck, U. & Kerschbamer, R. 2006: On Doctors, Mechanics and Computer Specialists: The Economics of Credence Goods, *Journal of Economic Literature* 44, 5–42.

Edwards, R. 2012: *Archaeological Archives and Museums 2012*, Society of Museum Archaeologists.

English Heritage, 1991: *Management of Archaeological Projects*, English Heritage, London.

English Heritage, 2006: *Management of Research Projects in the Historic Environment*, English Heritage, London.

Evans, C. 2012: Archaeology and the Repeatable Experiment: A comparative agenda, in A.M. Jones, J. Pollard, M.J. Allen & J. Gardiner (eds) *Image, Memory and Monumentality: Archaeological engagements with the material world*, Prehistoric Society Research Paper 5, Oxbow Books/The Prehistoric Society, Oxford, 295–306.

Everill, P. 2007: British Commercial Archaeology: Antiquarians and Labourers; Developers and Diggers, in Y. Hamilakis & P. Duke (eds) *Archaeology and Capitalism: from Ethics to Politics*, Left Coast Press, Walnut Creek CA, 119–36.

Everill, P. 2009: *The Invisible Diggers: A study of commercial archaeology in the UK*, Heritage Research Series 1, Oxford.

Fitzpatrick, A. 2012: Development-led archaeology in the United Kingdom: a view from AD 2010, in L. Webley, M. Vander Linden, C. Haselgrove & R. Bradley (eds) *Development-led Archaeology in Northwest Europe*, Oxbow, Oxford, 139–56.

Fulford, M. 2010: The impact of commercial archaeology on the UK heritage, in J. Curtis, M. Fulford, A. Harding & F. Reynolds (eds) *History for the taking? Perspectives on material heritage*, British Academy, London, 33–53.

Giddens, A. 1984: *The Constitution of Society: Outline of the Theory of Structuration*, Polity Press, Cambridge.

Hamilakis, Y. 2007: From Ethics to Politics, in Y. Hamilakis & P. Duke (eds) *Archaeology and Capitalism: from Ethics to Politics*, Left Coast Press, Walnut Creek CA, 15–40.

Hardman, C. 2009: The Online Access to the Index of Archaeological Excavations (OASIS) project: facilitating access to archaeological grey literature in England and Scotland, *The Grey Journal* 5, 76–82.

Heaton, M. 2014: Constructing Archaeology: The Application of Construction Management Practices to Commercial Archaeology in Britain, *The Historic Environment* 5.3, 245–57.

HEFCE, 2011: *REF 02.2011. Assessment Framework and Guidance on Submissions* http://www.ref.ac.uk/media/ref/content/pub/assessmentframeworkandguidanceonsubmissions/GOS%20including%20addendum.pdf (accessed 06.11.2015).

Jones, B. 1984: *Past Imperfect: The Story of Rescue Archaeology*, Heinemann, London.

King, T.F. 2004: *Cultural Resource Laws and Practice* (2nd edition), Left Coast Press, Walnut Creek CA.

Kristiansen, K. 2009: Contract archaeology in Europe: an experiment in diversity, *World Archaeology* 41.4, 641–8.

Lipe, W.D. 1996: In Defense of Digging: Archeological Preservation as a Means, Not an End, *Cultural Resource Management* 19.7, 23–7.

Marciniak, A. & Pawleta M. 2010: Archaeology in crisis: the case of Poland, in N. Schlanger & K. Aitchison (eds) *Archaeology and the global economic crisis. Multiple impacts, possible solutions*, ACE / Culture Lab Editions, Tervuren, 87–96.

Olivier, A. 1996: *Frameworks for our Past a review of research frameworks, strategies and perceptions*, English Heritage, London.

Perring, D. 2010 (for 2009): Archaeology and the Postwar Reconstruction of Beirut, *Conservation and Management of Archaeological Sites* 11.3–4, 296–314.

Perring, D. 2014: Involving the public in archaeological fieldwork: how heritage protection policies do not always serve public interests, in P. Stone & Z. Hui (eds) *Sharing Archaeology: Academe, Practice and the Public*, Taylor & Francis, New York & London, 167–79.

Perring, D. 2015: Recent advances in the understanding of Roman London, in M. Fulford & N. Holbrook (eds), *The Towns of Roman Britain: The Contribution of Commercial Archaeology since 1990*, Britannia Monograph 27, London, 20–43.

Pitta, D.A., Fung H-G. & Isberg, S. 1999: Ethical issues across cultures: managing the differing perspectives of China and the USA, *Journal of Consumer Marketing* 16. 3, 240–56.

Rahtz, P. (ed.) 1974: *Rescue Archaeology*, Penguin, Harmondsworth.

Sayer, F. 2014: Politics and the Development of Community Archaeology in the UK, *The Historic Environment* 5.1, 55–73.

Schofield, J., Carmen J. & Belford P. 2011: *Archaeological Practice in Great Britain: A Heritage Handbook*, Springer, New York.

Sinclair, A. 2010: The end of a golden age? The impending effects of the economic collapse on archaeology in higher education in the United Kingdom, in N. Schlanger & K. Aitchison (eds) *Archaeology and the global economic crisis. Multiple impacts, possible solutions*, ACE / Culture Lab Editions, Tervuren, 31–44.

Smith, L. 2004: *Archaeological Theory and the Politics of Cultural Heritage*, Routledge, London.

Society for American Archaeology, 1996: *Principles of Archaeological Ethics*, http://www.saa.org/publicftp/public/resources/ethics.html (accessed 06.11.2015).

Southport Group, 2011: *Realising the benefits of planning-led investigation in the Historic Environment: A framework for delivery*, Institute for Archaeology, London.

Thomas, R. 1991: Drowning in data? – Publication and Rescue Archaeology in the 1990s, *Antiquity* 65, 822–8.

Thomas, R. 2007: Development-led Archaeology in England, in K. Bozoki-Ernyey (ed.) *European Preventive Archaeology*, Papers of the EPAC meeting, Vilnius 2004, https://www.coe.int/t/dg4/cultureheritage/heritage/Archeologie/EPreventiveArchwebversion.pdf (accessed 06.11.2015).

Vander Linden, M. & Webley L. 2013: Introduction: development-led archaeology in northwest Europe. Frameworks, practices and outcomes, in L. Webley, M. Vander Linden, C. Haselgrove & R. Bradley (eds) *Development-led Archaeology in Northwest Europe*, Oxbow, Oxford, 1–8.

Willems, W.J.H. 2012: Problems with preservation in situ, in C. Bakels & H. Kamermans (eds) *The End of our Fifth Decade: Analecta Praehistorica Leidensia* 43/44, 1–8.

Zorzan, N. 2010: The Political Economy of a Commercial Archaeology: A Quebec Case-Study, PhD thesis, University of Southampton.

13 | Balancing stakeholders in the Netherlands. A plea for high-quality municipal archaeology

Dieke Wesselingh

Abstract: In the Netherlands the implementation of the Valletta Convention has led to archaeology being fully integrated into spatial development. Local governments take the majority of decisions, as it is they who draw up zoning plans and issue relevant permits. Dutch 'Malta archaeology' is a scientific endeavour as well as a pre-construction service. These two do not necessarily exclude one another, as is illustrated by the approach used in Rotterdam. Spatial development without destruction of valuable archaeological heritage and, no less importantly, without unnecessary excavations, is crucial for gaining and retaining public and political support. Archaeologists need to be selective and take care to explain their choices, in order to meet the expectations of all other stakeholders.

Keywords: preventive archaeology, the Netherlands, municipal archaeology, stakeholders, evaluation work

Introduction: the Dutch system

In the Netherlands, the Valletta Convention was incorporated in national legislation in 2007, but several years earlier archaeology had already become part of the spatial planning process. At the same time, a market for archaeological services had been tentatively introduced, followed (in 2007) by a system of quality assurance. Nowadays archaeology is fully integrated into spatial development. This means that local (municipal) authorities make the majority of decisions about which sites to protect or to excavate and how to do this. The idea behind this decentralisation was that most decisions on spatial development are taken at a local level. Thus the role of the competent authority as far as archaeology is concerned is directly derived from its responsibility for issuing the relevant permits. The revised Monuments Act (2007) requires municipalities to seriously take account of archaeological values. Zoning plans are important tools with which local government can attribute archaeological value to areas and thus oblige developers (and other 'initiators') to apply for building permits and, if necessary, to have archaeological research carried out.

Municipal archaeology

Of the 403 municipalities in the Netherlands, 268 have archaeologists working for them on a permanent basis, often regionally organised (situation at the end of 2014, data based on Vonk & Berkvens 2014 and Buitelaar 2015, see also Table 1). There is a difference between so-called regional archaeologists, working for a group of municipalities and often primarily involved in the issuing of permits, and municipal archaeologists who occasionally work for neighbouring municipalities too. The latter are civil servants and their job consists of various tasks deriving from the legal role of municipalities concerning archaeological heritage. Seriously taking account of archaeological values means writing archaeological paragraphs for zoning plans, assessing building plans that involve earth removal, deciding if and how further archaeological research is necessary, composing design briefs for archaeological fieldwork and approving site reports. This work is also carried out by contract archaeologists, mostly on an ad-hoc basis, and even by other advisers or civil servants not trained as archaeologists. Next to the abovementioned tasks, municipal archaeologists often also maintain a data system (digital archives and maps) and engage in public outreach activities. The 7% shown in Table 1 (municipalities employing their own

Table 13.1: Archaeological expertise for policy tasks in Dutch municipalities.

	number of municipalities	percentage
municipal archaeologist	29	7%
municipal archaeologist from neighbouring municipality	75	19%
regional archaeologist	164	41%
unknown (either hiring contract archaeologists on ad-hoc basis, or no expertise at all)	135	33%
total number of municipalities in the Netherlands (in 2014)	403	100%

archaeologist) may seem relatively low, but these 29 municipalities cover all large and most of the medium-sized towns in the Netherlands.

Fieldwork is carried out by archaeological companies or by municipal archaeologists themselves. In the current system, 25 municipalities have an excavation permit, which is also needed for borehole surveys. This permit covers only the territory of the municipality, unless there is a form of collaboration that allows the permit to be valid for adjacent municipalities. In exceptional cases, a university or the Cultural Heritage Agency may undertake fieldwork. Sites that have been classified as 'not worth preserving' are sometimes investigated by avocational archaeologists (who do not qualify for an excavation permit).

Stakeholders and their interests

The main archaeology stakeholders in the Dutch planning process probably do not differ a lot from those elsewhere in Europe (see also Van den Dries, this volume). They are:

- The government / competent authorities. As stated above, these are usually local authorities, but in larger infrastructural projects regional and national authorities may play an important role too.
- The so-called initiators: private persons, companies (developers) or other organisations that instigate a building project or infrastructural project and, in line with Valletta, have to pay for any archaeological research involved. Note that governments can also be initiators.
- The archaeologists (commercial, governmental or otherwise) that give policy advice and / or carry out the fieldwork. Note that policy advice can be given by other advisers too.
- The public in the broadest sense of the word: this may include the initiators, but also local residents, schoolchildren or the general public.

Their goals and expectations, or rather their interests, can be summarised as follows, in reverse order:

- The public expect to hear an interesting story about the past: what did the archaeologists find, how did they reconstruct the past, what happened here? What can we (not just 'they') learn from this project; why is this important, and does it justify the time and money spent? Some members of the public want to be involved during the project, not just afterwards.
- The archaeologists aim to carry out meaningful research, they want enough time and resources to document what is being destroyed and to gather the evidence they need in order to answer the research questions formulated. If they work for a commercial company they have additional interests, such as cost recovery and other regular business goals. In any case, policy-advising archaeologists want to be involved at the earliest possible stage of a project.
- The initiators want to be able to realise their projects without unnecessary or (even more importantly) unforeseen loss of time and money. If archaeological research is deemed necessary, they want to hear convincing arguments as to why. If archaeology provides added value to their project, they may consider this an extra.
- And, last but not least, the authorities. Local governments especially must take into account a range of goals and concerns. Municipalities want to realise a high-quality living environment, solve parking problems and air-pollution issues, and at the same time safeguard their cultural heritage: so, do we build an underground car park or not? And if we do, how much money can (or has to) be spent on archaeological research? They are constantly balancing various local and regional interests, and the choices they make depend on many factors. The extent to which archaeology is embedded in the organisation (e.g. the presence of a municipal archaeologist) can make a large difference. But even then the choices made may depend on the person of the alderman, or the archaeologist for that matter. Sometimes other values prevail, which is the prerogative of local governments. They are free to weigh and choose, as long as it is in a justified and well-founded manner. If the authorities decide that archaeological research is necessary, their interests overlap with those of the other stakeholders: they want the work to be carried out well, without unnecessary loss of time and money, and preferably yielding a good story that will enhance the identity of their town or province.

The matter of how to balance these different interests will be addressed further on in this paper. First, there is another question to be answered: is the Dutch system for preventive archaeology primarily a scientific endeavour or a pre-construction service? I would say it is both, as they need not exclude one another. This I will illustrate by outlining the approach adopted in the City of Rotterdam and its neighbouring municipalities.

The Rotterdam approach

Rotterdam is the second largest city in the Netherlands. Its historical centre was destroyed during the Second World War. Because of its continuous (re)development activities, Rotterdam was the first Dutch municipality to avail itself of its own archaeological expertise. When the City of Rotterdam Archaeological Service (BOOR) was founded in 1960, its principal aim was to safeguard the archaeological heritage, and at the same time to facilitate the planning process. This combined responsibility has not changed in the 55 years of BOOR's existence. It means that the archaeologists must demonstrate that their job is more than just clearing away obstacles (and doing so as fast and cheaply as possible) and at the same time that it generates more than just some data of interest only to a small group of academics. The Valletta Convention has only emphasised the necessity of proving both these points, since political and public support is indispensable.

The Rotterdam Archaeological Service also issues policy advice (based on archaeological expertise and knowledge of the area) to 8 neighbouring municipalities. Its fieldwork is mainly limited to

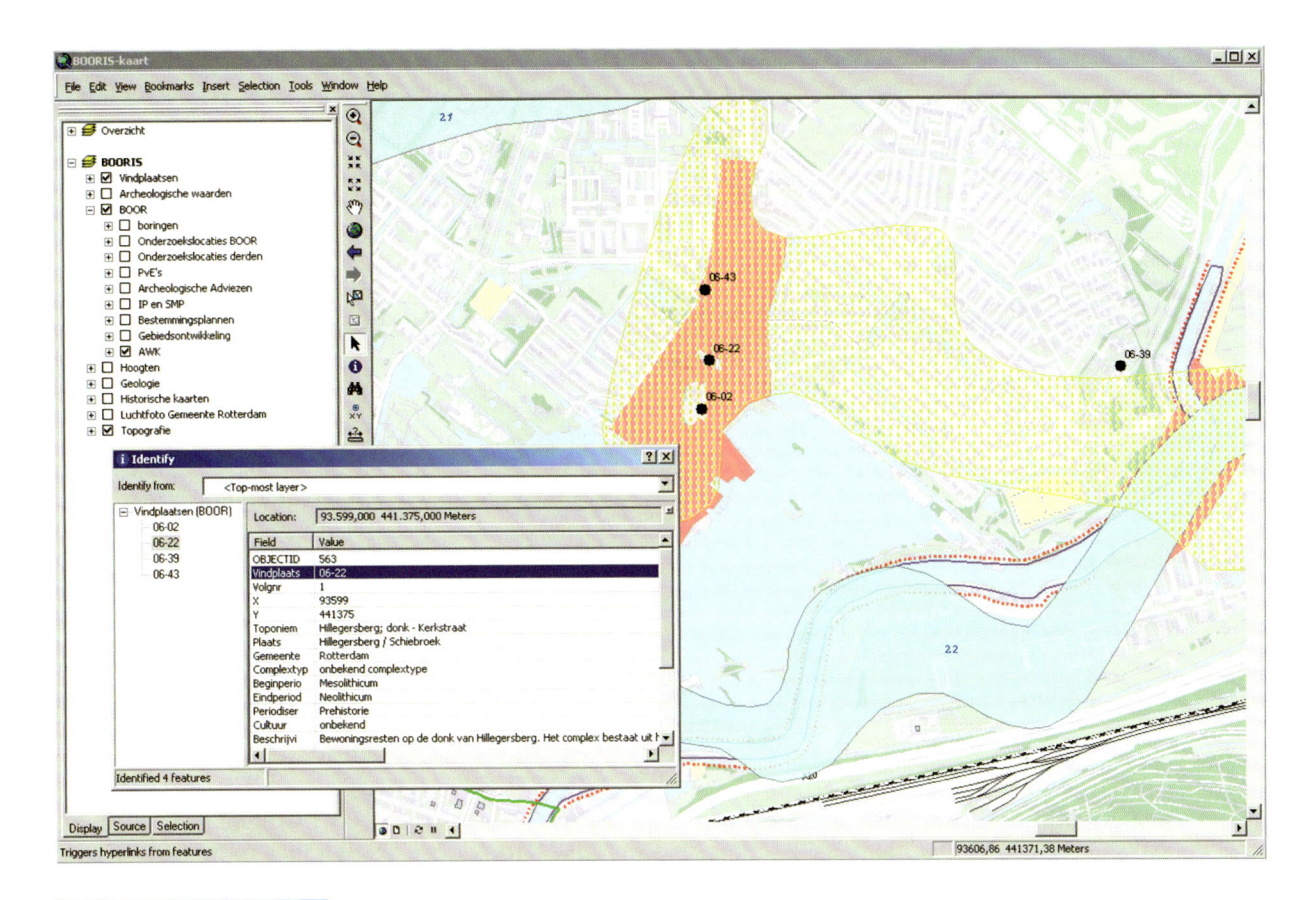

Figure 13.1: A screenshot of the GIS-based information system BOORIS.

Rotterdam itself. The backbone of the advisory practice is BOORIS, which stands for BOOR Information System (Figure 13.1). This system basically consists of GIS-maps connected to several large databases, allowing rapid access to information about locations in the area. This ranges from archaeological data (reports, finds database), geological and historical maps and results from borehole surveys down to individual cores, to municipal data on landfills, contour plots, depths of riverbeds and ports. Also incorporated are all kinds of policy data: zoning plans, design briefs, reports by contract archaeologists and all earlier decisions on building plans. In contrast to a static map showing archaeological values, this system is updated daily and ensures that the archaeologists at BOOR can base their advice on the latest data. Thus the first step, an assessment (or 'quick-scan') of any building project through desk research, is the most important.

The basic statistics for Rotterdam and its neighbours show that a great deal of evaluation work (mainly desk-based research and borehole surveys) is carried out, as opposed to a relatively small number of excavations. Out of every 100 building plans involving earth removal or pile driving that are assessed by (quick-scan) desk research, roughly 80 get a building permit straight away. It is important to note that an initial selection will already have taken place before the plans are handed in to be checked for archaeological consequences: zoning plans define the critical surface area and, more importantly, the depth of any earth removal. Anything smaller or shallower than the defined margins for that particular area will not need a permit.

In the 20 remaining cases evaluative fieldwork is needed, usually a borehole survey. Six out of these 20 get a follow-up with an extra borehole survey or trial trenches. And finally, one out of 100 assessed plans will lead to an excavation. This may be anything between a three-day campaign and a full-scale project lasting several months, although the latter is obviously an exception.

So is this percentage a poor outcome? Do we 'need' more excavations? The answer is no: all other building plans can go ahead without archaeological values being destroyed, either because there are none at all or none classified as 'worth preserving', or because they are not threatened by the building activities in question. The desk research (or quick-scan) and the borehole surveys often lead to this conclusion – this in fact is their purpose, not to generate new insights about the past (see also Van den Dries & Van der Linde 2012, 9). The one site to be excavated, however, is carefully selected and is expected to answer important research questions. This is a project that will be able to inform the public, to tell the story, and to add to our knowledge of the past.

No excavation, no relevance?

Critics – among them archaeologists as well as other stakeholders – have stated that this kind of archaeology (borehole surveys or even basic site reports) produces nothing relevant (Willems 2014, 152–153) and provides no valuable new insights (Raemaekers 2008). Another view is that the low proportion of excavations indicates that too much evaluation work is done, as a result of

Figure 13.2: Municipal archaeologist carrying out a hand-coring survey at the Rotterdam Markthal site. Photo BOOR.

municipalities attributing (too) high archaeological values to most of their territory. Desk research and borehole surveys without subsequent excavation should not have been carried out in the first place (Breimer & Sueur 2015 and Link 1). I strongly disagree, as this certainly does not apply to Rotterdam.

This kind of evaluation work makes possible a building process in which no valuable archaeology is destroyed and, no less importantly, no unnecessary (follow-up) research is carried out. To gain and retain a solid basis of political and public support, the latter is crucial. This was also one of the conclusions in the evaluation of the revised Dutch Monuments Act (Van der Reijden et al. 2011).

To put it boldly, the inclination of archaeologists (as voiced by Willems) is to want more excavations, since only through excavating and using the results in synthetic analyses can more knowledge about the past be generated, whilst initiators and authorities want less evaluative research, since it rarely leads to excavation and hence does not generate new insights about the past that appeal to the public. But obviously the two are connected, as archaeological heritage management is a cyclical process. In order to make choices, to be selective and to spend precious time and money on an excavation that does provide new knowledge, archaeologists need good desk research, borehole surveys and other focused evaluation work.

Making choices

Balancing the various stakeholders' interests also requires making well-founded choices. To this end it is essential to ensure continuity of local and regional knowledge. A handkerchief-sized excavation, or even six boreholes, may yield valuable information if placed in a larger context. Archaeologists with in-depth knowledge of a region or city can do this through stating the research questions in a specification or design brief, but preferably by carrying out the research themselves and using the results in a wider analysis. This requires a local research agenda, linked to regional

Figure 13.3: Mechanical borehole surveying at the Rotterdam Markthal site. Photo BOOR.

Figure 13.4: Describing and analysing cores. Photo BOOR.

and national research programmes, and up-to-date information. In Rotterdam the crucial choice is made as early as possible in the process, i.e. when assessing a building plan, preferably even before a permit is applied for. Since the soil of Rotterdam may harbour archaeological sites at depths of several metres, it is often the depth and specific type of earth removal or pile driving that determines whether or not further research is necessary. This decision can be made quickly and on the basis of tailored desk research. Since the well-argued decisions based on such quick-scans will be directly incorporated in the BOORIS GIS-system, they will be readily available to play a part in future evaluations and policy decisions.

Next to archaeological knowledge, municipal archaeologists also employ another kind of expertise: being able to take into account other aspects and interests. Obviously their job is to assess and value the archaeology at stake and, on the basis of the outcome, to advise on whether to preserve or investigate. But, especially when deciding on the extent of an investigation, it is important to consider time, budget, planning and public benefit as well. Municipal archaeologists work in a public and political context, and creating support is part of what they do. This also means they should be able to explain their choices to other stakeholders, which requires more than just archaeological expertise.

Local expertise combined with high-quality information and data management will allow well-founded choices. Archaeologists have to make and account for these choices, even though they are often suspected of wanting to excavate as much as possible (and in some cases, this might be true). For all other stakeholders, however, being selective in what to investigate and being quick about it is the number-one priority. While this attitude may be prompted by economic considerations, it can still allow meaningful research if the choices are made by archaeologists, rather than by others or by circumstance.

To be improved…

The above can be read as a plea for high-quality municipal archaeology, or a comparable system in which sufficient local knowledge (-management) is employed, in order to make the right choices. Rotterdam may be presented as an example of good practice, but there are still many improvements to be made and problems to be tackled before the Dutch system can work in the same effective way.

Figure 13.5: Carefully selected: the Rotterdam Markthal excavation (2009). Photo BOOR.

Figure 13.6: The 'Time Stairs' in the underground car park at Rotterdam Markthal: archaeological finds displayed at the levels at which they were excavated. Photo Bas Czerwinski.

As outlined at the start of this article, a third of all Dutch municipalities have not structurally organised their archaeological expertise. Some of these even lack any archaeology policy. Obviously not all 400 can or should employ their own archaeologist(s), but there are various good examples of sharing expertise within a region. This should be encouraged and facilitated. The evaluation of the Monuments Act has resulted in a programme through which the Cultural Heritage Agency offers knowledge to municipalities and promotes good practice. However, local authorities that do not value archaeological heritage will probably not take this up, and without locally-based expertise it is impossible to apply a made-to-measure approach.

Some municipalities hire archaeological advice from contract archaeologists. If this is done on a permanent basis it has a better chance of working out, since the archaeologist in question works in the area (and within the local organisation) at least several days a week. If advice is sought per project (e.g. assessing a building plan, writing a design brief, checking a site report) it will not be effective in the long run, since little local expertise is being accumulated. The same goes for fieldwork: contract archaeologists working all over the country will not have an opportunity to gather insights into the particular archaeological, historical and soil characteristics of a specific area.

The economic crisis combined with the free market for fieldwork has sparked fierce competition on price among archaeological companies in the Netherlands. In some cases this has led to poor-quality research, which has nevertheless been accepted by the commissioning parties (the initiators), who have other priorities (see also Van den Dries, this volume). Unfortunately, the current system of quality assurance has no way of preventing this from happening. Even worse are archaeologists advising to carry out (further) research when this is actually pointless – either because the evaluative research was not conducted properly and the conclusion was drawn without due deliberation, or perhaps even as a means of creating work for themselves. This will result in dwindling support from stakeholders, politicians and the public.

There definitely is benefit to be gained from closer collaboration between municipal and regional archaeologists, universities, museums and archaeological companies. In several cases this has proved to work well, such as the Ancestral Mounds Project (Link 2), but more often these parties misjudge each other or simply are unaware of what others are working on. Another problem with development-led archaeology is the lack of time and funding for synthetic analyses to collate the evidence from site reports, although there have been initiatives to address this issue, such as the 'Oogst van Malta' programme.

I should like to conclude this article with some thoughts on the goals of the Faro Convention. Whereas Valletta focused on the need to conserve archaeological heritage and the search for ways in which to protect it, Faro is about why heritage is protected and for whom, and how to involve these communities. 'Heritage communities', involving the public rather than just informing them, and 'community archaeology': these are concepts that are only tentatively being tested by Dutch archaeologists (or other heritage managers, for that matter). More traditional ways of informing the public, such as exhibitions, guided tours, books, lectures or websites, are plentiful and there are many excellent examples of these. Fully involving the public, especially in making choices, is usually considered a bridge too far. Still, archaeologists with in-depth knowledge of their region or city, and by this I mean a close acquaintance with both the archaeology and the present-day inhabitants, should be able to involve the public in new ways.

References

Breimer, J.N.W. & Sueur, C. 2015: *Archeologie in de ruimtelijke ordening: een schot hagel!*

Buitelaar, S. 2015: Bodemschat is kostenpost. *Binnenlands Bestuur* 02, 24–6.

Raemaekers, D.C.M. 2008: Het einde van Malta. *Archeobrief* 12 (1), 17–20.

Van den Dries, M. & Van der Linde, S. 2012: Twenty years after Malta: archaeological heritage as a source of collective memory and scientific study anno 2012. *Analecta Praehistorica Leidensia* 43/44, 9–19.

Van der Reijden, H., Keers, G. & Van Rossum, H. 2011: *Ruimte voor archeologie. Synthese van de themaveldrapportages*, RIGO report P18090, Amsterdam.

Vonk, S. & Berkvens, R. 2014: Heeft regionale archeologie de toekomst?, *Archeobrief* 18 (4), 18–23.

Willems, W.J.H. 2014: Malta and its consequences: a mixed blessing, in V. van der Haas & P. Schut (eds): *The Valletta Convention: Twenty Years After Benefits, Problems, Challenges*. EAC Occasional Paper No. 9, Brussels, 151–6.

Websites

Link 1: http://www.a-is-m.nl/?pid=9 (accessed 08.09.2015).

Link 2: http://www.grafheuvels.nl (accessed 08.09.2015).

14 | The legal basis and organisation of rescue archaeology in Poland

Michał Grabowski

Abstract: At the beginning of the 1990s, Poland underwent not only a transformation of its political system, preceded by the fall of communism, but also witnessed an ensuing period of unprecedented development of its national infrastructure and construction industry. Simultaneously, a substantial debate commenced about the role of rescue archaeology in the advancement of science. The debate is still ongoing. Recent changes made to the regulation of archaeological work undertaken at development sites reduced rescue archaeology to a mere service subordinate to the construction industry. This shows that the introduction of excessively liberal legal regulations in a sector which, by its very nature, requires careful control and management, has a negative impact and makes the protection and guardianship of archaeological heritage a very difficult task. And even though most of the aforementioned changes were fortunately revoked only a few months later, this incident has demonstrated that there is a lack of concept at government level for a coherent conservation policy that will define standards for archaeological work and the subsequent study and storage of finds.

Keywords: professional standards, rescue archaeology, large-scale archaeological projects

Beginnings

In the late 1980s and early 1990s, the reborn Poland witnessed more than the long-expected memorable change of its political system. Communism had fallen, followed by the unprecedented development of the country's infrastructure and construction industry. Along with large-scale infrastructural projects, numerous archaeological rescue operations commenced. This immediately triggered discussion about the role of rescue archaeology in the advancement of science and the process of recording and studying historical monuments (Gąssowski 2000). A similar debate had taken place just after Poland had regained independence after the First World War. It concerned the overarching role of archaeology as a branch of science intrinsic to the development of research into historical monuments.

Eventually, the debates resulted, in 1929, in the creation of the National Board of Archaeological Inspectors, followed by the Act of 6 March 1928 on the Protection of Monuments, and the ordinance on the newly created national registry of archaeological

Figure 14.1: Smólsk, Site 2/10 (Włocławski District, Łódzkie Voivodeship). Bird's-eye view of the archaeological site during excavation in June 2009; red lines mark the projected course of the A1 motorway (© Przemysław Muzolf).

Figure 14.2: Smólsk, Site. 2/10 (Włocławski District, Łódzkie Voivodeship), A1 motorway. Plan of pits and postholes recorded during archaeological excavations (© Przemysław Muzolf).

monuments. Coming just before the outbreak of the Second World War, this was one of the most modern and innovative legislative solutions in Europe. In the 1930s, numerous archaeological sites were recorded, secured and catalogued. The galloping development and industrialisation of the Republic of Poland favoured such actions. The Second World War brought an end to all works; many archaeologists were killed, and numerous museum collections were stolen, destroyed and lost.

The post-war rebuilding of Poland, undertaken in the 1950s and 1960s, initiated a new series of archaeological rescue projects: the rebuilding of Warsaw and Gdansk, and the construction of the Tadeusz Sendzimir Steelworks in Cracow were the best examples of these initiatives. Sadly, archaeology simultaneously

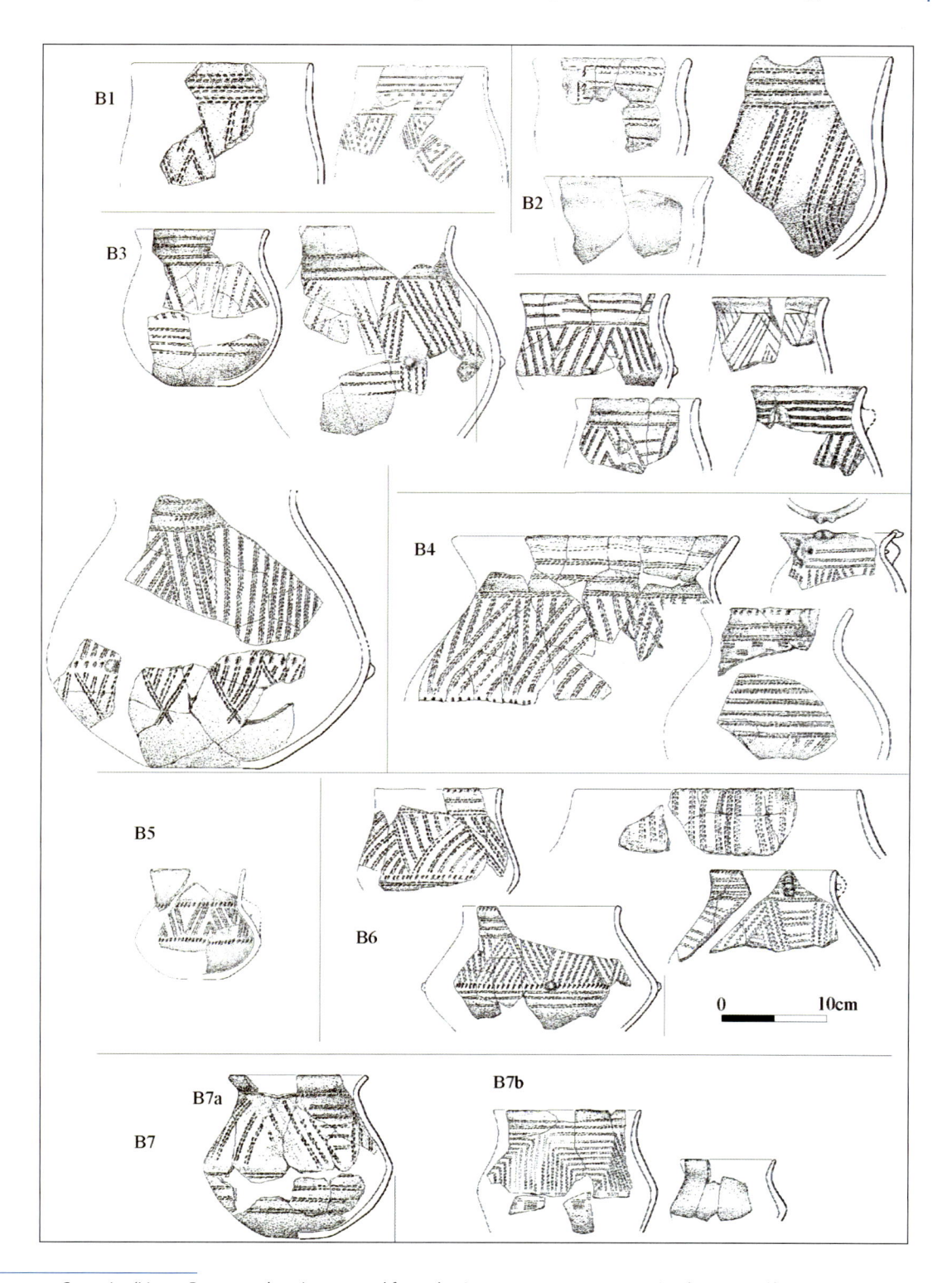

Figure 14.3: Ceramics (Linear Pottery culture) recovered from the A1 motorway excavations (© Błażej Muzolf).

became a quasi-tool used in ideological warfare by the communist regime.

During the communist era there seems to have been a simple division of roles within the archaeological community in Poland: it was generally assumed that universities were to educate students, the Polish Academy of Sciences fostered research activities, archaeological museums collected artefacts, and the Polish Studios for Conservation of Cultural Property carried out survey and rescue works, the latter being considered by many to be of little scientific significance.

The most important task for archaeologists was to unearth artefacts and examine them quietly behind the closed doors of their offices.

A new chance

After 1989 large investments and developments took place, such as the Yamal-Europe pipeline, followed by the construction of north-south and east-west motorways (the A1, A2 and A4) and expressways. All of the above construction projects were preceded by archaeological excavations.

Figure 14.4: Mediaeval wells during excavation on theA2 motorway (© Grzegorz Kałwak).

Figure 14.5: Bronze pin (Przeworsk culture) excavated during work on the A2 motorway (© Grzegorz Kałwak).

Numerous archaeological sites, rich in artefacts and historical evidence, discovered during pre-construction surveys made the researchers, conservation services and politicians realise how complex and important the issues associated with rescue archaeology were: the problems and challenges arising from the exploration of vast areas, the conservation and storage of recovered finds, as well as the writing up and, finally, publication of research outcomes. It became evident that archaeological rescue excavations had become an enormous enterprise. Such large-scale projects involved financial, logistical, organisational and scientific operations, the cooperation of specialists representing various disciplines and the accessibility of specific storage facilities, conservation studios and laboratories (Bukowski 2001). The Act of 27 October 1994 on Toll Roads, as adopted by the Polish parliament, followed by the establishment of a national programme of motorway construction in Poland, became a turning point and crucial factor in the matter under discussion. The planned construction of new motorways and two-lane expressways (2,300 kilometres in total) made the exploration of several thousand archaeological sites imperative. According to Polish law and regulations, each archaeological site (as specified in the Act of 23 July 2003 on the Protection of Monuments and the Guardianship of Monuments, and treated *expressis verbis* as a historical monument) must not be destroyed or abandoned without being previously explored and documented.

Organisation and legal basis

In the face of new challenges for archaeology, in 1995 the Minister of Culture and Art established a new institution – the Centre for Archaeological Rescue Research (Grabowski 2012). The Centre was established as a result of a long-term dispute about the role of rescue archaeology in Poland, shortly after which Poland ratified the Valletta Treaty (formally the European Convention on the Protection of the Archaeological Heritage). The Centre's goals were precisely specified: protection and documentation of endangered cultural heritage monuments located within areas affected by the planned construction of motorways and expressways. This assignment was implemented in cooperation with the National Service for the Protection of Monuments, the General Directorate of the National Roads and Motorways as well as other national agencies, local authorities and NGOs. The Centre was also responsible for the organisation and supervision of archaeological rescue excavations.

Unfortunately, the lack of qualified staff, unsatisfactory level of government funding as well as personal antagonisms made it impossible for the Centre to carry out its own independent works with the cooperation of its permanent employees and other individual researchers. The Centre, fully financed from the state budget, merely entrusted other contractors and entities with the task of carrying out the rescue excavations. They were supervised by the Centre and the Provincial Inspectors of Monuments.

In my opinion, this was a missed ideal opportunity to build a central institution for rescue archaeology in Poland – a place with its own storage facilities, laboratories and, above all, a never-created central institution which would set professional standards for fieldwork and documentation.

Instead, a specific mechanism emerged in the mid-1990s: most major archaeological institutions, museums, the Polish Academy of Sciences and university institutes were involved in the rescue excavation process. Private archaeological companies functioned as subcontractors for fieldwork and post-processing. The selection of a particular firm was based on the single-source procurement procedure, the determining factors being the company's staff capacity, its experience within the given region and the proposed costs of the work required.

In 2002 the Minister of Culture transformed the Centre for Archaeological Rescue Research, whose work focused mainly on rescue excavations along the routes of the planned motorways, into the Archaeological Heritage Preservation Centre. Its goals included the implementation of national policy on the protection of monuments, the organisation of rescue excavations, development and implementation of new methods of protection, as well as raising social awareness of the necessity of national heritage protection. The selection of archaeological contractors was based on a partly regulated system which acknowledged the primacy of major regional archaeological institutions, so that excavation and research in particular regions of Poland would be carried out by teams representing the highest level of regional archaeological knowledge and experience. Smaller institutions and private archaeological business entities worked under their supervision. This system was approved by the Agency for the Construction of Motorways (the later National Directorate). The above principles were to be taken into consideration when selecting the main contractor. The Centre functioned as the chief supervisory body under the Ministry of Culture, but was managed by an independent director as a legal entity separate from the government administration. This system remained in operation for several years.

In 2006, the Centre's director and his deputy, along with an employee of the National Directorate were charged with corruption relating to the process of selecting potential contractors, which initiated a series of perturbing political decisions, perceived as worrisome for the archaeological community. Some well-known and hitherto respected archaeologists were also charged with the offence of offering bribes in order to secure contracts. After a lengthy legal battle, they were all convicted, with the former Centre's Director serving a several-year-long jail sentence. This scandal, undermining all trust in archaeologists, both from the perspective of government and public opinion, started an avalanche of damaging political decisions. The uneasy results are still evident today. At first it caused problems in managing archaeological heritage in Poland and gave rise to a complete lack of trust in archaeologists on the part of decision makers. Archaeological issues have fallen significantly down the government agenda. That is why a decision was made at the Ministry to merge the Archaeological Heritage Protection Centre with the National Heritage Board of Poland as of 2007.

After the corruption scandal, the public sector entities (local government bodies), in accordance with the new recommendation issued in 2007 by the Director of the Public Procurement Office, began organising contracts for archaeological works based on the public procurement law. Trying to be as transparent as possible, it soon became evident that the only decisive factor was now the cost. This made those bidding for contracts compete solely on the basis of proposed rates for archaeological rescue excavations, which fell quickly and dramatically.

Paradoxically, and as a result of this situation, smaller archaeological companies with lower costs started to win tenders for large rescue excavations conducted before the commencement of large-scale construction projects. It turned out that they were able to adapt to the free market and decreasing rates perfectly, mainly due to low labour costs. Today, it is estimated that approximately 650 privately-run small companies actively participate in the market of archaeological services.

According to the report for the years 2008-2010 (Link 1) created by the National Heritage Board of Poland, there was a steady rise in the presence of commercial archaeology within all types of archaeological fieldwork (Czopek & Pelisiak 2014).

Competing solely on the basis of price has inevitably led to a reduction in the scope and quality of the archaeological services offered. This is reflected both at the fieldwork stage and during further research, when analytical methods are limited to a bare minimum. The dissemination of results is an even more difficult issue, as the law only requires a very general report, which does not have to be published. It is therefore very rarely that investors or contractors decide to disseminate the results of their rescue operations using their own means, and promotion of excavation results to the general public is simply non-existent.

Since 2011 the National Heritage Board of Poland has made several attempts to change this situation, both at policy level and in terms of promoting rescue excavation results. Unfortunately, 'the lowest price wins' approach in rescue archaeology contracts is satisfactory for all stakeholders apart from archaeologists themselves. Both investors and the government save money, since major infrastructure construction projects like motorways are financed through public funds. The general public is a potential ally for archaeologists; but with non-existent programmes promoting the value of archaeological heritage to communities, this remains an abstract idea.

Conclusions

Unluckily, any actions initiated by different representatives of the archaeology sector over the past few years (National Heritage Board of Poland, Scientific Association of Polish Archaeologists and other professional NGOs, researchers, private archaeological companies) in the field of creating coherent conservation policy which would ensure the adequate quality of work and the relevant post-processing and storage of artefacts do not go hand-in-hand with the government's approach, visible both at policy level and in the new legislation.

Recent changes to the regulations governing archaeological work undertaken at development sites, aimed at simplifying the work of developers, reduced rescue archaeology to a mere service subordinate to the construction industry. This shows that the introduction of excessively liberal legal regulations in a sector which, by its very nature, requires careful control and management, has a negative impact and makes the protection and guardianship of archaeological heritage a very difficult task. And even though most of these changes were fortunately revoked only few months later, the incident has demonstrated there is a lack of concept at government level for a coherent conservation policy that will define standards for archaeological work and the subsequent study and storage of finds.

Total outlays for archaeological rescue work in Poland over the last 20 years amount to more than 600 million Polish zlotys, constituting the equivalent of around €150 million. Thousands of archaeological sites have been explored within this budget. The decrease in the prices paid for archaeological services in recent years is considered beneficial by the public, government and private constructors. Archaeology in general is not considered a priority by the Ministry of Culture and National Heritage and, therefore, archaeologists, who are neither obliged nor encouraged to publish or promote their results have lost an opportunity to present how significant their research of the last 20 years has been.

Some discoveries made in Poland during archaeological rescue excavations within the last 20 years are indeed quite spectacular, and it is a great pity that they have not become common public knowledge. This is, however, due more to insufficient access to information than to lack of interest (Florjanowicz 2015). Moreover, these two decades of unique experience can constitute the basis for formulating new standards of research and documentation, and for establishing a brand new rescue archaeology system, which is currently an essential requirement in Poland. I believe, that with the support of the archaeological community such a new system will be created in the future.

References

Bukowski, Z. 2001: Od Redakcji, in Z. Bukowski (ed.): *Raport 96-99*, ORBA, Warsaw, 58.

Czopek, S. & Pelisiak, A. 2014: Autostrady i co dalej?, in: S. Kadrow (ed.): *Raport 9*, Narodowy Instytut Dziedzictwa, Warsaw, 429-30.

Florjanowicz, P. 2015: How should we manage the archaeological sources in Poland?, in P.A.C Schut, D. Scharff & L. C. de Wit (eds): *Setting the Agenda: Giving New Meaning to the European Archaeological Heritage*, EAC Occasional Paper No. 10, 137-42.

Gąssowski, J. 2000: *The Great Archaeological Dig. Pipeline of Archaeological Treasures,* Catalogue of Exhibition, RuRoPol GAZ S.A., Poznań Prehistoric Society, Poznań.

Grabowski, M. 2012: Od Ośrodka Dokumentacji Zabytków do Narodowego Instytutu Dziedzictwa. Rewolucyjna zmiana czy procesowa kontynuacja zadań z zakresu archeologii?, *Ochrona Zabytków* 1-2, 73-80.

Links

Link 1: http://www.nid.pl/pl/Dla_specjalistow/Badania_i_dokumentacja/zabytki-archeologiczne/badania_archeologiczne_zestawienia (accessed 28.08.2015).

15 | Preventive archaeology in Wallonia: perspectives

Alain Guillot-Pingue

Abstract: In this article, the author summarises the evolution of preventive archaeology in Wallonia during the last 25 years, after the transformation of Belgium into a federal state. In 1989, Walloon archaeology was incorporated into the remit of the General Directorate of Spatial Planning, Housing and Heritage. The author also depicts the future structures and tools that will progressively improve the dialogue between stakeholders, enable rational choices and answer citizens' expectations.

Keywords: planning, new codex, funding, structural changes, operational tools

Past history

By the end of the 1970s, Belgium was still one of the few countries in Europe to have no archaeological legislation. Excavations were primarily planned and conducted by universities, scientific institutions and the National Service for Excavations.

Universities, alerted by the frequent destructions of archaeological remains, set up an interventions cell: SOS Excavations. It was later renamed Rescue Archaeology.

Ten years later, still under the leadership of universities, atlases of the archaeological subsoil in Wallonia were gradually released, accompanied by a ministerial circular. These were the first steps towards a planned preventive archaeology and closer links with the regional planning services. Since 1989, archaeology has been incorporated into the spatial planning process. Since 1991, it has benefited from an archaeological decree, anticipating the terms of the Valletta Convention (revised in 1999).

Present

The Directorate of Archaeology of the Public Service of Wallonia is divided between a central Directorate of the Department for Heritage situated in the regional capital, Namur, and five decentralised hubs in each province, led by a Director of Spatial Planning. The central Directorate is mainly composed of experts (by historical periods and specialties, including ceramology, restoration, geomorphology, etc.). The funding of archaeological operations in Wallonia is mainly provided by the Public Service.

The decentralised hubs are in charge of the archaeological operations and are mainly composed of field archaeologists, technicians and diggers.

Like all countries/regions of Europe, Wallonia has suffered from the effects of the 2008 crisis with significant budget cuts, the temporary non-replacement of staff, other economic and social priorities taking precedence, etc. Heritage has suffered for some years from this crisis, and it is not a priority in terms of budget and staff resources.

It is therefore urgent to review the work and the funding of preventive archaeology operations.

Added to this is another factor, which is the major overhaul of the Code for Spatial Planning (CWATUP 2015). A new code was adopted by the Walloon Government in March 2014 but it is not in force yet (CoDt). The new code reinforces the power of local authorities, which implies a dilution of the legal initiative into 262 entities and means that the state archaeology service should take action as early as possible in order to be efficient. Thus, two tasks are necessary for the management of archaeology. The first is to integrate archaeology into the new Spatial Planning Code; the second is to ensure that it is given full consideration in the Heritage Code, which is at present being drafted and should be adopted during the current legislature (20142019). A great opportunity thus stands before Walloon archaeology.

Future

Structural changes

Walloon archaeology operates with some decentralisation: central management and five local hubs in the provinces. Even if the two bodies confer monthly in order to achieve a comprehensive policy, the fact remains that some discrepancies exist. It is therefore urgent to review the current administrative organisation chart.

The idea is not new; but it will take time for it to happen. Recently, the Walloon Government has recognised this and allowed the remodelling that is currently underway. Since 2015 the so-called 'provincial' archaeologists have been subordinate to the Director of Archaeology. However, a partnership has been maintained between the 'provincial' spatial planning directors and archaeologists so that they remain largely integrated into the spatial planning process. This reorganisation has streamlined expenses, the pooling of human and material resources and the development of a common action policy. Among the range of improvements, emphasis has been placed on the need to work on a project-by-project basis and thus avoid costly and sometimes unnecessary interventions. This task, given the randomness of this hidden heritage, will not be easy and will require detailed desk-based assessments, non-intrusive surveys and perhaps a final diagnosis. This approach, particularly in urban archaeology, will connect links that have been unclear up to now. In terms of rural archaeology, it also seems efficient to move the teams of archaeologists between the provinces depending on the operations which need to be carried out. This will improve both efficiency and the quality of data recording, as well as subsequent interpretation. Finally, the Public Service of Wallonia, should foster closer links with enthusiasts and amateurs who, through their knowledge of the field, can also fuel our knowledge and protect our heritage.

Implementation of operational tools

Preventive interventions must also be provided with tools that operate fairly and uniformly in the Walloon territory. Priorities must be set in terms of information, operational choices, traceability of interventions and new technologies in non-intrusive surveys.

Information

As elsewhere, the emphasis is on information that is based on the earliest possible moment in the planning process. On the other hand, a new archaeological operation dossier (DOA) has recently been implemented, which aims to formalise the traceability of the archaeological work.

Inventory

For nearly 25 years, the Spatial Planning Code has mentioned the keeping of an archaeological inventory. This idea, however interesting, was never considered acceptable by archaeologists since it would have been tantamount to producing a distribution map of known sites, when it seems clear that it is in the unknown sites that the archaeological potential is most threatened. Also, these maps, if they were to be published, might tempt some malicious looters.

Archaeological mapping

It was essential to produce a map of the archaeological potential of Wallonia a map to inform the various stakeholders (decision takers, applicants, municipalities, consultants, notaries, etc.) about it. This led to the creation of a map-based information tool which uses four colour-coded zones. In the blue zone, a notice is required for any application for a planning permit, urbanisation permit or 'single' permit. In the green zone, the notice is required when the total area of the permit is at least 5,000 m². In the yellow zone, a notice is required when the total area of the permit is at least 10,000 m². In the grey zone, no notice is required.

The mapping is considered as an administrative and guidance document that leads to the identification of the required services. Unlike France, it will not be enforceable against third parties. The map is not yet officially online because it is in a test phase.

Operational choices

Alerted by a planner discovering the archaeological potential on the map, the archaeologist will decide whether or not to intervene.

Archaeological Operation Dossier (DOA)

Any archaeological field operation involves a significant number of steps and stakeholders. In Wallonia, there is a single document that ensures the necessary coordination: the DOA. It is a management tool that is initiated and maintained by the archaeologist in charge of the operation. It gathers all the data relevant to the implementation of the project, from the preparation of the intervention, to post-excavation research publication and archiving of the site. As a communication tool, it is the interface between archaeologists, scientific staff, administration and developers. In this way, it coordinates the priorities of the Directorate of Archaeology through objective and sound management.

LiDAR

The entire territory of Wallonia is covered by a LiDAR operation conducted by all branches of the combined Public Service of Wallonia. The survey resolution (one point per metre) enables details to be mapped at the level of macro-relief. This remote sensing level is insufficient to detect most archaeological features. The aim is to achieve a finer resolution of 520 cm. In the coming months it is planned to buy a drone, or UAV equipped with a more precise LiDAR system. In the meantime, a terrestrial LiDAR unit (3D scanner) is used which has a resolution of 1 to 2 mm.

The financing of archaeology and relationships between planners and archaeologists

Following the recent regional elections, the new Walloon Government has included in its Regional Policy Statement the possibility of creating a fund for preventive archaeology. The goal is to keep the public service in full control, while managing funds coming from the private sector, thus avoiding a switch to a commercial archaeology system.

Publications

Publication of scientific results remains a priority and a duty *vis-a-vis* the academic and scientific world, and also the wider public. A very tight schedule of upcoming publications has been drawn up. Although regularity has had its ups and downs, the 34th volume of Walloon scientific documents (Études et documents) has just been published. Thirty-four studies in 25 years.

In addition, there is also the annual output of the *Chronical of Archaeology in Wallonia* (*Chronique de l'Archéologie Wallonne*), which provides an overview of the interventions of a given year carried out by all the actors in the archaeological sphere in Wallonia (universities, museums, associations).

Valletta and Faro

In the spirit of the Valletta and Faro Conventions, the Directorate of Archaeology marked the milestone of its 25th year of existence in 2014 by organising a public awareness campaign. No less than 190 events were organised with 115 partners: seminars, conferences, publications, exhibitions, site visits, activities, excursions and other festivities. This 'Archaeology, everywhere for everybody' has been very successful and has helped to affirm the visibility of the work of archaeologists and its importance for every citizen.

Conclusions

The sound management of preventive archaeology is a quest which is not always easy to fulfil and must constantly be adapted to the prevailing socio-economic climate and political changes. One must learn to negotiate with stakeholders, policy makers and planners by keeping in mind the principle of proportionality. Moreover, one must never lose sight of the responsibility to nurture science. Finally, one must communicate to the wider public in order to justify that mission and its cost to the tax payer. Nothing is ever guaranteed, so one must always be utterly professional. It is the only way to keep Walloon preventive archaeology in Wallonia in the public realm, and to avoid a commercial archaeology system.

References

CWATUP 2015: *Code wallon de l'Aménagement du territoire, de l'Urbanisme et du Patrimoine* (CWATUP), The archaeological decree of 1991 (revised in 1999), TITRE IV. De l'archéologie art. 232 to 252 & 515 to 529, http://dgo4.spw.wallonie.be/DGATLP/DGATLP/pages/DGATLP/Dwnld/CWATUP.pdf (accessed 06.01.2016).

CoDt: *Code du Développement Territorial* adopted by the Walloon Government in 2014 to replace CWATUP in the coming months.

Session 3

Assuring quality

Examples of NRA exhibitions. Upper left: photo of 'Hidden Landscape: searching for the lost Kingdom of Mide' (© Studio Lab); lower left: photo of 'Migrants, Mariners, Merchants' exhibition (© Studio Lab); right: display board from 'ASI: Archaeological Scene Investigation in North Louth' (© County Museum Dundalk and NRA), (from the Irish case study, see Swan).

16 | Is everybody happy? User satisfaction after ten years of quality management in development-led archaeology in Europe

Monique H. van den Dries

Abstract: The topic of the last session during the symposium of the annual meeting of the Europae Archaeologiae Consilium (EAC), in Lisbon 2015, was assuring quality in development-led or preventive archaeology. It was stated in the conference announcement that one of the greatest challenges of development-led or preventive archaeology is to determine *how* to monitor quality – the quality of both the archaeological research process and the valorisation of the results. The latter includes the process of sharing with different target groups (researchers and the public) and of ensuring a lasting public benefit. The suggestion I will discuss in this article is to look at it from the perspective of the users or customers of development-led archaeology and to try to 'measure' how satisfied they are.

Keywords: development-led archaeology, quality management, user groups, stakeholders, user (customer) satisfaction

Introduction – user satisfaction

It is about ten years ago, at the start of the new millennium, that the archaeological sector in Europe started to seriously and intensively discuss issues of quality management. The main reason for this attention to the quality of archaeological work was the emergence and rapid expansion of a contract-based, and sometimes commercial, practice in development-led archaeology in Europe. Outside of Europe, like in the USA and Canada, quality management had been addressed for many years already, but not in Europe. Back then, we had only just started to explore and discuss suitable ways and instruments to implement and achieve quality assurance and quality management in the newly developing European archaeological practice.

In 2005 and 2006, the topic was on the agenda during two international conferences. The first took place in Rosas, Spain and was organised by the Europae Archaeologiae Consilium (EAC), the other was in Dublin, during the World Archaeological Congress (WAC), when Willem Willems and I organised a session on the issue of quality management. On the basis of the contributions to these two meetings we composed a volume on quality management in archaeology (Willems & Van den Dries 2007). It consisted of ten accounts – eight from European countries and two from North-America – on the approaches applied to maintain academic standards of work under the changing circumstances that had been initiated by development-led research and heritage management policy in archaeology, which took its course in the 1990s.

In 2015, the topic of assuring quality in development-led or preventive archaeology was again on the agenda. It was one of the main issues during the symposium of the annual meeting of the EAC in Lisbon.

The conference announcement stated that one of the greatest challenges of development-led or preventive archaeology is to determine *how* to monitor quality: the quality, on the one hand, of the archaeological research process and, on the other hand, of the valorisation of the results. The latter includes the process of sharing with different target groups (researchers and the public) and of ensuring a lasting public benefit. As the co-editor of *Quality Management in Archaeology* (2007) I was invited to provide an introduction to this session. Given the fact that this EAC meeting in Lisbon was taking place ten years after the first meeting on quality assurance in Rosas (Spain), it occurred as an appropriate moment to review what had happened in this past decade. But in order to be able to assess the state of affairs ten years later, it was essential to first ask ourselves what we mean when we talk about managing and assuring quality in our sector. We would need to know what quality we are looking for and what criteria we use to evaluate it.

For this article, I adopted the quality assessment approach that is customary in the professional domain of quality management. In this domain it is generally acknowledged that quality is relative; it is assessed by establishing whether a service or a product fulfils certain requirements. In order to assess quality, these requirements have to be defined, and they usually relate to the users of the products or services for which the requirements are defined. The International Organization for Standardization (ISO), for instance, mentions that the aim of standards for quality management is to meet the needs of customers (www.iso.org). Quality is defined as 'The totality of features and characteristics of a product or service that bear on its ability to satisfy stated or implied needs'. Consequently, when we talk about quality in our sector, and if we want to assess the level of quality we have achieved, we should evaluate whether we satisfy the needs and requirements of our specific customers.

This implies that we first of all need to define *who* our (main) customers or users are and *what* their quality requirements or needs are.

In defining who our user groups are, my point of departure is that the contemporary archaeological sector, which works primarily in a development-led archaeological context, cannot afford to act as if it is living on an island. We need to acknowledge that, in this context, archaeology is strongly tied with society and fully dependent on the goodwill of society because society needs to be willing to pay for what archaeologists do. It means that if the sector wants to have public support, it needs a positive public opinion. And only if there is an interest in its deliverables will there be continuing support to produce them.

Taking this perspective further, it follows that we should be interested to learn how happy the customers of our services and users of our products are with what we do and produce. During the EAC meeting, I therefore addressed the issue of quality management in relation to four main groups of customers or end-users: the authorities that commission research in the context of development-led archaeology; us, the sector, as we are the consumers of each other's knowledge products; the developers as the people and organisations who cause a lot of archaeological research and subsequently pay for most of it; and the public. The aim of this article is to repeat that exercise in a more elaborate way. I want to discuss some qualitative and quantitative indications that may provide insights into customer satisfaction across the board for development-led archaeology in Europe. Even though the heritage management systems in Europe continue to differ hugely from one nation state to the other (see for instance other contributions to this volume, and Bozóki-Ernyey 2007; Willems & Van den Dries 2007; Webley et al. 2012; Guermandi & Salas Rossenbach 2013; Van der Haas & Schut 2014), the majority of the archaeological work in Europe is prompted by land and town development, and the principle of developer funding is now established across most of Europe too (even if the Malta Convention has not been signed, such as in the case of Iceland). Consequently, archaeologists in most European countries do serve these four main categories of clients or customers.

Authorities

Requirements
The first group we should look at if we want to evaluate user satisfaction, are the national authorities, in particular where it concerns their role as decision takers and policy makers. As in most countries the authorities have the responsibility to take care of the archaeological resource as a public task, this usually means that their requirements are laid down in laws, regulations and policies. Within the context of development-led archaeology, I take it that their main requirement or need is that the sector complies with the national law(s) on antiquities/archaeology/monuments/heritage, i.e. that the policies are being followed and applied. In Europe this generally means that the safeguarding of the archaeological record is taken care of along the principles set out by the Malta Convention, as nearly all states have at least signed and ratified it, and many are in the process of implementation. In most European countries the national heritage laws aim to prevent unauthorised excavations, to register archaeological sites, to define conditions for carrying out research, to arrange the financing of research, to designate and protect scheduled monuments, and to determine the ownership of finds, etc. This means, in terms of the three levels to quality management that are usually distinguished (i.e. the management of the process of archaeological work, of the products of this work, and of the people that carry it out), that most often the third level is covered through regulations. A direct reference to this requirement can be found in Article 3 of the Malta Convention, which states that 'To preserve the archaeological heritage and guarantee the scientific significance of archaeological research work, each Party undertakes: (…) to ensure that excavations and other potentially destructive techniques are carried out only by qualified, specially authorised persons'. The first and second levels are usually not included in legislation. National governments usually leave it to the sector, through training and education and through self-regulation, like codes of practice, codes of conduct and other guidelines, to establish the professional quality requirements regarding the process of the work and its results (the products). As the quality of the deliverables usually is not defined through legislation, my assumption for this article is that most authorities do not have specific requirements for the quality and scientific value of the archaeological sector's reports, services and other deliverables. In the case of authorities that commission development-induced research, their most important requirement probably is that archaeologists deliver in time and according to the project outlines as defined, to make sure development and construction work can continue without unexpected delays and costs.

A second main requirement of authorities is that *knowledge* – in whatever form – is given in return for their investment in archaeological research. Any mayor, member of parliament, or minister of culture, when talking about archaeology, always underlines its values for society. In particular cultural education, tourism and active participation in cultural activities are seen as high-impact factors for socio-economic development. And increasingly authorities talk about inclusion, social cohesion, personal wellbeing and quality of life as objectives to be achieved in the context of cultural heritage management. This is not merely national policy; it is highly influenced by the policies of the European authorities too, as they actively propagate the idea that every person has the right to engage with the cultural heritage of their choice. It is, for instance, laid down in the Council of Europe's Faro Convention on the Value of Cultural Heritage for Society (2005), which aims to involve everyone who so wishes in the process of defining and managing cultural heritage. The message is also actively disseminated through the cultural policy of the European Commission, which says that cultural heritage enriches the life of citizens and that it is an important resource for economic growth, employment and social cohesion (for an overview of the EU policies on heritage see Florjanowicz, this volume). So, taking the audience into account and producing knowledge for society has become a moral obligation, and, as such, a user requirement. We should therefore include it in an evaluation of customer satisfaction, despite the fact that knowledge dissemination and public outreach is usually not explicitly included in national legislation on archaeological heritage management. In fact, in most European regulations on archaeological heritage protection, the commissioning and tendering processes do not include scientific analysis and publication beyond the production of basic site reports. Only Sweden seems to have another approach to this. In the implementation regulation of the Swedish Heritage Act, the concept of 'good scientific quality' is characterised as 'the use of scientific methods to acquire meaningful knowledge of relevance to authorities, research, and the general public. This requires that the results be made available and useful to the various interested parties.' (KRFS 2007:2, in Andersson et al. 2010, 18).

User satisfaction

In order to evaluate across the board for Europe the level of customer satisfaction of authorities, ideally we would need to have dedicated studies. To my knowledge there are, however, no surveys or other studies conducted to verify with the decision makers how satisfied they are as a customer group. It is only in some countries, like in Ireland, Sweden and the Netherlands, that the authorities – together with the sector and other stakeholders – conducted in the past decade a review of, or a national debate on, the effectiveness of the archaeological legislation.

In Ireland the Minister for the Environment, Heritage and Local Government undertook a major review of archaeological policy and practice and of the National Monuments Act in 2007. Based on a briefing document with key issues to be discussed (Department of the Environment, Heritage and Local Government 2007), public consultation meetings and a conference organised by the Institute of Archaeologists of Ireland (Bolger 2008), the main conclusion reached was that a major 'value-shift' was needed within archaeological practice and management, with a move away from the concepts of an acceptable level of 'record' to a more focused concept of knowledge creation and knowledge gain (pers. comm. M. Gowen 2015). It was recognised that new (higher) standards in Irish archaeological practice were needed, as were more efficient and suitable methods of dissemination. Suggestions were made to the Department to define and to implement those in the national legislation (Gowen 2009), but these have not yet been followed.

An evaluation of the development-led archaeological practice that results from the Swedish Heritage Conservation Act (1988), which was carried out by the Swedish National Heritage Board in 2007, led to some similar conclusions; the scientific results of this practice should no longer be the aim but the means (pers. comm. C. Andersson 2015). As the objective was 'to transform and present the results of an excavation for different target groups in an interesting and relevant manner' (Andersson et al. 2010, 19), the focus of the new orientation of the regulations was to make sure preventive archaeology would include both scientific documentation and the dissemination of the results, as it was explained by C. Andersson during her presentation at the EAC Symposium in Lisbon.

In the Netherlands, an extensive evaluation of the effectiveness of the revised Monument Act of 2007 was commissioned by the Dutch Ministry of Education, Culture and Science and carried out in 2011 by RIGO Research en Advies. It showed the Ministry that the act does what it is supposed to do on most aspects (Van der Reijden et al. 2011; for an English version see Guermandi & Salas Rossenbach 2013, 179). It was, for instance, concluded that archaeology is sufficiently taken into account in development and planning activities; that the preservation of archaeological remains *in situ* has become more common, and that there are sufficient financial resources generated by the disturber-pays system to carry out necessary research. Moreover, the Dutch State Inspectorate showed in 2013 that almost all local authorities (93%) have integrated archaeology in their development plans (Erfgoedinspectie 2013). So, even though the sector might be critical because there is sometimes another reality in daily practice – with many exemption rules being applied and an actual shortage of the staff at municipalities that should take care of these tasks (Figure 16.1) – the Ministry was overall rather satisfied with the effectiveness of our legislation on the protection of the archaeological heritage.

Figure 16.1: Part of a press article in *Binnenlands Bestuur* (30 January, 2015) – a magazine of (local) authorities – in which municipal archaeologists warn of the lack of time and money that is available for them to do a proper job. This could lead to archaeology being seen as a burden rather than a source of inspiration by local authorities.
(http://www.binnenlandsbestuur.nl/ruimte-en-milieu/achtergrond/achtergrond/bodemschat-is-kostenpost.9462527.lynkx, accessed 22.09.2015)

Nevertheless, RIGO Research en Advies also reported a lack of quality management by the sector and an insufficient level of quality monitoring (Van der Reijden et al. 2011). The Minister of Education, Culture and Science connected the two and initiated measurements to improve the level of self-regulation by abolishing the state-run licensing system for excavation work and replacing it with a system of certification that would have to be run fully (both issuance of certificates and quality monitoring) by the private sector.

Next to these evaluation reports, we could use the intelligence available from members of the Europae Archaeologiae Consilium (EAC), whom we may consider as reliable representatives of the national authorities. In particular the 2013 EAC survey on the implementation of the Malta Convention may serve as an indication of the level of satisfaction with this group. Although the results of this survey show that there are many things that need to be improved, the general impression is that respondents are quite positive about the level of implementation in their country of the articles on the protection and preservation of the archaeological record (Articles 2 and 5) (Olivier & Van Lindt 2014, 168). The state respondents are also rather positive with regard to the implementation of Article 3 of the Malta Convention. In fact, the most significant achievements were reported on this article (Olivier & Van Lindt 2014, 168). This suggests that it is quite well-guaranteed that archaeology in Europe is carried out by people whom the authorities consider qualified.

These observations seem to suggest that we may assume that authorities are probably sufficiently satisfied with the way the law is being carried out, as far as this concerns the role of archaeologists and their compliance with these regulations, although there may be exceptions in some European countries. This assumption may not necessarily be valid for the overall effectiveness of the legislation, as there are testimonies from several European countries that many issues are subject to improvement, like illegal digging and looting, the extent to which preservation *in situ* has been accomplished, the integration of archaeology in the planning process, etc. These are however issues that do not relate to the quality of the performance of the archaeological sector, so they will not be discussed further in the context of this article.

This high level of satisfaction with the performance of our sector might not be reached for the other main requirement that authorities have, i.e. the sector's production and dissemination of knowledge for society. As far as I know there are, again, no evidence-based studies available that assess the satisfaction of the authorities on this aspect, but if we look again at the EAC survey, it is clear that the respondents to this evaluation are less optimistic about issues relating to public outreach. Almost 60% of them (20 out of the 34), indicated that Article 9 of the Malta Convention has not yet been successfully implemented. In fact, of all articles, number 9 emerged as the least implemented (Olivier & Van Lindt 2014, fig. 22.4).

This seemingly low level of involvement of the archaeological sector with society is also the overall picture that emerges from an analysis of the level of engagement our sector has with the public (Van den Dries 2015), which was based on the Discovering the Archaeologists of Europe (DISCO) surveys from 2013–2014. The bottom line of that analysis is that in many European countries public engagement activities are only a minor part of the professional activities. When in a particular country visitor services do look like a substantial field of work, which involves many people, then it turns out that the amount of time spent on such activities is quite low. If expressed in figures, engagement with the public does not seem to be one of the priorities of the profession.

To summarise the above, we do not have many figures, but those we have do not seem to be in our sector's favour. The indications are that the authorities may be satisfied with how the sector serves the aims of archaeological regulations, but may be less happy with the level of valorisation it receives in comparison with the volume of research that is being facilitated.

Archaeologists

User requirements

The second group of end-users for which we could review customer satisfaction is our sector itself. The main objective of archaeologists usually is to conduct research in order to gain knowledge about the past. In order to be able to conduct research the first requirement is to have sufficient and good quality education. In order to be able to build upon the state of affairs and to enhance the sector's level of knowledge, we secondly need to have good quality products from our peers and colleagues, like site reports, documentation and synthesised work. The importance of these requirements is acknowledged through the Malta Convention; Article 7 concerns the dissemination of knowledge, and Article 8 the (international) exchange of knowledge. In order to be able to live up to our own professional standards and to produce the required good-quality products, the sector thirdly needs the right conditions, such as sufficient time and money to gather data and conduct analysis and interpretations. Moreover, it needs to define its precise requirements. This can be done by means of quality standards, but most often only the basic principles of behaviour are defined by codes of ethics or codes of conduct.

User satisfaction

Also in the case of our sector, we lack studies and data across the board on our satisfaction level regarding the quality we (are able to) deliver. However, it was recently evaluated through the DISCO surveys how satisfied our sector is with its training and education. In 21 countries assessments were made of what both archaeologists and employers think of the level of training archaeologists get and whether they have particular skills gaps. The results revealed that archaeology in Europe has a highly qualified workforce, with 94% being graduates (Aitchison et al. 2014, 36), but that a lot of employers are not yet satisfied with the knowledge and skills of their employees. In almost all the participating countries, the majority of both the employers and employees indicated they have a demand for additional training, for instance in computer-based skills, interdisciplinary research methods, new field methods, project management and public outreach (Van den Dries 2015, 49–50). Some countries (e.g. Latvia, Romania, Bosnia and Herzegovina) do not have a specialised or full range of study programmes for archaeology (bachelor's, master's and PhD degrees), and they see this as a disadvantage. In these countries archaeology is, for instance, part of the education programme in history or it is taught only up to undergraduate level.

Furthermore, employers were asked whether any (annual) budget is set aside for training and development of employees, and whether there is a training programme for the employees. Strikingly, a lot of organisations reserve a budget for training, but lack training programmes for their employees. Moreover, in most countries the sector is not satisfied with the availability and quality of the vocational training programmes (idem). Another concern in many of the 21 DISCO reports is that most employers provide training only, or mainly, for their permanent staff, not for their temporary staff. This is a problem as only 63% of the archaeologists held permanent contracts at the time of the DISCO research; the remaining 37% had temporary contracts (Aitchison et al. 2014, 7).

Unfortunately, issues relating to education are not the only problem the sector identifies in relation to quality management. Another and probably even more pressing issue concerns the second requirement, i.e. the availability of good quality knowledge products and the dissemination of this knowledge. When Willem Willems and I edited the volume on quality management in archaeology, back in 2007, nearly all authors indicated that they were not satisfied with the quality of the research that is conducted in the context of development-led archaeology projects. The main problems they mentioned were a lack of reports, a lack of accessibility of reports (grey literature) and a lack of synthesised research; they also mentioned the need for standards for recording data and for reporting. Moreover, in the proceedings of a colloquium that was held on preventive archaeology in 2004 (Bozóki-Ernyey 2007), which covered the institutional and legislative background in 16 countries (Belgium, the Czech Republic, England, Estonia, Finland, France, Greece, Hungary, Iceland, Ireland, Italy, Luxembourg, Poland, Romania, Slovenia and Spain), the contributors provided some information on the particular strengths and problems in their country. When taken together, the two volumes show that 11 authors – out of the 17 countries that provided an analysis of the weaknesses of development-led archaeology – explicitly talk about issues with quality. No less than 9 of these 17 testimonies reported major difficulties in relation to reporting and/or dissemination among peers. Thomas, for instance, complained that for many of the thousands of excavations taking place in England every year, the reports were produced in limited numbers and were hard to get hold of (Thomas 2007, 41). Of these 17, 6 also mention a lack of professional standards as an important gap in the system or a need to improve the standards applied. Only 4 authors consider standards to be in place in their country (France, United Kingdom, the Netherlands and Germany).

In some states, the concerns relating to the availability of good quality knowledge products were heard by the authorities, who, in response, launched special programmes to stimulate the development of both research agendas and syntheses. In Ireland, for example, a research-needs assessment (The Heritage Council 2007) was followed by the Irish National Strategic Archaeological Research (INSTAR) programme. It was launched (in 2008) to transform three decades of archaeological data into new knowledge on Irish archaeology. In the Netherlands the state and its organisation for scientific research (NWO) financed in the past decade the *Oogst van Malta* programme and the *Odyssee* programme. The first is a research programme ('Malta's harvest') to produce scientific syntheses of development-led excavation results; the second finances the analysis of unpublished excavations that were conducted between 1900 and 2000.

However, many complaints concerning reporting and dissemination continue to be heard throughout Europe. They are, for instance, documented in the EAC's volume on 20 years of archaeology based on the Malta Convention (Van der Haas & Schut 2014), and other volumes with evaluations of the effects of the Malta Convention (e.g. Guermandi & Salas Rossenbach 2013). Moreover, most respondents in the EAC survey reported a lack of significant achievements on the implementation of Articles 7 and 8 of the Malta Convention (Olivier & Van Lindt 2014, 168). It was also apparent during the recent session on quality at the EAC 2015 conference in Lisbon that reporting and dissemination are still an issue. The representative of Slovakia, for example, stated that the work of private companies usually does not include publication activities (see Bednár et al., this volume). Also for Sweden a need for thematic and geographical syntheses on a national level was reported both in 2010 and again during the EAC Symposium in Lisbon (Andersson et al. 2010, 26). According to Vander Linden & Webley (2012) there is no funding model yet in use in Northwest Europe that consistently provides resources at a level that archaeologists might wish. So, it presumably will remain a persistent problem as long as most European countries exclude reporting and dissemination as one of the obligatory deliverables of development-led projects.

What, however, in regard to quality management can also be noticed from these recent evaluations and from the DISCO surveys is an increased awareness of and attention to developing and implementing quality standards for archaeological research. Across Europe, many more instruments are being put in place to manage quality issues. Ten years after the discussions in Rosas and Dublin, it can, for instance, be seen that an increasing number of national authorities obliges organisations and/or people to have a licence for conducting archaeological research, both with the state-run and the commercial archaeology enterprises, and that more bodies have been installed to verify the quality of the work performed and of the final site reports. Moreover, the sector itself has put quite some effort into discussing quality management too. Professional associations in particular play an important role in this. They are usually active in establishing codes of conduct and guidelines for various activities and work processes, from desk-based assessments and consulting to the compilation of archives. Sometimes they play a role in setting specific membership criteria and providing a register of professional archaeologists. According to the information that is available with the European Association of Archaeologists' (EAA) Committee on Professional Associations, at least 20 professional associations for archaeologists can be counted across Europe today (pers. comm. Gerald Wait 2015). It is also interesting to notice the emergence of special associations for contractors in archaeology, for instance in Italy, the Netherlands and Poland.

But despite this growing awareness towards quality management, it also emerges from some of the national DISCO reports that only a tiny minority of the archaeological organisations use quality instruments such as ISO certifications: in Romania 4%, in the Netherlands 7%, in Ireland 10%, in the United Kingdom 12%, in Cyprus 16% and in Italy 17% (Van den Dries 2015, 48–49). This does not mean of course that without such an ISO standard certificate there is no quality, but it does show that to this day little attention is paid to such agreed-upon instruments for achieving quality assurance.

To summarise the above, it seems we have to conclude that the many quality issues that are mentioned suggest that several of our sector's needs are not yet fully met, especially not in the areas of (post-graduate) training and knowledge dissemination within the profession.

Developers

User requirements

The third main group of consumers that we should take into account when evaluating the quality of our services and products are the developers, i.e. the building and construction sector or those who in many countries are contributing financially to archaeological research as a means to mitigate their building actions. Clearly, developers must have specific requirements, but also in this case, the archaeological sector does not know much about them. In fact, this is probably the group of customers about which the archaeological sector knows least. For example, we hardly talk about them in our publications and evaluations of development-led archaeology. In two volumes from 2007 (Bozóki-Ernyey; Willems & Van den Dries) that together consist of testimonies from 19 countries, only two contributions take the interests of developers a little bit into account and mention, for example, that the sector needs to be more selective and transparent for this stakeholder group. None of the other contributions pay attention to the developers while discussing the strengths and weaknesses of the development-led practice.

This lack of attention to this customer group is striking if you look at it from a quality management point of view. Why is it that we are the least familiar with those we are most dependent on, in the sense that in many countries developers provide the prime reason for archaeological research and as such support our work financially most? It suggests that many archaeologists are not that interested in them and do not actually consider them partners in a joint venture. In contrast, developers do talk about archaeology, at least in the Netherlands. Branch magazines and newspapers for the building sector (e.g. *Cobouw*) regularly report on the experience developers have with archaeology. Such sources are very interesting and informative for our sector with regards to user requirements and user satisfaction. They show, for instance, that developers and constructors most of all need to know what to expect. They do not want any surprises that may cause unforeseen delays or additional costs during the research process. They also show that if they need to pay for archaeological research, they at least want to get something in return for their investments – something which demonstrates that the research they paid for was actually worth their investment.

User satisfaction

With regard to this user group there is, again, little data available regarding its level of satisfaction. Individual organisations sometimes assess user satisfaction of their own organisation or services, but hardly any larger-scale studies are available. There seems to be only one – not very recent – study from London (Corporation of London 2001). It was held among developers who had all together carried out 36 large-scale projects in the City of London. In this study it was found that 'whilst developers generally do not express enthusiasm about paying for archaeological work, there is little opposition to the view that it is a legitimate factor for consideration in the City of London.' (Corporation of London 2001, 25). Many developers regarded the sums they had to pay for archaeological research as relatively insignificant. They were less concerned about the direct costs of a project than about the effects of unexpected delays on their projects (idem, 1). In terms of risks, the developers in London mentioned that their greatest concern is about letting risks, associated with delays to completion dates and potential loss of tenants. There also seemed to be serious concern about the effects of the possible loss of floor space as a result of archaeological considerations. They were mainly afraid of losing basements due to obligations to preserve archaeological remains *in situ* (idem, 18).

In the same study, there were hardly any complaints about the professionalism of archaeologists (idem, 1). If there were complaints, these concerned the insignificance of the archaeological results for the cost involved (idem, 41). There was, for instance, a lot of grumbling about excavations and watching briefs which cost a lot and do not provide evidence of unexpected archaeology. The report also says that a significant number of interviewees raised the question of who actually benefits from archaeology. Most of the developers argued that they did not. Several developers also believed that the balance of considerations was weighted too heavily in favour of archaeological interests, especially in the matter of whether remains should be preserved *in situ* or not (idem, 1). Moreover, none of them indicated that archaeology added a direct commercial value to their developments (idem, 24), and preserving archaeological remains as a feature in commercial developments was not regarded as a useful objective.

In the Netherlands the indications as to user satisfaction are similar. It was, for instance, said in the evaluation that the Ministry commissioned in 2011 that, according to the majority of the local authorities involved in the evaluation, the new policy on development-led archaeology and its funding system did not have a negative impact on the volume of land purchases by developers (Van der Reijden et al. 2011, 18). Moreover, the 'disturber pays' principle was found to enjoy wide support among the major 'disturbers' (Van der Reijden et al. 2011, 5). Some Dutch developers who were interviewed by one of our students from the Faculty of Archaeology suggested this as well (Van Donkersgoed 2011). They said they consider taking care of archaeological remains an intrinsic part of their duties as a developer.

What was, however, identified as a major issue in the Netherlands, like in London, is the lack of transparency with regard to the site evaluation and selection process. In a meeting with developers and planners from municipalities in the Netherlands, the archaeological sector was criticised for not being transparent in its criteria for making decisions regarding what research is needed (Caspers et al. 2011, 36). It was stated that for the other domains which developers have to take into account, like safety, nature and environment, it is usually perfectly clear what the criteria, requirements and standards are, but not for archaeology. This complaint was confirmed by interviews with developers which formed the basis of an appraisal of the process of site evaluation applied in the Netherlands. Both planners and developers seem to consider the selection process of the archaeological sector too much of a black box (Vestigia 2013). It is, for instance, not clear on the basis of monument maps and archaeological policy plans what kind of research will be demanded by authorities and archaeologists. Decisions are not based on objective and transparent criteria and values but rather on the subjective judgement of experts. This makes it very difficult for developers to make a well-educated guess prior to development as to the risks and costs they can expect regarding the archaeology (Vestigia 2013, 87). This results in insecure planning and sometimes development projects that are not profitable or even cost-effective.

Another problem also identified in my country is the lack of valorisation for society. Developers are often not convinced the research they have to pay for is useful and will make a difference to society, because in most cases they see little or no return on investment in the sense of results given back to society. They accuse archaeologists of almost exclusively being interested in the scientific results (e.g. Silvester 2015). Moreover, developers do not see a link between the costs and the benefits (Vestigia 2013, 91) because in their experience very expensive research does not yield a return on their investment.

None of the studies mentioned above are statistically representative, but they do give an indication as to the level of satisfaction developers have as a customer group. If we were to do a survey among developers, they may conclude that due to these shortcomings the archaeological sector does not have customer satisfaction high on its agenda. Developers are not really involved in the decisions and do not benefit much from the valorisation of the results. We could reply by saying that in practice the archaeological sector is very much selective – in the case of the Netherlands the number of excavations that result from field inventories is only 1 to every 16 evaluations – and that it does produce many new insights in important scientific questions. But whether their image of us is true if we look at the statistics is not really relevant. If developers feel it that way, then it is their truth, and that *is* relevant. We then need to try to change that negative image by satisfying their needs much better.

To better meet these requirements it has been suggested by Dutch planners and developers to integrate societal values in the valuation of archaeological sites (e.g. Caspers 2011, 36; Vestigia 2013, 87) and to include them in the decisions on what should happen with the results and finds from research (Silvester 2012; Figure 16.2). One of my master's students at the Faculty of Archaeology actually proposed an expansion of the current site valuation method with a societal value assessment, which includes a site's potential for education and for the local community (Elemans 2013). However, the main direction followed by the sector is to define knowledge lacuna, to include more transparent criteria in the selection process and to provide better explanations and motivations for the choices archaeologists make, rather than focusing on participatory governance and on including the stakeholders (and society) in the valuation and selection process.

If we made a serious effort in Europe to consider developers an important stakeholder – an ally instead of an enemy – we might all benefit from such a partnership. In the Dutch evaluation report of 2011 it was, for instance, said that developers show an interest

Figure 16.2: Part of a press article in *Cobouw* (30 January, 2012) – a newspaper for developers – indicating 'Developers want influence on archaeology'. It says that developers want to have a say in what happens with finds.
(http://www.cobouw.nl/nieuws/algemeen/2012/01/30/projectontwikkelaar-wil-invloed-op-archeologie, accessed 22.09.2015)

in anticipating the public interest in archaeology by organising events or producing output for the public. In the case of large research projects they seem to be willing to offer a budget for public outreach (Van der Reijden et al. 2011, 91). The developers whom our Leiden student interviewed also indicated they would like to utilise the archaeological research results much more for public relations purposes (Van Donkersgoed 2011). As the sector indicates in many countries in Europe that it lacks sufficient skills, for instance to organise dissemination activities (see Van den Dries 2015), it could be advantageous for both parties to collaborate on this aspect.

The public

User requirements

Finally we have to satisfy the fourth group, the public, or 'society' at large, on behalf of whom authorities safeguard the archaeological heritage. Although this stakeholder group is discussed last in this article, they should actually be our sector's main concern, because the public's satisfaction influences that of the authorities. Without any public support, there will be no political support.

But of all the customer groups discussed in this article, it is probably for this one that it is most difficult to evaluate whether they are satisfied as a customer group. First of all because, so far, no large-scale studies have been conducted among the public in Europe to find out what its requirements are. While writing this article, the author is participating in a large-scale survey on the public's perception of archaeology that is being conducted within the context of the European research project NEARCH (www.nearch.eu), but the final results are not yet available. Secondly, unlike authorities and developers, 'the public' is much less an entity with shared interests. It consists of little subgroups which we as a sector engage with. Across Europe some visitor satisfaction studies have been done in relation to a number of museums and archaeological site parks, but almost by definition they usually relate to little sub-groups of the public only. Thirdly, the public hardly ever voices its satisfaction or dissatisfaction in the way developers or authorities may do, for instance, in Parliament or in the media. There are usually no protests by the public either, even if they are unhappy with what we are doing, except perhaps for some pagan groups and indigenous communities, although public complaints from the latter are very unusual in Europe. As a consequence, we know for a very few local subgroups what their needs or requirements are, but not for the public at large.

User satisfaction

With regard to user satisfaction for the public, there are some indirect indications, like visitor numbers to museums and monuments and participation levels for heritage activities. Across the board in Europe, these figures are known to be rather low, and the latest Eurobarometer on cultural access and participation (TNS Opinion & Social 2013) even indicates that the interest in cultural heritage has diminished since it was measured last, in 2007. The percentage of people not having visited one single historical monument or site in the last 12 months prior to the survey had increased from 45% in 2007 to 48% in 2013 (TNS Opinion & Social 2013, 8). For 37% of the respondents lack of time was in 2013 the main reason for not visiting a historical monument or site, while for 28% it was a lack of interest. Luckily, expense was not the main obstacle (9%), nor was it the poor quality of the activities where people live or the limited choice (10%) (TNS Opinion & Social 2013, 21). However, it is worrisome that for the whole of Europe the *lack of interest* in cultural activities went up from 27% in 2007 (Eurobarometer 2007) to 33% in 2013, while the *lack of time* was mentioned less as a barrier (falling from 42% to 28%).

Small local participation studies, for instance from England, also suggest that archaeological projects are not always accessible or appealing to potential visitors (e.g. Treble et al. 2007). This perhaps has to do in the first place with the limited opportunities on offer rather than the attractiveness of the individual activities, as open door days and school programmes at excavations are usually very well-attended. Moreover, the first and very preliminary results of the European survey of the NEARCH project seem to suggest that many European citizens rarely think of archaeology in terms of leisure and amusement. They do not consider it to contribute a lot to their quality of life (www.nearch.eu).

Besides these figures, our own experiences may provide some indications as to user satisfaction as well. Ten years ago it was said by several authors of both the EPAC volume on development-led archaeology from the European Preventive Archaeology Project meeting of the Council of Europe (Bozoki-Ernyey 2007) and the one on quality management (Willems & Van den Dries 2007) that much more should be done to generate new knowledge from this new practice and to share it with society. In fact, the archaeological sector in Europe has been repeating this for several decades now. It is striking to notice the large number of authors, both in the EAC occasional papers and other Malta evaluations, all from various European countries, that still mention the need for more dialogue with the public and a need to better show the public benefit of archaeology to society. Even for Sweden an insufficient level of dissemination to society was reported at the 2015 EAC meeting.

In our sector's defence, I would say that what makes it extremely difficult to successfully reach a sufficient level of valorisation for society is that there is no clear consensus on the meaning of public benefit and how we can best make it operational. We have some agreed-upon general principles and rules of behaviour, but we have not established standards of public engagement in the context of development-led archaeological research and guidelines for achieving them. For instance, the ICOMOS Charter for the Interpretation and Presentation of Cultural Heritage Sites (Ename Charter 2008) defines the seven basic principles of interpretation and presentation as essential components of heritage conservation efforts. The EAA Code of Practice even has 'archaeology and society' as its first main chapter. It states that 'archaeologists will take active steps to inform the general public at all levels of the objectives and methods of archaeology in general and of individual projects in particular, using all the communication techniques at their disposal.' (EAA 2009, article 1.3). However, no guidelines for archaeologists can be found that address valorisation and provide best practices to meet the needs of society. This is a rather remarkable hiatus, as we do have guidelines that serve our own (economic) interests, such as best practices for archaeological tourism. Moreover, some countries, like Ireland, have even defined guidelines for developers, to make sure archaeologists are involved in early phases of planning (The Heritage Council, 2000).

Another complicating factor is that in Europe valorisation demands are hardly ever explicitly included in legislation on archaeological heritage, except in Sweden, as was discussed above. Consequently, in our primarily development-led archaeological practice there are no legal obligations, no facilities and very few funding opportunities. It is mostly left to the initiative and creativity of the sector to solve the problem. Even though the sector may have an abundance of opportunities and possibilities to handle the issue and to find alternative sources to finance community involvement, it can be considered unfair and unrealistic of the authorities to have valorisation requirements but not to facilitate them and to mainly abdicate the issue to the sector.

But despite the fact that the problem seems to be persistent, one can nonetheless observe a clear difference between the current situation and that of 10 years ago. In many countries an outward-looking attitude can nowadays be observed. While a decade ago merely 7 out of 17 contributions mentioned the public or society in relation to development-led archaeology and quality management, today the public is on the radar throughout Europe when we discuss such issues. It is, for instance, illustrated by the huge amount of attention given to the public in the last three volumes in the EAC Occasional Paper series, and it was noticeable during the 2015 meeting in Lisbon too. For example, in the contribution from Estonia it was stated that their point of departure is to assure the quality of preventive archaeology without forgetting the interests of society (Pillak, this volume). The representative from Slovakia added to the complaint concerning the lack of site reports that this also means that a lot of knowledge is not accessible to laypersons (see Bednár et al., this volume).

But before we get too optimistic and tempted to downsize the problem, we should keep in mind that this attention was mainly given by people working in the heritage management sector. It is not obvious in all domains of the archaeological profession that the public is on the priority list. If we consider the number of sessions during the annual conferences of the EAA on 'archaeology and the public/society' (not including tourism) as an indication of the level of attention within the academic sector, we have less reason to be very optimistic. Although the number of sessions went from 1 in 1995 to 5 in 2014, their share in relation to the total number of sessions remained the same, from 3.7% in 1995 (EAA meeting Santiago de Compostela) to 3.6% in 2014 (EAA meeting in Istanbul). There was even a decrease in the percentage of papers on topics relating to the public, from 6% (10 papers out of 27) in 1995 to 4.4% (82 of 1,849) in 2014.

We should in any case remain sharp regarding user satisfaction and its development towards the future, as the demands for inclusiveness and accessibility are still getting stronger at the European policy level. In its search for inclusiveness, the Council of the European Union adopted in November 2014 a Work Plan on Culture 2015–2018 in which the first priority in the area of heritage is *participatory governance* (Council of the European Union 2014, 11). The council states that 'participatory governance of cultural heritage offers opportunities to foster democratic participation, sustainability and social cohesion and to face the social, political and demographic challenges of today' (Permanent Representatives Committee 2013). As so far the developments in our sector do not seem to be keeping pace with the political ambitions, the gap between the expectations of society and what we provide may widen.

Discussion

When we compare the present-day situation concerning quality management in Europe with that of 10 years ago, some important developments can be observed. In particular within the archaeological sector there has been a growing attention for the development and implementation of quality management instruments and for defining our own standards. But when we consider quality from the perspective of other users of our services and products, which is customary in the domain of professional quality management, it turns out there are some issues left that need more attention.

This exercise has also shown that it is, however, difficult to answer the question of whether everybody is happy for each of the four customer groups that were distinguished in this article. In the domain of professional quality management it is commonly known that one cannot evaluate and account for quality if one lacks the requirements that define this quality. It is only after requirements are defined and assessed that it can be decided if a particular service or product is satisfactory or if and what corrective actions need to be taken. In our case, we do not yet clearly understand the quality requirements of our various users. We would need to adopt an interest in our users in order to learn about and anticipate their needs.

The second factor that complicates an evaluation of customer satisfaction is that we lack evidence-based data. We have large-scale studies that focus on our own sector, like the DISCO surveys on employment and training satisfaction, but no studies that measure the contentment of, for instance, developers and other user groups within society. In this sense, we should adopt both a customer focus and a more factual approach to decision making.

On the basis of the signals we *do* have and the observations we can make, it seems we may conclude that some user groups may be happier than others but that a recurring and consistent issue with all user groups is our sector's lack of valorising research results. All seem to be limitedly served with regard to the need for new knowledge as a return on investment. In terms of quality management, this implies that our procedures fall short. If we want to aim for a higher level of user satisfaction, we would also in this respect need to adopt a more outward-looking attitude, a customer focus. In my opinion, this suggests that we should not only work on standards for the knowledge *production* process, but that we may think about implementing quality procedures for the subsequent phase of the *valorisation* process as well.

A crucial question that will always immediately be asked next is who will pay the valorisation bill? It is nowadays beyond doubt that the final objective of development-led archaeology is to provide society with new knowledge. But if society does not get what it is paying for (either through taxes or purchasing the result of development projects), there can only be one of two answers: either society is not yet paying enough or the available resources should be redistributed. Logically, the choice should be made by those that have set the valorisation guidance and requirements, i.e. the (inter) national authorities. The European Commission seems to recommend addressing the issue at all three levels; in a reaction to the decreased participation levels for culture in 2013 (TNS Opinion & Social 2013), Androulla Vassiliou, the European Commissioner for Education, Culture, Multilingualism and Youth, stated 'This survey shows that governments need to re-think how they support culture to stimulate public participation and culture's potential as an engine for jobs and growth. The cultural and creative sectors also need to adapt to reach new audiences and explore new funding models. *The Commission will continue to support cultural access and participation through our new Creative Europe programme and other EU funding sources*' (Link 1). Most national authorities, however, are not at the fore yet in taking responsibility or in initiating the discussion on taking responsibility for the huge task of valorising the sheer volume of knowledge that development-led archaeology in Europe is creating. As this debate should definitely be on the agenda, and should include participants from *all* our customer groups, this could perhaps be a suitable topic for a future EAC conference.

Acknowledgements

This article has been conducted in the context of the NEARCH research project on developing new scenarios for a community-involved archaeology (www.nearch.eu), which is funded by the European Commission. I am thankful to Gerald Wait, Carolina Andersson and Margaret Gowen for providing information, to the EAC for facilitating my contribution to the conference, and to Paulina Florjanowicz for organising most of it, for chasing me to write this article and for editing this volume.

References

Aitchison, K. , Alphas, E., Ameels, V., Bentz, M., Borş, C., Cella, E., Cleary, K., Costa, C., Damian, P., Diniz, M., Duarte, C., Frolík, J., Grilo, C., Initiative for Heritage Conservancy, Kangert, N., Karl, R., Kjærulf Andersen, A., Kobrusepp, V., Kompare, T., Kreković, E., Lago da Silva, M., Lawler, A., Lazar, I., Liibert, K., Lima, A., MacGregor, G., McCullagh, N., Mácalová, M., Mäesalu, A., Malińska, M., Marciniak, A., Mintaurs, M., Möller, K., Odgaard, U., Parga-Dans, E., Pavlov, D., Pintarič Kocuvan, V., Rocks-Macqueen, D., Rostock, J., Pedro Tereso, J., Pintucci, A., Prokopiou, E.S., Raposo, J., Scharringhausen, K., Schenck, T., Schlaman, M., Skaarup, J., Šnē, A., Staššíková-Štukovská, D., Ulst, I., Van den Dries, M.H., Van Londen, H., Varela-Pousa,R., Viegas, C., Vijups, A., Vossen, N., Wachter, T. & Wachowicz, L. 2014: *Discovering the Archaeologists of Europe 2012–14: Transnational Report*. York Archaeological Trust, York.

Andersson, C., Lagerlöf, A. & Skyllberg, E. 2010: Assessing and measuring: on quality in development-led archaeology. *Current Swedish Archaeology* 18, 11–28.

Bolger, T. 2008: *Rethinking Irish Archaeology: Old Ground, New Ideas. A summary report of the proceedings of the IAI Autumn Conference 2007*. Institute of Archaeologists Ireland, Dublin.

Bozóki-Ernyey, K. (ed.) 2007: *European preventive archaeology. Papers of the EPAC Meeting, Vilnius 2004*. Council of Europe & National Office of Cultural Heritage Hungary, Budapest, https://www.coe.int/t/dg4/cultureheritage/heritage/Archeologie/EPreventiveArchwebversion.pdf (accessed 21.09.2015).

Caspers, S., Knol, W. & Kars, H. 2011: *Richtlijnen voor Maatwerk. Onderzoeksrapport project Archeologievriendelijk bouwen & fysiek behoud*. Instituut voor Geo- and Bioarchaeologie, Vrije Universiteit Amsterdam, Amsterdam.

Corporation of London 2001: *The Impact of Archaeology on Property Developments in the City of London*. London, http://www.cityoflondon.gov.uk/business/economic-research-and-information/research-publications/Documents/2007-2000/The-Impact-of-Archaeology-on-Property-Development.pdf (accessed 08.09.2015).

Department of the Environment, Heritage and Local Government 2007: *Review of Archaeological Policy and Practice in Ireland: Identifying the issues*. http://www.ahg.gov.ie/ie/Foilseachain/FoilseachainOidhreachta/SeadchomharthaiNaisiunta/45.%20Review%20of%20Archaeological%20Policy%20and%20Practice.pdf (accessed 08.09.2015).

Council of the European Union 2014: *Draft Conclusions of the Council and of the Representatives of the Governments of the Member States, meeting within the Council, on a Work Plan for Culture (2015–2018)* (Document 16094/14, 26 November 2014). Brussels, http://data.consilium.europa.eu/doc/document/ST-16094-2014-INIT/en/pdf (accessed 08.09.2015).

Elemans, L. 2013: *The heritage of World War II in the Netherlands. The development of new criteria to value the traces of World War II in the Netherlands*. Leiden University (Unpublished master's thesis, Faculty of Archaeology), Leiden.

Erfgoedinspectie 2013: *Monitor Erfgoedinspectie. Staat van de naleving 2011–2012*. Erfgoedinspectie, The Hague.

Eurobarometer 2007: *European Cultural Values, Special Eurobarometer #278* (September). The European Commission, Brussels.

European Association of Archaeologists 2009: *The EAA Code of practice*. http://e-a-a.org/EAA_Code_of_Practice.pdf (accessed 08.09.2015).

Gowen, M. 2009: Ministerial Review of Archaeological Policy and Practice October 2007 – 2009/10. *IAI News Series* 2, Issue 1, 2–3, http://iai.ie/wp-content/uploads/2014/09/IAINews_Ser2Vol1low.pdf (accessed 08.09.2015).

Guermandi, M.P. & Salas Rossenbach K. (eds) 2013: *Twenty years after Malta: preventive archaeology in Europe and in Italy*. Istituto per i Beni Artistici, Culturali e Naturali della Regione Emilia Romagna, Bologna, http://online.ibc.regione.emilia-romagna.it/I/libri/pdf/twenty_years_after_malta_26_09_2013web.pdf (accessed 08.09.2015).

KRFS 2007:2. *Riksantikvarieämbetets föreskrifter och allmänna råd avseende verkställigheten av 2 kap. 10–13 §§ (1988:950) lagen om kulturminnen m.m.* Stockholm.

Olivier, A. & Van Lindt, P. 2014: Valletta Convention perspectives: an EAC survey, in V.M. Van der Haas & P.A.C. Schut (eds): *The Valletta Convention: Twenty Years After – Benefits, Problems, Challenges*. EAC Occasional Paper 9, EAC, Brussels, 165–76.

Permanent Representatives Committee 2013: *Draft Council conclusions on participatory governance of cultural heritage* (Document 15320/14, 13 November 2014). Council of the European Union, Brussels, http://data.consilium.europa.eu/doc/document/ST-15320-2014-INIT/en/pdf (accessed 08.09.2015).

Silvester, M. 2012: Projectontwikkelaar wil invloed op archeologie, *Cobouw* 30-01-2012, http://www.cobouw.nl/nieuws/algemeen/2012/01/30/projectontwikkelaar-wil-invloed-op-archeologie (accessed 08.09.2015).

Silvester, M. 2015: Archeoloog vindt zijn weg in de bouw, *Cobouw* 19-01-2015, http://www.cobouw.nl/nieuws/algemeen/2015/01/17/archeoloog-vindt-zijn-weg-in-de-bouw (accessed 08.09.2015).

The Heritage Council 2000: *Archaeology & Development. Guidelines for good practice for developers*. The Heritage Council, Dublin, http://www.heritagecouncil.ie/fileadmin/user_upload/Publications/Archaeology/Guidelines_for_Good_Practices_for_Developers.pdf (accessed 08.09.2015).

The Heritage Council 2007: *A Review of Research Needs in Irish Archaeology*, http://www.heritagecouncil.ie/fileadmin/user_upload/Publications/Archaeology/Research_Needs_in_Irish_Archaeology.pdf (accessed 08.09.2015).

Thomas, R.M. 2007: Development-led Archaeology in England, in K. Bozóki-Ernyey (ed.): *European preventive archaeology. Papers of the EPAC Meeting, Vilnius 2004*. National Office of Cultural Heritage Hungary, Budapest & Council of Europe, Brussels, 33–42.

TNS Opinion & Social 2013: *Special Eurobarometer 399 – Cultural Access and Participation*. European Commission, Directorate-General for Education and Culture, Brussels, http://ec.europa.eu/public_opinion/archives/ebs/ebs_399_en.pdf (accessed 08.09.2015).

Treble, A., Smithies, G. & Clipson, H. 2007: *The Wider Community's Perception of Archaeology – Elitist or Accessible? Evaluating the Grosvenor Park Excavation*. Council for British Archaeology, http://www.britarch.ac.uk/caf/wikka.php?wakka=GrosvenorParkResearch (accessed 08.09.2015).

Van den Dries, M.H. 2015: From Malta to Faro, how far have we come? Some facts and figures on public engagement in the archaeological heritage sector in Europe, in P.A.C. Schut, D. Scharff & L.C. de Wit (eds): *"Setting the Agenda": Giving new meaning to the European archaeological heritage*. EAC (EAC Occasional Paper 10), Brussels, 45–55.

Van der Haas, V.M. & Schut, P.A.C. (eds) 2014: *The Valletta Convention: Twenty Years After Benefits, Problems, Challenges*. EAC (EAC Occasional Paper 9), Brussels.

Vander Linden, M. & Webley, L. 2012: Introduction: development-led archaeology in northwest Europe. Frameworks, practices and outcomes, in L. Webley, M. Vander Linden, C. Haselgrove & R. Bradley (eds): *Development-led Archaeology in Northwest Europe*. Oxbow, Oxford, 1–8.

Van der Reijden, H., Keers, G. & Van Rossum, H. 2011: *Ruimte voor archeologie. Synthese van de themaveldrapportages*. RIGO Research en Advies BV, Amsterdam, http://cultureelerfgoed.nl/sites/default/files/downloads/dossiers/rapport-ruimte-voor-archeologie-synthese-van-de-themaveldrapportages.pdf (accessed 08.09.2015).

Van Donkersgoed, J. 2011: *Archeologie; bron van inspiratie of bodemloze put? Onderzoek naar het draagvlak voor archeologisch onderzoek bij projectontwikkelaars*. Leiden University (unpublished bachelor's thesis, Faculty of Archaeology), Leiden.

Vestigia 2013: *Evaluatie en optimalisatie waarderingssystematiek Kwaliteitsnorm Nederlandse Archeologie*. Vestigia (Report nr. V1107), Amersfoort, http://www.sikb.nl/upload/documents/archeo/Vestigia-rapport%20V1107%20SIKB%20Evaluatie%20Waarderingssystematiek%20KNA%20def-versie%2017-12-2013.pdf (accessed 08.09.2015).

Webley, L., Vander Linden, M., Haselgrove, C. & Bradley, R. (eds) 2012: *Development-led Archaeology in Northwest Europe*. Oxbow Books, Oxford.

Willems, W.J.H. & Van den Dries, M.H. (eds) 2007: *Quality Management in Archaeology*. Oxbow Books, Oxford.

Websites:
Link 1 http://europa.eu/rapid/press-release_IP-13-1023_en.htm (accessed 08.09.2015).

17 | Challenges and opportunities for disseminating archaeology in Portugal: different scenarios, different problems

Ana Catarina Sousa

Abstract: From Valletta to Faro, much has changed in Portuguese archaeology: legislation, archaeologists, heritage administration and communication with society. Several archaeological stakeholders recognise that dissemination is still one of the major gaps in post-Valletta Portuguese archaeology. This article will separately analyse the main problems and opportunities in disseminating archaeological knowledge in Portugal, using case studies and crossing data with some personal views. For different actors and contexts there are different challenges and opportunities many lost, others rediscovered.

The following scenarios will be retrospectively analysed:
1. Urban archaeology (Lisbon),
2. Rescue archaeology in major projects (EDIA – Alqueva Development and Infrastructure Company),
3. Archaeology in the municipalities (Mafra),
4. Archaeology in universities and research centres (UNIARQ – Centre of Archaeology at the University of Lisbon),
5. Archaeology by the cultural heritage authorities (IPA – *Instituto Português de Arqueologia*, IPPAR – *Instituto Português do Património Arqueológico e Arquitectónico*), IGESPAR – *Instituto de Gestão do Património Arquitectónico*, DGPC – *Direcção Geral do Património*),
6. Community and associative archaeology.

This review will cover the period between 1997 and 2014, beginning with the date of ratification of the Valletta Convention in Portugal.

Keywords: Portugal, archaeology, dissemination, public archaeology, Valletta

1. From Valletta to Faro, making a stop at Lisbon: a retrospective of Portuguese archaeology

In Portugal, during recent decades there has been an almost 'uncontrolled' rise in archaeological activity: a sharp increase in the number of archaeological excavations and in the number of public and private archaeologists, the emergence of archaeology companies and an increasing number of universities offering degrees in archaeology. This growth was exponential until 2009, when it experienced a decline related to the financial crisis that led to the Portuguese financial rescue between 2011 and 2014 (Sousa 2013; Bugalhão 2011).

This quick growth has caused some discrepancies, particularly in the field of dissemination, which was clearly left behind, a fact recognised by the archaeological community locally and at a European level, according to the DISCO project (Discovering the Archaeology of Europe).

Portuguese archaeology has been losing part of the main role it had attained at the start of this growth process. With the actions taken by institutions within the public administration (Portuguese Institute of Archaeology – 1997 to 2006) and the subsequent implementation of a legal framework following the principles of the Valletta Convention, the conditions were laid for a growing assertion of archaeology in Portugal. However, the last decade has witnessed a reversal in the visibility of archaeology in the public sphere, as it has become obscured within other more general categories, resulting in a clear decline in its media presence. This situation can be explained by economic, organisational and social factors. However, in contrast to the public's concern for other sectors, such as museums and libraries, there has been hardly any public reaction regarding archaeology.

Are we therefore condemned to archaeology merely *for* and *from* archaeologists? To assess this issue, we focused on the promotion of archaeological activity in Portugal, which is affected by a complex web of contexts, agents, processes and means. In terms of context, there are differences in the types of measures used to publicise archaeological issues: promoting archaeology in urban areas, in large enterprises and at local level are very different propositions. There are several means of disseminating information about archaeological activity. We should differentiate any actions targeting the archaeological/scientific community (databases, scientific publications, conferences) from the promotion of initiatives aimed at the general public (media disclosure, publications, public presentations, exhibitions, musealization

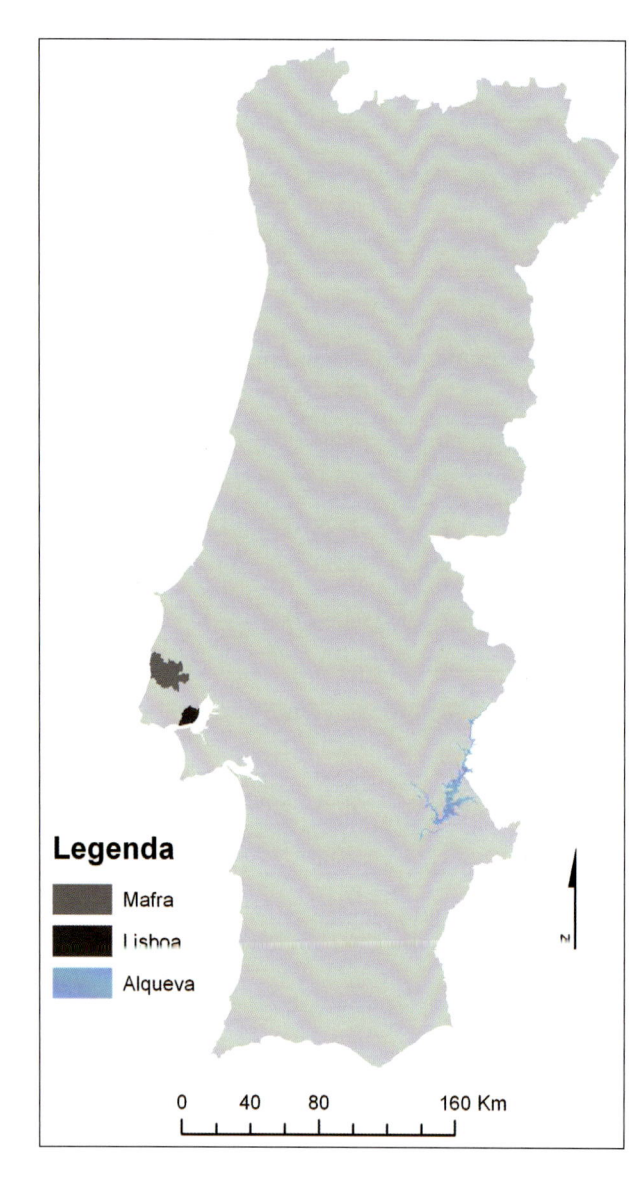

Figure 17.1: Location of the case studies referenced in this paper.

and enhancing of archaeological remains, heritage education, and new technologies).

A wide range of agents are directly involved in archaeological activity: we, the archaeologists (administration, companies, universities, associations), and others (developers, local government, the media, and the education system or tourism agents).

This article aims to examine the broad scope and perspectives for the development of archaeology in Portugal, somehow reflecting my own personal journey as an archaeologist. Declaration of interest: many of the reflections listed here are drawn from my own experience in the Municipality of Mafra (1997–2011), Directorate-General for Cultural Heritage – DGPC (2011–2013) and the Faculty of Letters, University of Lisbon (since 2008).

The analysis begins in 1997, when Portugal signed the Valletta Convention, and extends until 2014. The nature of this study will necessarily be broad and short, with references to particularly relevant case studies.

2. Different scenarios, different problems, different opportunities

2.1. Urban archaeology in Lisbon

Performing archaeology of cities is quite different from performing archaeology in cities (Martins & Ribeiro, 2009–2010), treating a metropolitan area as a sole archaeological document in spatial and temporal terms and with a technical and scientific specificity concerning intervention and interpretation.

The promotion of archaeological activity in urban areas is probably one of the greatest challenges that developers, archaeologists and public authorities face nowadays. Despite the existence of international conventions such as the Venice Charter (1964), the International Charter for the Protection of Historic Towns (Washington 1987) or the European code of good practice for urban archaeology (Archaeology and the Urban Project – a European code of good practice, European Council 2010), there are no specific guidelines for this discipline in Portugal (Lemos 2004; 2006).

The outlook of archaeology of cities in Portugal is very unequal as very different approaches to this topic coexist. Urban areas like Braga, Mértola and Beja have taken up an integrated management of archaeological activity, understood as a global research project. This is particularly relevant in Braga (*Bracara Augusta*), where, since 1977, a model of focused intervention has been developed by the Archaeology Unit of the University of Minho, with the collaboration of the Municipal Archaeology Office (Martins et al. 2013) from 1992. In the overall national scenario there is no integrated management of archaeological excavations, which are carried out by various parties: private companies, municipal archaeology centres, central

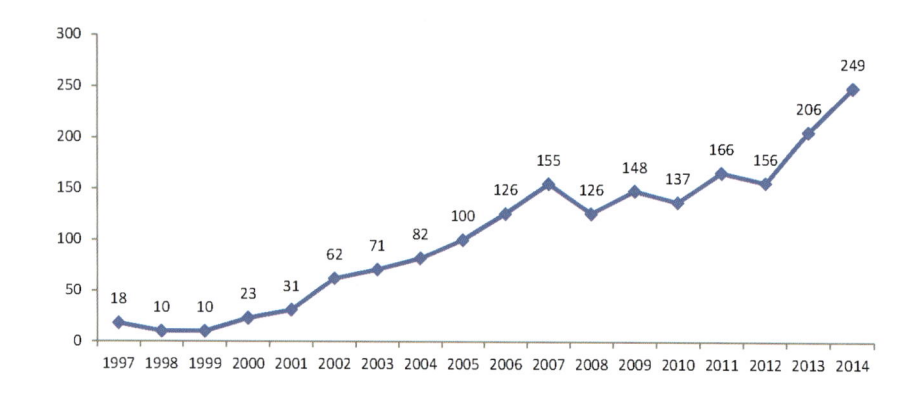

Figure 17.2: Archaeological interventions in Lisbon between 1997 and 2014. Source: Endovelico (DGPC).

Figure 17.3: Archaeological excavation in Praça D. Luis I (Lisbon, Portugal) in 2012. Lisbon's riverfront yielded numerous archaeological contexts relevant to the history of the city's ports. In the plaza D. Luis I, the excavation of an underground car park enabled the identification of port structures of the 16th/17th century (dockyard tide gauge) overlying a Roman anchorage. Excavations directed by Alexandre Sarrazola, Era Arqueologia.
Photo José Paulo Ruas/DGPC.

administration. Elsewhere, actions directly related to archaeological research are almost non-existent, corresponding almost exclusively to preventive interventions.

Lisbon archaeology, naturally, assumes an unparalleled scale in terms of the extent and chronological spectrum of excavations undertaken when compared with other Portuguese urban centres (Bugalhão 2007). In this city, there is no integrated management, as archaeological work is developed independently by different public and private teams. This work fragmentation greatly affects the interpretation of data collected, particularly because the dissemination of technical and scientific documentation is time-consuming and in many cases non-existent. Since 2001, archaeological excavations have been carried out almost exclusively by private companies, with over 15 of these operating in Lisbon (Bugalhão 2007).

The competent cultural administration decides on a case-by-case basis what constraints are to be applied; it assesses work plans and defines minimisation measures. It seldom includes specific guidelines for enhancing and promoting archaeological assets.

Spatial planning instruments (municipal master plans, detailed plans) are often generalist and inadequate for furthering knowledge about the archaeological resources in Lisbon's subsoil. Examples of this inadequacy are the recent interventions on the riverfront, which led to constant underground works at important port-related sites.

Between 1997 and 2014 the number of archaeological excavations grew by 1,283%, reflecting the overall

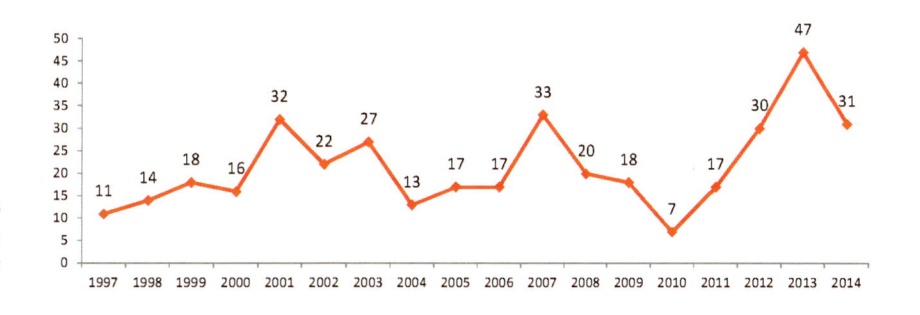

Figure 17.4: Bibliographic references relating to archaeology in Lisbon between 1997 and 2014. Source: Bugalhão 2014.

	CSJ - A	MNA	MAC	NARC	ML	CB	MTR	BdP	MG	TOTAL
1997		37821		4918						42739
1998		59653		4430						64083
1999		54166		6000						60166
2000		55465		9671						65136
2001		50324	24 180	9754						84258
2002		72394	48 701	8202						129297
2003		75129	54 052	9917						139098
2004		46 358	70263	5300						121 921
2005		61756	45 589	6538						113883
2006		102026	63 803	5000						170829
2007		129104	69 990	4153						203247
2008		125594	74 852	4063	70610		41882			317001
2009		126140	82 585	4700	60761		73087			347273
2010	59279	93374	79 009	8754	123192		39242			402850
2011	213213	85343	78 011	9733	76853		44902			508055
2012	244212	79210	89 000	8126	71828		47756			540132
2013	269347	80139	103 000	9087	70552		13721			545846
2014	348955	103068	130000	11830	77674	18480		30250		720257

Table 17.1: Visitors to archaeological monuments and museums in Lisbon (CJS – Castle of St George; MNA – National Museum of Archaeology; MAC – Carmo Archaeological Museum; NARC – Archaeological Centre of Rua dos Correeiros; ML –Museum of Lisbon; CB – Casa dos Bicos; MTR – Roman Theatre Museum; MG – Geology Museum).

Figure 17.5: Núcleo Arqueológico da Rua dos Correeiros (NARC), archaeological museum with musealized archaeological structures. This museum is managed by Millenium bcp, a private bank located near the arch in Rua Augusta, occupying almost an entire block in the Pombaline historical centre of Lisbon. Between 1991 and 1995, the renovation works carried out there revealed 2,500 years of Lisbon's history. Photo Jacinta Bugalhão.

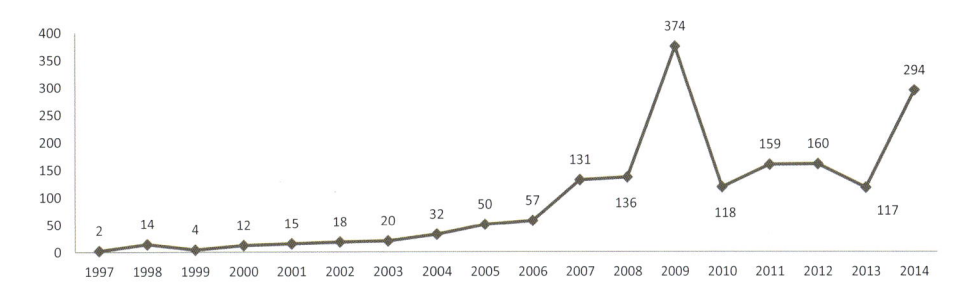

Figure 17.6: Archaeological interventions in Alqueva between 1997 and 2014. Source: Endovelico – DGPC.

trend of the national archaeological scenario. The type of intervention is dominated by archaeological monitoring of construction areas with underground impact in order to evaluate archaeological potential, with emphasis on large-scale underground projects, namely the construction of car parks or architectural remodelling, including construction of basements.

This type of intervention, usually carried out by private companies, is essentially influenced by the cost-speed trade off, and its principal aim is to comply with legal restrictions that only cater for 'rescue by registration'.

The evolution of urban archaeology in Lisbon has been remarkable. Up to the late 1980s, emergency rescue and inadequate urban policy tools were the rule. Today, preventive archaeology is deployed (Bugalhão 2007), but despite positive developments, the overall picture is still very unsatisfactory (Fabião 2014).

While field activities are legally secured (excavation and monitoring), study, publication and dissemination are often postponed until better financial conditions arise. However, postponing the dissemination of urban archaeology is probably the worst scenario. From this point of view, the opaque nature of archaeological activity during the fieldwork stage jeopardises interest and awareness from the general public.

Archaeology often makes the news when a certain street is closed to traffic for months as it waits for completion of an archaeological work. On the other hand, there is rarely an option for *in situ* conservation, even when ongoing projects are considered. It seems that 'rescuing by recording data' is the only 'reasonable' solution for the interests of the developer and the community, as it ensures that sites are 'unpolluted by ruins after the passage of archaeologists' (Martins & Ribeiro 2009–2010).

The disclosure of information in scientific circles is dispersed in different media. As for technical and scientific data, only the Endovelico Information System, managed by the cultural heritage administration (IPA – *Instituto Português de Arqueologia*, IGESPAR – *Instituto de Gestão do Património Arquitectónico*, DGPC – *Direcção*

Figure 17.7: Xarez 12 (Reguengos de Monsaraz, Évora), prehistoric habitat (Mesolithic, Early Neolithic) excavated between 1998 and 2002 in the Alqueva reservoir. The excavations took place under the direction of Victor S. Gonçalves (UNIARQ, University of Lisbon) and co-direction of the author. Photo by Victor S. Gonçalves.

Geral do Património), brings together all the information and makes it available on its website (Gomes et al. 2012; Link 1). It is expected that the scope of this tool will be increased in the near future with online reports and detailed georeferencing of interventions.

There is still room for improvement as regards the regularity and quality of the technical and scientific information produced, with long delays in reporting often complicating the interpretation of the archaeological remains found by different teams.

In terms of (scientific) publications, there is a general trend for growth, although the percentage increase is much smaller than the number of archaeological works actually carried out. Monographic studies have seldom been published, with preliminary reports predominating. A noteworthy exception is the case of the Archaeological Centre of Rua dos Correeiros, which has already published 48 titles (Bugalhão et al. 2012–2013).

Scientific meetings (congresses, conferences) are scarce. In this respect, the role played by the Association of Portuguese Archaeologists should be emphasised due to the regular conferences it promotes, where archaeological work and specific themes are presented. Archaeology companies also organise annual presentations of archaeological work, where archaeological work in Lisbon plays a prominent role.

In addition to these actions focused on the scientific community, some interventions have registered extensive impact in the media, especially during the construction phase. In most cases, the disclosure of information comes from outsiders, since developers tend to fear releasing information about the archaeological discoveries made on their sites.

The list of public spaces related to archaeology in Lisbon is relatively small, but it has a long history:

1. Museums: Geology Museum (1859), Carmo Archaeological Museum (1864), National Museum of Archaeology (1893), Museum of Lisbon – Pimenta Palace (1979);
2. Visitable archaeological sites including a museum: Museum of Lisbon – Casa dos Bicos (1987), Museum of Lisbon – Roman Theatre (1988), NARC (1995), St George's Castle (2008), Bank of Portugal (2014), cloister of the Cathedral (1993);
3. 'Memory' Spaces: Praça Luís De Camões Car Park (2000), Parking Plaza Don Luis (2014).
4. Other archaeological remains: Cryptoporticus (1986), Ribeira das Naus Dock (1990), Napoleon Shop (1994), Chinese Mandarin (1998), Academy of Sciences of Lisbon (2005).

With a total of 1,876 archaeological interventions at 340 sites, it would be expected that the increase in archaeological activity between 1997 and 2014 would translate into more visitable archaeological sites or *in situ* structures integrated into rehabilitation works. However, the list of such sites is very scant, with only two having been created following post-1997 interventions and subsequent enhancement projects: the Bank of Portugal and the archaeological museum at St George's Castle. Furthermore, memory spaces in car parks have been registered.

Even in construction projects financed by public administration, such as Centro Cultural de Belém (1992) or the National Coach Museum (2015), contemporary architecture was chosen at the expense of preserving *in situ* archaeological remains relevant to the history of the city (port structures).

This approach is clearly divergent from the aforementioned international conventions, including the Valletta Convention. Attempts to reconcile new rehabilitation projects with pre-existent structures (underground or in the built environment) have not often been successful. This is currently a major threat to Portuguese historic centres: 'the absence of knowledge acquisition and of the diachronic evolution of a site leads to the subordination of cultural and heritage values in favour of more aesthetic options devoid of historical context.' (Martins 2012, 252).

The Museums of Lisbon have different institutional frameworks, bearing modest relation to the archaeology carried out in Lisbon in recent decades. Exhibitions related to preventive archaeology are very scarce. The first exhibition of this type, dating back to 1966, was organised by Irisalva Moita at Rossio metro station. At irregular intervals, some exhibitions have been held at museums managed by the city of Lisbon, such as *Town Square – the archaeology of a location* (City Museum, 1999) or *The Archaeology of Lisbon – Sessions at the City Museum* (2007). Despite their very limited number, a general reading of exhibition attendance allows us to verify the relevance of these cultural spaces in terms of visitor numbers.

Within this context, there are several challenges and opportunities as regards the promotion of archaeology and its accessibility to society, with various agents having different responsibilities. The increasing number of visitors to museums and heritage-related sites, coupled with the media impact of some findings

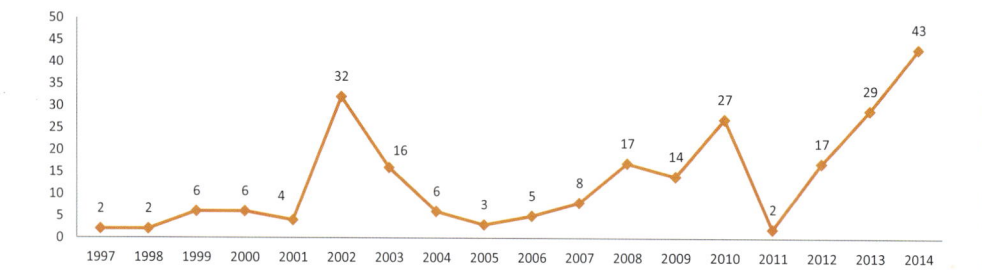

Figure 17.8: Bibliographic references relating to archaeology in Alqueva between 1997 and 2014. Source: Apud bibliography published in Silva 2014.

Figure 17.9: Torre Velha 3 (Serpa, Beja), Bronze Age site identified under the minimisation of archaeological impact of the Alqueva irrigation canals. The opening of the Alqueva irrigation canals enabled the identification of a completely unknown reality for this period, with numerous ditches, pits and hypogea. Excavation directed by Eduardo Porfirio, Miguel Serra, Catarina Costeira and Catarina Alves. Photo Eduardo Porfirio (Palimpsesto).

in Lisbon, confirm the interest in these issues from the local community and visitors to Lisbon in general.

Promotion and social returns need to be addressed at all stages of the archaeological process: during planning and land use planning, when determining archaeological constraints, and during the implementation of archaeological work and its integration in multidisciplinary research projects with the participation of universities and research centres. Furthermore, there is a need to strengthen resources and expertise in urban archaeological management, both in terms of the heritage authorities and at the level of local administration.

It is anticipated that the newly created Archaeology Centre in Lisbon (CAL – Lisbon City Hall) will ensure the collation of all dispersed documentation concerning heritage and detailed georeferencing.

Despite the fact that improvements are required in urban archaeological management, in recent years there has been some good progress, with a number of initiatives that appear to demonstrate the commitment of various stakeholders in dissemination.

2.2. Rescue Archaeology in major projects: the Alqueva Dam
The Alqueva Dam project is to date the largest carried out in the country. This dam is located in southern Portugal in the Guadiana River basin, affecting a large area of Alentejo and the Spanish Extremadura. It is the largest artificial water reservoir in Western Europe, extending for 250 square kilometres. In addition, the reservoir involves a series of irrigation canals, still under construction. The total investment of the project amounts to €1 billion, of which 14 million is related to mitigating its impact on cultural heritage (Martins 2012, 40).

The construction of the project was phased under the management of the Alqueva Development and Infrastructure Company (EDIA). Construction of the heritage and economic framework dates back to 1985 (Silva 2002, 57), with amendments in 1996. An archaeological survey formed the basis for devising a heritage minimisation plan for the backwater area of the Alqueva Dam, involving the definition of 16 thematic/chronological blocks and 200 interventions developed between 1998 and 2001. The minimisation plan was supervised by a monitoring committee and also involved experts and representatives of municipalities and heritage associations.

The development of a minimisation plan for the Alqueva backwater between 1998 and 2001 was relevant in the national archaeological scenario, as it coincided with the beginning of so-called 'contract archaeology', at a time when archaeology companies were still embryonic and when archaeologists started

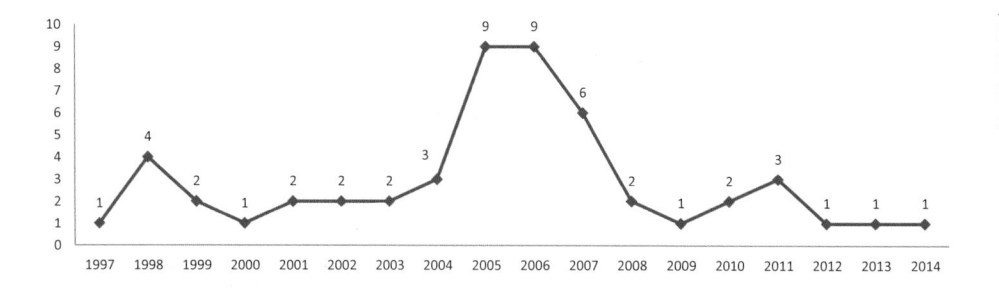

Figure 17.10: Archaeological interventions in Mafra between 1997 and 2014. Source: Endovelico (DGPC).

being professionally recognised. The works were organised in blocks and awarded to universities, heritage associations, individual archaeologists and some private companies.

A second phase began in 2007 mainly corresponding to the construction of the overall Alqueva irrigation system extending for 120,000 hectares (Melro & Deus 2014). The new heritage monitoring committee became exclusively bilateral (EDIA/IGESPAR), with the administration ensuring the coordination process, according to a protocol signed in October 2007 (Melro & Deus 2014). Execution of the archaeological work focused exclusively on the business perspective, with payments for completed excavations being awarded by the cubic metre.

In comparison to the first phase, the second had a higher number of archaeological interventions. It should be noted that the type of intervention was very different from the earlier ones. In the 1998–2001 phase, the work focused on sites following a sample global intervention plan. In the subsequent phase, from 2007, the intervention was geared towards minimising impacts on linear channels, making site interpretation more difficult. The visibility of archaeological remains was higher in the second phase because it involved land mobilisation. For example, it was noted that in the area examined from 2007 onwards, ditches, enclosures and negative structures (pre- and proto-historic) proliferated, but these did not occur only in the backwater area: differentiated visibility or distinct land-use dynamics?

These are two completely different perspectives with regard to the management of archaeological work, research and promotion.

In terms of management, the first phase was monitored by a joint committee including various disciplines and organisations and a scientific committee. In the second phase, monitoring was carried out exclusively by the heritage authority and the developer along with the archaeological contractor. As the fieldwork was carried out exclusively by private companies, usually with confidentiality agreements, a blanket of silence covered the Alqueva Dam's archaeology, only interrupted by occasional news of spectacular discoveries.

As regards research, the intent was completely different. The first phase of archaeological work at the dam site involved some teams that had previously run research projects in the area, and thus they viewed the 'Alqueva period' as an extension of an integrated action. Other teams were formed by companies without research experience in the region, so that the Alqueva project provided leverage for start-up companies. After 2007, interventions were performed exclusively by private companies, with little or no coordination with academia.

Disclosure of information has always been the biggest obstacle of the whole project. In the first phase there was a plan and an agreement for producing monographic studies and setting up a regional museum. This museum was never built, assets were scattered and consequently there was a loss of an integrated view of the entire cultural heritage under study.

With regard to monographic studies, 80% of the teams concluded them, which was a rather time-consuming process (Silva 2014). Unfortunately, after completion of the monographs in 2007, they were not published until 2013/2014. Publication was made possible thanks to the patronage of the regional heritage body (Alentejo Region Directorate for Culture), benefitting from European funding.

Despite the delay in publication and problems of distribution, emphasis should be drawn to the enormous volume of published information from the Alqueva Dam project and its importance to Portuguese archaeology: 23 monographs and a special edition in an archaeological journal.

Regarding the second stage concerning Alqueva's irrigation channels, there is less public information. Contracting model studies are unclear, with the whole process focused on the duality of excavation/monitoring and the production of technical and scientific reports. However, some information has been presented at congresses, but there is no known plan concerning the publication of monographs.

In terms of published material, there was an initial peak in 2002, corresponding to the end of the first phase of the Alqueva project. The publishing rate remained relatively stable (though lower than the level previously reached) up to 2010, when the first preliminary studies of irrigation channel interventions began to be published, as mentioned above (Silva et al. 2014). Public presentations at specially organised conferences were also irregular (1996, 1999, 2001, 2010). A lot of archaeological documentation remains unpublished, but the greatest weakness lies in delivering information to the general public. Some promotional material was published (CD-ROMs, DVDs, brochures and articles in special-interest magazines) but the impact on communities was not very significant.

The lack of a specifically dedicated museum was never overcome, despite the creation of a local museum in the new Aldeia da Luz, which plays an important role at local level and where some themed exhibitions have been held (*Vinha das Caliças – The slow awakening*, 24 February 2010, *O Touro de Cinco Reis* 8–27 April 2012, *Barca do Xerez de Baixo: a testimony rescued from history*: inaugurated 23 September 2013).

An evaluation of the whole 'mega-operation' in terms of disclosing information about the Alqueva project since 1998 shows some considerable changes in management models and communication strategy. The project promoter has disclosed the total budgets of the archaeological activity (Martinho 2002; 2014), but apparently delivering information to the general public is minimal. The full potential of delivering information for heritage education in the region and ensuring its socio-economic exploitation is yet to be achieved. Dissemination cannot once again be the end of the line after the all the fieldwork, writing of reports and preparation of monographic studies. This issue should be very well outlined from the beginning of the project and integrated in a heritage conservation plan.

This problem is common to most environmental impact statements in Portugal, as a strategy to communicate findings to the public is usually omitted or too vague. A contrast to this reality, of which the Alqueva Dam is an example, is the situation in Brazil, where the funding of heritage education projects has been mandatory since 2002 (IPHAN Ordinance No. 230,2002), (Almeida et al., 2009, 37).

The participation of the archaeology authorities in environmental impact assessment committees has progressively increased since 1997, today reaching almost all of the regions of Portugal (Branco 2014, 247). The effort undertaken in the implementation of heritage protection measures was complemented by an increasing volume of projects which showed a shortage in human resources to supervise works, manage information, create methodology guidelines and to promote disclosure of information.

2.3. Act local, think global:
Mafra and archaeology in the municipalities

After the Portuguese revolution of 25 April 1974, the local administration in Portugal took over an important role in land management, culture, education and social development. The existing 308 municipalities are characterised by their diversity, making it difficult to generalise about them.

With regard to archaeology, it is important to refer to the activity developed at municipal level in several areas: research, land management, enhancement and disclosure of information.

Figure 17.11: Zambujal fortress (Mafra, Lisbon): fortification inserted in the Lines of Torres Vedras – a system of defences created between 1809 and 1811 during the Napoleonic Wars. The regional project included excavation, restoration and creation of interpretive centres. Grants were funded by the EAA (Norway, Iceland and Liechtenstein), and the project received an Europa Nostra award nomination. Excavation directed by Ana Catarina Sousa and Marta Miranda (Câmara Municipal de Mafra).

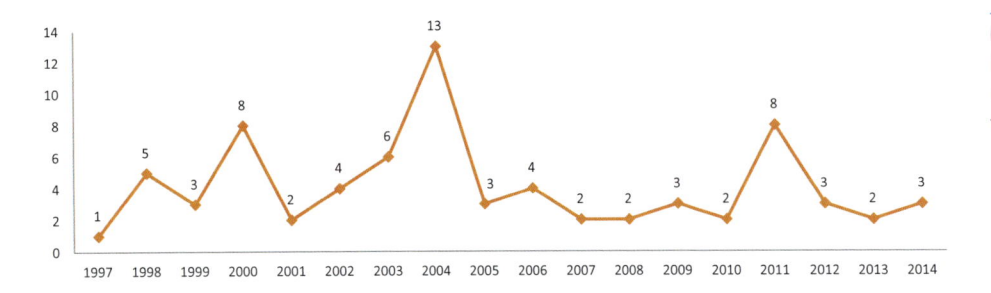

Figure 17.12: Archaeological bibliographic references relating to Mafra between 1997 and 2014.

It is very difficult to characterise archaeological activity in the municipalities, as there is no significant data available. There were several attempts to assess the pattern of archaeology in Portuguese municipalities, especially by the Professional Association of Archaeologists (APA), which conducted surveys in 2002 and 2006 (Almeida, 2006, 2007). Recently, under the DISCO programme, new surveys were carried out (Costa et al. 2014). It should be noted that these surveys were not exhaustive: in the 2006 survey 107 municipalities participated (Almeida 2007, 130), while in the 2014 survey only 53 did (Costa et al. 2014, 92).

From the data available in the Endovelico Information System (DGPC), Jacinta Bugalhão concludes that in 2010, about 12% of Portuguese archaeologists were working in local municipalities (Bugalhão 2011, 35). In 2014, the DISCO project estimates that 27.2% of archaeologists were working in local administration (Costa et al. 2014, 93).

It is difficult to perform diachronic readings very accurately. The 2006 survey revealed two moments of growth in the number of municipalities employing archaeologists: the 1980s and the period between 2000 and 2005 (Almeida 2007, 135). In the 2014 survey, there appears to be a reduction in archaeological activity in the municipalities (Costa et al. 2014, 17).

The first rise in archaeological activity in the municipalities, in the early 1980s, relates to the growing importance of the municipalities after the April 25 revolution, when they benefited from remarkable financial and administrative autonomy. Growth in the early 21st century seems to reflect the influence of the post-Valletta legal framework (Basic Law for Cultural Heritage – DL 107/2001 and Archaeological Works Regulation – DL 270/99) and the action of an independent authority for archaeological heritage – the Portuguese Institute of Archaeology. Economic recession in Portugal, culminating in the 2011 financial rescue, contributed to the reduction in municipal archaeological activity, as a result of an organisational reshuffle and budget constraints.

There are no accurate data on municipal funding for archaeological purposes, but it is generally agreed that municipalities were the major funders of research projects, development and promotion of archaeology.

Archaeology in the municipalities has very different organisational structures. In most cases it is included in the culture department, though it can also come under construction and planning (Almeida 2007). Archaeological work is carried out directly by municipal teams, through private companies or universities and research centres.

The general trend of archaeological activity seems to indicate a decline from 2009, coinciding with the financial crisis (Sousa 2013). However, in recent years there seems to have been an increase in heritage promotion and education, according to data available in the DISCO 2014 project (Costa et al. 2014, 102). This trend may also follow the evolution process of archaeological activity at local level: an initial phase of surveys and research studies, bridging century-old gaps until the early 21st century, followed by the last decade, where projects were aimed at generating social and economic return.

The municipalities' proximity to communities makes them a privileged vehicle for public disclosure. Since the 1990s, municipalities have played an increasingly important role in the education and social sectors, with a growing autonomy and responsibilities in these areas. The exponential growth of tourism in Portugal has also raised awareness among municipalities of the importance of developing archaeological sites, whilst working with local partners for the protection and management of archaeological sites has also shown positive results.

Nevertheless, the reality still points to weaknesses in this model, since many municipalities have promoted their own projects without liaising with other agents at regional level, thereby hindering the development of itineraries with national and international visibility.

Faced with so many variables, I chose to analyse a specific case: the municipality of Mafra, both a personal choice and one representing the national outlook.

The municipality of Mafra is located in the metropolitan area of Lisbon, just 40 kilometres from the capital. With a surface area of 291.66 square kilometres and 76,685 inhabitants (2011 census), Mafra is still essentially a rural landscape. The history of archaeological activity in this region dates back to the 19th century, but archaeological research was minimal until 1997, totalling only 4 excavated archaeological sites. From 1997, following the establishment of a municipal archaeology office, the situation changed significantly with the creation of a small technical team, laboratory infrastructure and backup, exhibition and educational spaces, and the enhancement of archaeological sites.

Excavation and archaeological monitoring at 35 sites was carried out for a total of 104 archaeological works, between 1997 and 2014, of which 57% were directly run by the Municipal Office. The progress of this activity, in a way, followed the national trend. In the first phase (1998–2004), the archaeological work was aimed at conducting specific research projects and site prospection. From 2004 until 2008, preventive archaeology intensified, in particular the works on Highway A21 – the only motorway sponsored by a municipality in Portugal. From 2008 preventive and research works slowed down and heritage enhancement projects increased.

Efforts to publicise this archaeological activity sought to target a range of audiences and various forms of communication: scientific and popular publications, exhibitions, guided tours, education services including schools, teacher training, historical re-enactment with local community participation, enhancement of archaeological sites (Miranda 2009).

In terms of publications, two types of work were published: general interest and scientific publications. Only two titles were published in the first category: one for children and youth and one for heritage site visitors. The remaining 74 published titles took the form of books (4), chapters, scientific articles and academic theses.

During this period, 9 exhibitions were held at various locations, including an exhibition area staged in association with an educational workshop and a long-term exhibition programme. Education services included a programme for various levels of education, teacher training and family workshops.

Preventive archaeology has also been regularly promoted by the local press and in themed exhibitions; a noteworthy example was the *A21 exhibition – Archaeology on the Highway*, which was launched in 2009 and generated considerable impact in the national media given the rarity of such initiatives.

Despite this intensive activity, it was only possible to implement enhancement measures for the inter-municipal project Historic Route of the Lines of Torres Vedras, with funding from EAA grants – a project that won a Europa Nostra award.

In addition, the local population of Mafra was further encouraged to participate in heritage activities such as archaeological excavations and historical re-enactments.

The case of Mafra highlights the importance of maintaining a balance in archaeological heritage management, research and dissemination, and of developing a long-term plan. Unexpected funding cuts have led to the closure of exhibition spaces and to a reduction in staff – a trend that can be seen in many other municipalities. Unfortunately the picture is very unbalanced at national level, depending more on personal initiatives and executives than on national policies.

Figure 17.13: Historical re-enactment in Zambujal Fortress. Publicity about the Historical Route of the Lines of Torres has a strong local impact on the communities involved in the maintenance and animation of this heritage. Photo Marta Miranda.

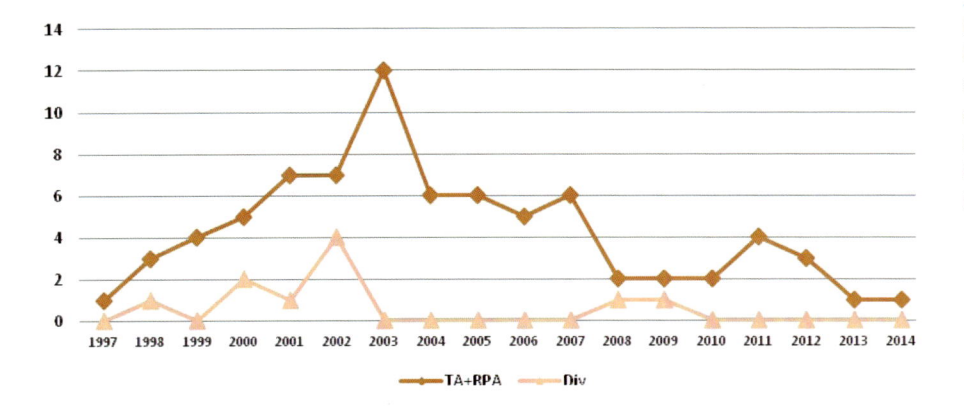

Figure 17.14: Publications by the central heritage administration between 1997 and 2014. TA+RPA: Trabalhos de Arqueologia and Revista Portuguesa de Arqueologia. DIV – Diverses.

3. We and the others: archaeological promotion agents in Portugal

Despite the long history of archaeology in Portugal (Fabião 2011) it is only in recent decades that there has been widespread dissemination of archaeological activity. Several interconnected factors can be mentioned in this respect: the Côa valley findings (1995–1997), the establishment of an autonomous archaeological authority (Portuguese Institute of Archaeology 1997) and specific archaeology legislation, the emergence of the first university degrees in archaeology (during the 1990s) and the professionalisation of archaeology. Together with the abovementioned circumstances, the economic contribution from European Community funds for the implementation of major archaeological projects has to be mentioned.

Currently, the majority of the Portuguese population is aware of archaeologists and archaeological activity. Even though the research process is widely recognised, the public tends to find it more difficult to understand and interpret the work details involved.

Communication in archaeology is primarily carried out by archaeologists and for archaeologists, which may make it less clear for large segments of the population. It is therefore important to broadly examine the promotion agents (us): the cultural administration (central and regional), museums, universities and research centres and businesses.

3.1. Promotion of archaeology by the cultural heritage authority

Currently, the Portuguese cultural heritage has a centralised administration bringing together architectural, movable, intangible and archaeological heritage. Management of these areas also has a regional component as regards museums, monuments and sites management (Decree Law 114/2012, Decree Law 115/2012).

Since 1997 the protection of the archaeological heritage has fluctuated between various organisations, as part of a major administrative reorganisation of the entire sphere of culture. As it is impossible to critically analyse the whole process, I will provide a brief overview of the main approaches to the promotion of archaeology during the study period (1997–2014).

Broadly speaking, promotion strategies are much more effective in the technical-scientific area, when targeting archaeologists or heritage technicians.

Implementation of the Endovelico Information System in 1995 (Bugalhão & Lucena 2006) was a milestone in archaeological heritage management, as it enables the inventorying, geo-referencing and publicising of land and underwater archaeological heritage, which currently amounts to more than 30,000 occurrences (Gomes et al. 2012). Its database is accessible via the Archaeologist's Portal – an online platform that provides e-services to professionals and information about archaeological sites for users in general. It has proved to be an effective tool for heritage promotion and protection.

The Portuguese Institute of Archaeology (1997–2006) promoted an editorial plan for publishing archaeological work, thereby fulfilling the requirement set out in the Regulation on Archaeological Works (DL 170/99, Article 15, paragraph 3), including a monographic series (archaeological work) and a bi-annual magazine (Portuguese Journal of Archaeology – *Revista Portuguesa de Arqueologia*), open to all of the archaeological community.

The regularity of the publications and their wide dissemination through a European network of exchanges, apart from being available online, has made them a reference source for Portuguese archaeology.

Archaeological publications 'survived' the changes in the organisational structure of the archaeology authority, remaining under IGESPAR (2007–2011) and the DGPC (2012–), albeit with a substantial decline. The new Regulation on Archaeological Works (DL 140/2014) maintains a reference to the monographic series and the Portuguese Journal of Archaeology. Between the 1999 and the 2014 regulations there was a clear need to find other forms of promotion, particularly for rescue archaeology. The 2014 regulation also mentions the availability of online publications, in particular concerning rescue archaeology.

Although the editorial overview of scientific publications is positive, there is no strategy for promotional publications. These are restricted to itineraries of visitable sites, including the "Archaeological Route" Collection. This gap has not been filled by other sectors such as museums or commercial publishers.

Figure 17.15: Canada do Inferno, Archaeological Park of the Coa Valley. The process of safeguarding Coa Art triggered a radical change in the legal framework of archaeological activity. Photo José Paulo Ruas / DGPC.

Under the national heritage agency, 35 congresses took place (20 organised by the national agency and 15 co-organised with other partners). The enhancement and management policy for archaeological sites has changed over time according to their different governing bodies. Management of archaeological sites requires direct monitoring and significant investment in conservation planning. It has led to short lifecycles in various archaeological enhancement projects developed by the governing authorities, such as in the Antas de Belas circuit. The recent shift in the responsibility for archaeological heritage protection to the regional directorates tried to bridge the gap between managers and the territory within their remit. As a result, several regional directorates have established collaboration protocols with the municipalities.

Currently the DGPC has a very limited number of archaeological sites under its jurisdiction. They include sites located in the area of Lisbon and the Tagus Valley and those inscribed on UNESCO's World Heritage List. A foundation model with an exclusively financial contribution was chosen for management of the only world heritage archaeological site in Portugal: the Côa Valley.

The outlook is somewhat different in the remaining regional directorates, which have a total of 26 visitable archaeological sites under their direct management.

About 500 archaeological sites have been legally protected by classification, but only a small number have been targeted for conservation, evaluation and interpretation.

In 2001 the Archaeology Centre of Almada conducted a survey of municipalities and cultural administration, having gathered an extensive dossier of 300 visitable sites in Portugal (Raposo 2001). This exercise brought together a number of archaeological sites with very different visitor access conditions. It refers to some unevenness in their geographical distribution (by district and municipality) and highlights the lack of regional and national plans. In 2001 only 20 sites were integrated in museums or associated to museum structures. In most cases visits are free (Raposo 2001, 104). Despite the undeniable economic impact of visitable archaeological sites through tourism, an investment by the public administration, namely by the heritage authority, will always be required.

3.2. Museums and archaeology

Archaeological museums are spread across the country with an estimated total of 208 museums in 2014 (Antas 2014, 226). This figure includes:

1. Archaeological museums, archaeological museums with musealized archaeological sites and multi-core archaeological museums;
2. Archaeology collection museums;
3. Interpretive centres (Antas 2014).

This multiplicity has been provided for by the Museum Framework Law (Law 47/2004 of 19 August), which stipulates that an archaeological site or ensemble can be considered a museum.

Within the restricted Portuguese Museum Network (of accredited museums), 52 archaeology museums or archaeology collections are referenced, representing 37% of all museums in this network. The establishment of this archaeological museum network progressed at a relatively steady pace until the 1990s, registering a peak in the first decade of the 21st century. Many of the new museums are musealized archaeological sites and interpretation centres, and 71% were created by the municipalities. There are two main explanations for this situation: on the one hand, the increase in archaeological activity and, on the other, the European funding of the last Community Support Framework. After a period of strong growth, there is a current downturn, and various sustainability issues in this network (Camacho 2008–2009). Some of the museums that emerged between 1990 and 2010 were closed down and others recorded downsizings in financial and human resources. The excess of local museums with very similar content clearly limits their attractiveness for non-local audiences; recently, there has been a tendency to create small thematic museums, such as the Southwest Script Museum (Almodôvar) or the Discovery Museum (Belmonte). There is also a growing tendency to establish integrated routes between museums, archaeological sites and other heritage sites, such as the Historic Route of the Lines of Torres Vedras or the Romanesque Route.

It should be emphasised that there was a gap between the discourse of the museums and recent developments in archaeology after Valletta. Some of the main Portuguese archaeology museums were established in the 19th century and their collections were brought together between the late 19th century and the 1970s. These museums have become true repositories of Portuguese archaeological history and are in a way detached from the contemporary world. They have nothing to do with management policy regarding the holdings of preventive archaeology activity, which is one of the main difficulties of archaeological activity in Portugal. The Alqueva Dam, which does not have a regional museum or an integrated management of its assets, is an example. Besides, there is a limited perception of the concept of archaeological holdings, often perceived as works of art rather than scientific documents (Correia 2013–2014).

Against this background, it would seem clear that museums could develop a more active role in promoting archaeological activity (research and prevention). Mediators are needed to handle the technical and scientific findings from the fieldwork carried out in recent decades.

The work of these museums is particularly important for engagement with local communities in terms of identity and as a tourist development engine (see the paring identity/economy developed by Correia [2013–2014, 155]).

3.3. Research centres and universities

The growth of archaeological activity was followed by an increase in academic degrees (bachelor's, master's and doctoral) in archaeology. At the same time, there are new universities all across the country.

During the period from 1997 to 2014, archaeological academe expanded with the creation of new archaeology degree courses at the Universities of Minho (1998), Nova de Lisboa (1995), Évora (2000–2001) and Algarve (2008). However, most archaeologists graduated from the 'old' universities such as those in Lisbon, Coimbra and Porto, accounting for the number of entries and the results of the recent DISCO study (Costa et al. 2014).

Universities have a double impact on the promotion strategy. On the one hand they essentially have a training capacity (Diniz 2008). In addition to their skills in technical and scientific training, university studies also include social skills, such as promotion. Although there are no specific curriculum areas for science communication, these concepts are addressed across various disciplines. Students are also required to participate in promotional activities undertaken by universities/research centres.

The issue of publications (*publish or perish* ...), is naturally at the centre of university actions. The challenge of communicating science to the public (Public Understanding of Science) became important in the 1980s in the UK (Entradas 2015), as there was an attempt to find a relationship between scientific knowledge and the public's attitude towards science.

In Portugal, communication science was developed by Mariano Gago (particle physicist, responsible for a scientific research agency between 1986 and 1989, and science minister for 12 years: 1995–2002, 2005–2011). In 1996 the creation of the Life Science Agency (*Ciência Viva*) set off an intense science education program, including the establishment of a network of 14 centres

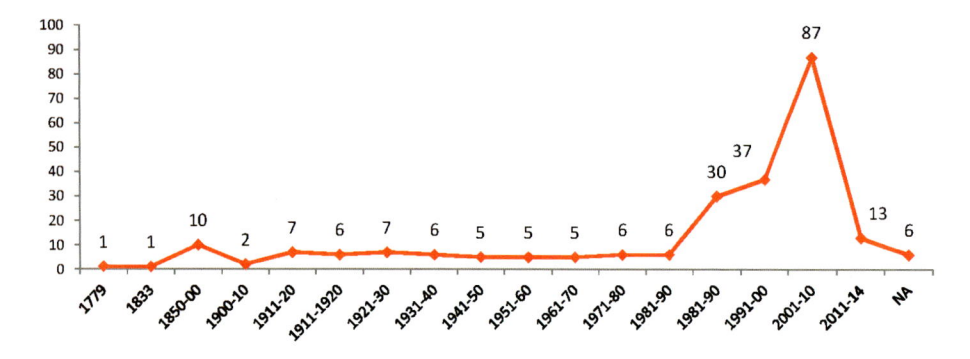

Figure 17.16: Foundation dates of archaeological museums in Portugal. Source: Antas, 2014.

Figure 17.17: Promotion actions by Portuguese archaeological companies. Source: Data from DISCO 2014.

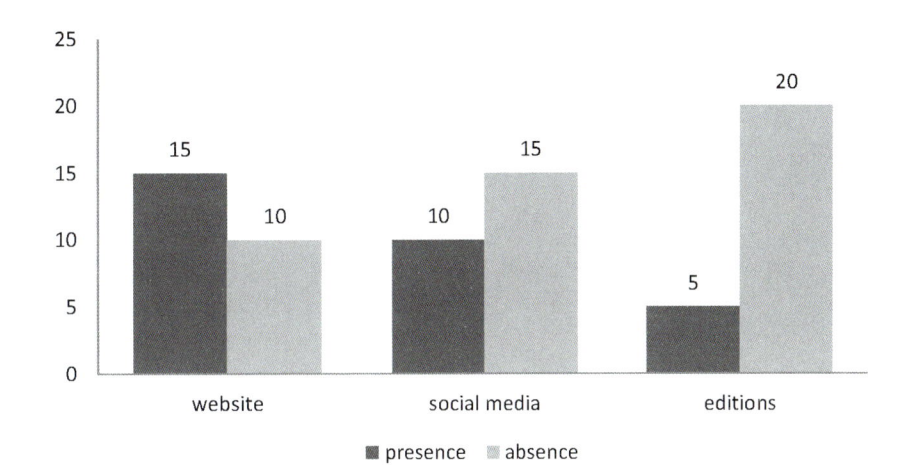

across the country. Even though social sciences were not a central core of these initiatives, some were dedicated to archaeology-related themes, thereby leading to archaeology communication beyond the sphere of cultural heritage. This trend was strengthened by specific guidelines for communication science developed by the Foundation for Science and Technology (FCT) during the process of evaluating projects and research centres.

These initiatives have recently spiked with projects such as the European Researchers' Night (*Researchers' Night*), promoted by the European Commission under the Marie Curie Actions since 2005 in order to celebrate science and engage in community outreach. This action, promoted by the Science Museums of Lisbon and Coimbra University included archaeological activities.

These actions were aimed at bringing together research archaeology and the local community. In addition, 'open days', mostly run by universities/research centres, have been developed in recent years to encourage visits to archaeological excavations. Beyond these occasional and seasonal initiatives, there is a strengthening of knowledge transfer in more permanent actions, such as scientific coordination of enhancement and musealization projects at several archaeological sites.

From this perspective, it seems that the coming years will register an increasing concern for communication science, a trend reinforced by Horizon 2020, – an EU programme aimed at capacitating European citizens, with specific funding lines (Reflective societies: transmission of European cultural heritage, uses of the past, 3D modelling for Accessing US cultural assets).

3.4. Archaeology companies

The free market model adopted by Portugal (Sousa 2013) led to the exponential growth of archaeology companies. In the absence of a permit or accreditation system, it is very difficult to quantify existing archaeological companies (Costa et al. 2014). DISCO 2014 reference frameworks are used suggesting that 'at the beginning of the financial crisis, Portugal had 39 active archaeology companies, and in 2014, that number dropped to 25.' (Costa et al. 2014, 79).

A total of 8 dozen archaeology companies have been active in Portugal, the majority being sole-trader companies, which have already closed down their business.

This scattering of micro-businesses naturally compromises promotion both in terms of organisational capacity and their financial capacity to invest in promotional activities. In most cases there is no communication strategy whatsoever.

Corporate communication has two main objectives: company promotion and social responsibility.

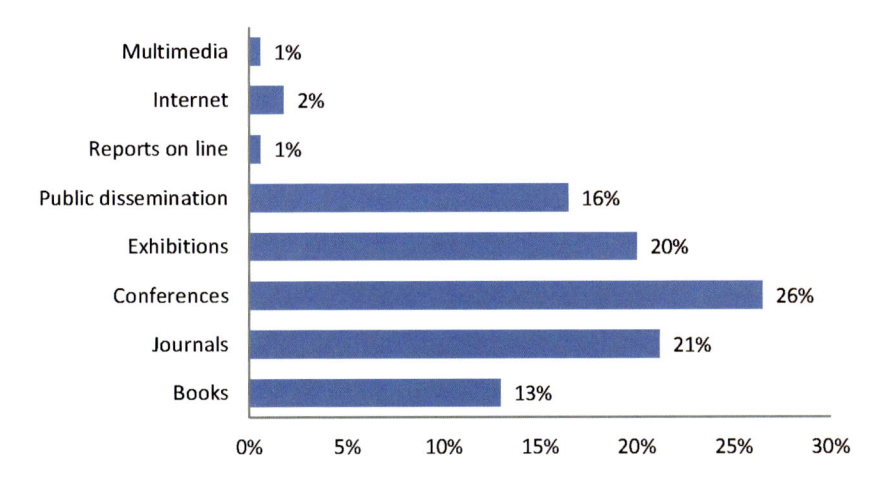

Figure 17.18: Dissemination media used by archaeological organisations (adapted from DISCO 2014 – Costa et al., 2014).

YEAR	TOTAL	ARCH	ORGANIZATIONS						
			OTHER	PRIVATE	ASSOCIATION	MUNICIPALITY	NATIONAL MUSEUMS	REGIONAL ADMINISTRATION	CENTRAL ADMINISTRATION
2014	575	60	5	5	7	28	10	5	0
2013	463	40	5	3	1	24	2	5	0
2012	525	57	0	3	5	39	2	8	0
2011	461	43	0	2	5	20	5	8	3
2010	444	15	0	6	1	4		3	1
2009	403	47	4	0	1	29	3	10	0
	2871	262	**14**	**19**	**20**	**144**	**22**	**39**	**4**

Table 17.2: Actions held on the International Day of Monuments and Sites in Portugal. Source: DGPC.

Using three simple indicators (website, social media, publications), it can be said that many archaeology companies do not pay particular attention to communication, even when investment is reduced, as is the case with virtual communication.

Websites and social media are currently the only vehicle for real-time dissemination of the biggest findings in preventive archaeology. As well as web promotion, community outreach has surfaced in recent years, particularly in the case of corporate research projects.

With regard to publications (online and print), the situation is more striking, as only 20% of companies have publications. As for printed editions, only two companies have published works for more than a decade.

In order to analyse scientific production in the corporate sector, especially as regards publications, it would be necessary to conduct a thorough literature inventory impossible in the current study. Antonio Valera attempted to make an analysis of the scientific production by Portuguese archaeological companies (Valera 2007), but inquiry-based surveys always have great representation issues.

This 'low-cost' archaeology (Almeida 2007) necessarily leads to a low social return rate, as repeatedly referred to by some Portuguese archaeological companies (Almeida & Neves 2006; Valera 2007; 2008).

Of course, the problem will always have to do with financial sustainability. If contracts make no mention of research and promotion, corporate archaeologists are not the only ones to be blamed.

3.5. Associations

Heritage protection associations, which are non-governmental organisations, played a major role in the post-revolution period (after 25 April 1974). Given the importance of these associations, the Law of Cultural Heritage (DL 107/2001) sets out the rights of those organisations in terms of the 'right of participation, information and popular action' and collaboration with public administration in promotion (art. 10). Nevertheless, they are not represented in the advisory bodies of the cultural heritage authority, including the National Council of Culture, section of Architectural and Archaeological Heritage.

In 1997, Jorge Raposo identified 45 heritage associations, most of them founded in the 1990s (Raposo 1997). Similar growth can also be observed in reference to Environment Protection Associations (Caninas 2011).

Concurrent to the associative movement of the 1990s and 1980s, local associations with direct impact on the archaeological heritage remain active to date.

The Portuguese Archaeologists Association (AAP) is the oldest heritage protection association in Portugal (established in 1863), being responsible for the Carmo Archaeological Museum (MAC) – the first art and archaeology museum in the country. In recent decades, the AAP has played an important role in disseminating information to professionals and the general public through lectures and seminars and by promoting initiatives such as the Festival of Archaeology.

The Mértola Archaeological Site and the Mértola Heritage Defence Association had a unique role in marking the boundary between academic and community areas. Established in 1978, they have implemented a research, enhancement and promotion plan. Contrary to what usually happens, their intervention in the territory is permanent, as researchers have settled down in Mértola. Their action programme includes rescue archaeology in the historic centre, enhancement and *in situ* conservation of archaeological assets, museum promotion and periodic publications.

The Archaeology Centre of Almada represents an exemplary case in Portugal as it combines research, training and promotion, with particular reference to periodic publications.

3.6. The others

Archaeology promotion is also ensured by other agents not belonging to the archaeological community.

In addition to Heritage Protection Associations (many of which have no archaeologists), there is the 'Groups of Friends' movement, connected mainly with museums and musealized sites. A notable example is the Group of Friends of the National Archaeological Museum, established in 1999, which has considerable powers of mobilisation.

Large impact digital media (facebook, websites) have recently emerged, as is the case with *Portugal Romano* (60,000 followers, about 200,000 weekly views) and many embryonic themed platforms. These platforms play an important role in promotion and awareness.

4. General trends

Generally speaking, the Portuguese (and European) archaeological community recognise(s) that it is absolutely necessary to reverse the current situation regarding archaeological promotion. There is a clear increase in initiatives undertaken by all agents. We do hope that this new trend may reverse the declining presence of archaeology in the media and in political agendas.

Based on the survey conducted by the Professional Association of Archaeologists for DISCO (Costa et al. 2014), it can be said that the current promotion model is still very focused on the archaeological community and on scientific knowledge production. The great challenge will undoubtedly be to develop communication and mediation skills targeting the general public by means of an interdisciplinary perspective and with the support of communication professionals.

Heritage enhancement must also be encouraged. An archaeological site is only perceived by the communities as 'their own' if the right mediation strategy is used. This is probably why archaeology ranks low in heritage promotion schemes such as the International Day on Monuments and Sites (ICOMOS) or the European Heritage Days (Council of Europe and European Commission). Our indicators are based on the last five years of the International Day on Monuments and Sites in Portugal, according to which archaeology represents only 9% of the total activities between 2009 and 2014.

The International Day on Monuments and Sites undoubtedly reflects the current situation in terms of promotional dynamics.

It is without any doubt the municipalities that are leading the initiatives with 55% of all activities. This percentage reflects the special attention paid by municipal archaeologists to promoting the archaeological heritage.

The regional culture directorates should also be mentioned as they represent 15% of total activities. This percentage reflects a dynamic promotion strategy as regards the archaeological sites under their protection.

The DGPC – the central body organising this initiative in partnership with ICOMOS – is virtually absent from the picture as far as archaeological data is concerned. This is due to the current cultural heritage management competencies in Portugal: the central heritage administration only manages sites located in the area of Lisbon and the Tagus Valley and those inscribed on UNESCO's World Heritage List. Foz Côa is the only archaeological site under this circumstance, but since 2011 it is managed by a foundation (included in the 'others' category).

Private companies are responsible for two major Portuguese archaeological sites that are under private management: the Roman ruins of Troia and the Archaeological Centre of Rua dos Correeiros.

5. Promotion by decree? Future prospects

Effective dissemination of information on archaeology cannot be ordered by law as it requires society to be convinced and get involved, and thus assimilate/appropriate the principles of the Faro Convention. Awareness of the need to change the current scenario led to the inclusion of promotion in the recently published Regulation of Archaeological Works DL 164/2014).

This concern can be found in the preamble of this decree-law:

According to the new Regulation of Archaeological Works, applicants for an archaeological work permit are required to submit a 'Plan for disseminating

> 'This decree-law hereby redefines and clarifies policies regarding management and disclosure of the results of archaeological works in the areas of scientific publication, heritage awareness and education. These should be designed to hold the archaeologist responsible and to move the scientific discipline closer to the citizens.'

archaeological work results to the community' (DL 164/2014, art. 7).

The future will evaluate the contribution of this legislation for the promotion of archaeology. But there is no doubt that the task of dissemination cannot be left exclusively to the goodwill of archaeologists. This responsibility also lies with policy-makers, cultural heritage administration, developers and the local communities.

Acknowledgements

Archaeological activity indicators are spread across multiple organisations. I would like to express my gratitude for the excellent collaboration of institutions and archaeologists who helped me develop this work by providing information from their archives.

Data was provided by the following institutions: Directorate General for Cultural Heritage (DGPC), Alqueva Development and Infrastructure Enterprise (EDIA), Municipality of Mafra (CMM), National Archaeological Museum (DGPC), Carmo Archaeological Museum (Association of Portuguese Archaeologists), Lisboa Museum (Lisbon City Hall), Millennium BCP Foundation (NARC), EGEAC (Lisbon City Hall), Geological Museum (LNEC), Bank of Portugal Museum.

My thanks go to Cintia Pereira de Sousa for the English translation.

I would also like to thank everyone who shared their own ideas and personal research, which made it possible for me to carry out this study in a timely fashion: Jacinta Bugalhão, who generously provided me with Lisbon data and with whom I exchanged a lot of ideas; Antonio Carlos Silva, for his collaboration regarding the Alqueva project, and the unfailing support of Filipa Neto, Filipa Bragança and Ana Sofia Gomes. I also extend my gratitude to Lidia Fernandes, Manuela de Deus, Gertrude Branco, João Marques and Maria Catarina Coelho. Any errors and omissions are, of course, my own responsibility.

References

Almeida, M. J. 2006: Colecções de Arqueologia em Autarquias: reflexões a partir de um inquérito promovido pela APA, *Praxis Archaeologica* 1, 37–57.

Almeida, M. J. 2007: Inquérito nacional à actividade arqueológica. Uma segunda leitura sobre a actividade arqueológica nas autarquias portuguesas, *Praxis Archaeologica* 2, 129–71.

Almeida, M. J., Almeida, M., Caldarelli, S. B., Cavalcanti, G. A., Costa, F. A, Dias, R. J., Lago, M., Malerbi. E., Ribeiro, J.P.C., Santos, M. C. M. M. & Tocchetto, F. 2009: Diálogos transatlânticos: Contribuições da arqueologia consultiva à pesquisa e proteção do patrim(ó)(ô)nio arqueológico no Brasil e em Portugal, *Praxis Archaeologica* 4, 27–43.

Almeida, M. & Neves, M. J. 2006: A Arqueologia Low-cost: fatalidade nacional ou opção de classe? O modelo empresarial.*Al-Madan*. IIª Série, 14, 86–91.

Antas, M. 2014: *A comunicação educativa como factor de (re)valorização do património arqueológico – boas práticas em Museus de Arqueologia portugueses* (unpublished PhD thesis).

Branco, G. 2014: *Avaliação de Impacte Ambiental: O Património Arqueológico no Alentejo Central* (unpublished PhD thesis).

Bugalhão, J. 2007: Lisboa e a sua Arqueologia: uma realidade em mudança, *ERA – Arqueologia* 8, Lisboa, 218–30.

Bugalhão, J. 2011: "A Arqueologia Portuguesa nas últimas décadas", *Arqueologia & História*, 6061, 19–43.

Bugalhão, J., 2014: Arqueologia de Lisboa balanço e perspectivas Conferência apresentada no *Seminário "Lisboa Subterrânea – Trajectos na Arqueologia Lisboeta Contemporânea"*. Lisboa, Sociedade Portuguesa de Geografia, Lisboa, em 21 de Maio de 2014.

Bugalhão, J., Gameiro, C., Martins, A. & Braz, A. F. 2012–2013: Núcleo arqueológico da Rua dos Correeiros: da intervenção à investigação, gestão e apresentação pública, *Arqueologia & História*, Lisboa, 191–201.

Bugalhão, J. & Lucena, A. 2006: As Novas Tecnologias como Instrumento de Gestão e de Divulgação do Património: o exemplo do Endovélico – Sistema de Gestão e Informação Arqueológica, *Encontros Culturais do Baixo Tâmega, Património. Actas*, Baião: Câmara Municipal, 175–92.

Camacho, C. 2008–2009: A rede portuguesa de museus e os museus com colecções de arqueologia – parâmetros de sustentabilidade. *Revista da Faculdade de Letras. Ciências E Técnicas Do Património*, I Série, Volume VII–VIII, 107–14.

Caninas, J. C. 2011: Associativismo e Defesa do Património (1980–2010) in: *100 anos de Património: memória e identidade*. Lisboa: IGESPAR, 2–12.

Correia, V. H. 2013–2014: Os Museus de Arqueologia e os seus Públicos, *Arqueologia & História*, Lisboa, 153–60.

Costa, C., Duarte, C., Tereso, J., Lago, M., Viegas, C., Grilo, C., Raposo, J., Diniz, M. & Lima, A. 2014: Discovering the archaeologists of Portugal 20122014, Associação Profissional de Arqueólogos.

Diniz, M. 2008: Arqueologia; Divulgação; Universidade: palavras-chave para um Novo Contrato Social, Praxis Archaeologica, Associação Profissional de Arqueólogos.

Entradas, M. 2015: Science and the public: The public understanding of science and its Measurements. *Portuguese Journal of Social Science* 14 (1), 71–85.

Fabião, C. 2011: *Uma Historia da Arqueologia Portuguesa. Das origens à descoberta da Arte do Côa*. Lisboa.

Fabião, C. 2014: Olhai esta Lisboa de outras Eras: Arqueologia urbana no sítio de Lisboa, *Rossio. estudos de Lisboa*, Lisboa, n. 3 maio 2014, 5.

Gomes, A. S., Leite, S., Neto, F., Oliveira, C. & Bragança, F. 2012: Inventariação e gestão do património imóvel na Direção-Geral do Património Cultural. *Documentazione e conservaziones del Patrimonio Arquitettonico ed Urbano*, Desegnarecom, Special edition.

Lemos, F. S. 2004: A salvaguarda do património arqueológico em contexto urbano, *Património e Estudos*, Lisboa, 33–40.

Lemos, F. S. 2006: A Lei e a Arqueologia Urbana, *Praxis Archaeologica*.

Martinho, M. 2002: Investigação arqueológica no Alqueva: a indispensável conclusão, *Al-madan*. Almada, 2nd Series, No. 11, 63–4.

Martinho, M. 2014: O património cultural no Empreendimento de Fins Múltiplos de Alqueva: caracterizar, avaliar, minimizar, valorizar in A. C. Silva,. F. T. Regala, & M. Martinho (eds) – *4º Colóquio de Arqueologia do Alqueva. O Plano de rega (20022010)*. Évora: EDIA/Direcção Regional de Cultura do Alentejo (Colecção Memórias de Odiana, 2ª Série), 34–44.

Martins, A. M. N. 2012: A salvaguarda do património arqueológico no âmbito dos processos de avaliação de impacte ambiental e de ordenamento territorial: reflexões a partir do direito do património cultural, do ambiente e da gestão do território, *Revista Portuguesa de Arqueologia*, vol. 15, 219–56.

Martins, M., Fontes, L. & Cunha, A. 2013: Arqueologia urbana em Braga: balanço de 37 anos de intervenções arqueológicas" in J. M. Arnaudet al. (eds): *Arqueologia em Portugal 150 anos, Associação dos arqueólogos Portugueses*, 81–7.

Martins, M. & Ribeiro, M. C. 20092010: A arqueologia urbana e a defesa do património das cidades, *Forum*, Ed Conselho Cultural da U. do Minho, http://hdl.handle.net/1822/13351 (accessed 13.09.2015).

Melro, S. & Deus, M. 2014: O acompanhamento no terreno do Projecto EFMAA na área da Extensão de Castro Verde in A. C. Silva, F. T. Regala & M. Martinho (eds): *4º Colóquio de Arqueologia do Alqueva. O Plano de rega (2002–2010)*, Évora: EDIA/Direcção Regional de Cultura do Alentejo (Colecção Memórias de Odiana, 2ª Série), 45–52.

Miranda, M. 2009: Viajar no Tempo através do Património histórico e arqueológico do Concelho de Mafra. *Revista Pedra e Cal. Anuário da Conservação do Património.*

Raposo, J. 2001: Sítios arqueológicos visitáveis em Portugal, *Al-madan*, 2nd series, 10, 100–5.

Raposo, J. (ed.) 1997: Património e Associativismo, *Almadan*, 2ª série, nº 6, Centro de Arqueologia de Almada.

Silva, A. C. 2002: Avaliação dos Impactos Arqueológicos em Alqueva: a formação do «Quadro Geral de Referência», *Al –madan*, Almada, 2ª série, nº 11 IIª S, nº 11, 56–65.

Silva, A. C. 2014: Alqueva – quatro encontros de Arqueologia depois. In A. C. Silva, F. T. Regala & M. Martinho (eds): *4º Colóquio de Arqueologia do Alqueva. O Plano de rega (20022010)*, Évora: EDIA/Direcção Regional de Cultura do Alentejo (Colecção Memórias de Odiana, 2ª Série), 19–33.

Silva, A. C., Regala, F. T. & Martinho, M. (eds): *4º Colóquio de Arqueologia do Alqueva. O Plano de rega (2002–2010)*, Évora: EDIA/Direcção Regional de Cultura do Alentejo (Colecção Memórias de Odiana, 2ª Série).

Sousa, A. C. 2013: A revisão do Regulamento de Trabalhos Arqueológicos e os contextos sociais da arqueologia portuguesa no século 21: uma breve reflexão. *Revista Património*. Lisboa, 1, 36–42.

Valera, A. C. 2007: Arqueologia empresarial e produção de conhecimento, *Al-madan*, Almada, 2nd series, No. 15, 75–82.

Valera, A. C. 2008: A divulgação do conhecimento em Arqueologia: reflexões em torno de fundamentos e experiências, *Praxis Archaeologica* 3.

Links

Link 1 http://arqueologia.patrimoniocultural.pt (accessed 29.12.2015).

# 18 | From Valletta to Faro – avoiding a false dichotomy and working towards implementing Faro in regard to archaeological heritage (reflections from an Irish perspective)

Margaret Keane and *Sean Kirwan*

Abstract: Despite not having ratified the Faro Convention (Council of Europe 2005), key aspects of heritage management in Ireland already reflect its values and principles. This reflects the fact the there is no conflict between Faro and Valletta. Faro is a framework convention which supports the sector-specific cultural heritage conventions such as Valletta. To present matters otherwise is to create a false dichotomy. Debates over issues such as partial versus total excavation in response to developmental impacts may well be necessary, but must not be presented as representing a conflict between Faro and Valletta. In this article the authors suggest that Faro joins with and supports Valletta in the continuing development of archaeological heritage management in Europe. This complementary rather than evolutionary relationship between the Conventions of Valletta and Faro is demonstrated in some particular programmes which have been implemented during the last decade in Ireland. Archaeology in the Classroom is a bespoke programme which enables children between the ages of 5 and 12 years to learn about and appreciate their heritage. This serves as a mechanism for the protection and conservation of that heritage into the future, achieving preservation through education. Arising from the implementation of the Convention of Valletta in Ireland, a collaborative grant programme – Irish Strategic National Research (INSTAR) – was established to foster the dual aims of advancing the vast quantities of new data into knowledge and to provide for collaboration across professional archaeological groups including the commercial, academic and public sector silos.

Keywords: false dichotomy, preservation by education, complementary conventions, collaboration

Introduction

This paper will respond to the theme of the issues raised for contemporary archaeological heritage management by the 2005 Council of Europe Framework Convention on the Value of Cultural Heritage for Society (the 'Faro Convention', Council of Europe 2005). It should be said at the outset that Ireland has neither signed nor ratified the Faro Convention, and there are no immediate plans to do so. Nevertheless, as the authors hope this paper will demonstrate, key aspects of archaeological heritage management in Ireland already reflect values and themes of the Faro Convention and have in fact done for some time. This could no doubt be said for other aspects of cultural heritage in Ireland apart from archaeological heritage. In that context, it is worth noting that the Faro Convention is not an instrument focused specifically on archaeological heritage. This is, perhaps, an obvious statement. Nevertheless, the authors, having attended the last two EAC annual meetings (2014 and 2015) and listened to the discussions and debates, would highlight the need to remember that the Faro Convention is exactly what its title describes it as – a framework convention. It seeks to provide an overarching framework for cultural heritage policy within which the pre-existing sectoral conventions (in the case of archaeology, the 1992 Revised European Convention on the Protection of the Archaeological Heritage, the 'Valletta Convention') remain vital and central. In that sense there can be no journey from Valletta to Faro – rather Faro joins with and supports Valletta in the continuing development of archaeological heritage management in Europe.

It should be noted that the authors are based in the National Monuments Service of the Department of Arts, Heritage and the Gaeltacht. Much of the work of the National Monuments Service is regulatory in nature (though not exclusively so, as will be clear from what follows), and in reflecting on themes in the Faro Convention it is necessary for the authors to refer to the role and work of other organisations and colleagues. This reflects the nature of archaeological heritage management in Ireland, with a number of bodies involved in varying roles.

Some reflections on the Faro Convention

As noted, Ireland is not a party (i.e. has not ratified) the Faro Convention. Ireland is, however, a party to the three sectoral cultural heritage conventions developed within the Council of Europe framework – the Valletta Convention, as referred to above (which for those states party to it replaced the 1969 European Convention on the Protection of the Archaeological Heritage), the 1985 Convention for the Protection of the Architectural Heritage of Europe (the 'Granada Convention') and the 2000 European Landscape Convention (the 'Florence Convention'). Furthermore, Ireland was one of the first states to ratify what might be seen as the original (and still in force) framework convention for cultural heritage in Europe – the 1954 European Cultural Convention. Ireland ratified this in 1955. The 1954 Convention is cited in the preambles to both the Valletta and Granada Conventions as a background to those conventions.

That being so, what has held Ireland back in regard to binding itself formally to (as opposed to implementing in practice, as outlined below in regard to archaeology) the principles of the Faro Convention? Here, the authors must stress that they do not express an official view. It might be noted first that Ireland is clearly not alone in treating the Faro Convention with a degree of caution. Ten years after its adoption the Convention has secured 17 ratifications. This is not an inconsiderable number, but the Faro Convention is certainly not yet at the level of near universal (within the Council of Europe framework) adherence achieved by the Valletta Convention, which currently has 45 ratifications (including one non-Council of Europe party). Clearly, the Valletta Convention has been open to ratification for a longer period than the Faro Convention, but by the end of 2002 (10 years after its adoption) Valletta had already secured 27 ratifications (including one non-Council of Europe party). (Data is derived from the Council of Europe website.) An analysis of the provisions of the Faro Convention is beyond the scope of the scope of this paper, but it can be safely said that it is an ambitious document – ambitious not just in terms of the wide scope of its subject matter (far beyond archaeology – a point archaeologists should remember) but also in terms of the range of governmental functions and actors which would be called into play to ensure its full implementation, or even to achieve agreement to proceed to ratification. Is there a danger that Faro is over ambitious, or perhaps that its terms lack the specificity needed to allow governments to have the necessary clarity as to what they are committing themselves to by ratifying?

On the other hand, at the risk of being provocative (and noting again that this is not an official view), when all is said and done, what does Faro really say which is additional to what was provided for with clarity and succinctness in Articles 1 and 5 of the 1954 European Cultural Convention ('Each Contracting Party shall take appropriate measures to safeguard and to encourage the development of its national contribution to the common cultural heritage of Europe' and 'Each Contracting Party shall regard the objects of European cultural value placed under its control as integral parts of the common cultural heritage of Europe, shall take appropriate measures to safeguard them and shall ensure reasonable access thereto'), as fleshed out and given at least some specificity in the succeeding sectoral conventions as referred to above? Insofar as it does add anything, it may be especially open to the charges of lack of specificity or over-ambition. Perhaps the most important aspect which Faro adds is the laudable focus in Articles 3, 4 and 7 on the need to respect cultural heritage diversity and, by implication, to avoid the perpetuation of narrow or exclusionary heritage narratives or views as to what constitutes cultural heritage of importance. Greater attention to this vitally important theme and the setting out of clearer aims and standards in that regard might, perhaps, have truly added more to the corpus of existing Council of Europe cultural heritage conventions. Examples of positive trends in regard to this theme in Ireland in recent years would include greater governmental and general public recognition of the role of Irish soldiers in the First World War and the development of an official visitor centre for, and access to, the site of the 1690 Battle of the Boyne – a pivotal battle in Irish history between forces led on the one side by King William III and on the other by King James II and which has had different resonances for different traditions on the island of Ireland right up to the present day.

However, the Faro Convention stands as, and must be respected as, the outcome of important work by the Council of Europe and its members. As noted, nothing said here is an official view and Ireland may well, in due course, decide to ratify the Faro Convention. There can be no objection to a body such as the EAC seeking to use the themes of the Faro Convention to map out future priorities for the progress of heritage management in Europe. Provided, that is, no false dichotomy is created between Faro and Valletta. The avoidance of this is explored below with respect to one aspect of discussions and debates in recent EAC annual meetings.

The Faro Convention and debates about selective archaeological recording

As noted, the authors have participated in the last two annual EAC meetings, including the discussions at Amersfoort in 2014 which provided the background to the Amersfoort Agenda document (Schut et al. 2015). The authors of this paper found that the issue of whether archaeological heritage management should move from a model based on the greatest level possible of archaeological recording of sites impacted on by development to a model of selective archaeological excavation was one which arose repeatedly, both in formal and informal conversation. There appeared, at least to the authors, to be at times a perception that Ireland was almost an extreme or ultimate example of an 'excavate everything' approach. Leaving aside the rights and wrongs of both this perception and the underlying policy issue (and for that see in particular Keane 2015), there is in fact nothing in the Faro Convention, either explicit or implicit, which the authors can see as speaking to the question of whether or not selective excavation is the appropriate policy for the future (indeed, there is very little if anything in the Faro Convention which speaks to the specifics of archaeological heritage management).

Equally, there is in fact little if anything in the Valletta Convention which explicitly demands on an absolute basis total archaeological excavation in advance of removal of sites to allow development. Articles 5 and 6 of the Convention require (in broad summary) the integration of archaeological concerns into the planning and development process and the provision of appropriate levels of funding for necessary archaeological work consequent on development, but it is clear that the text gives a significant measure of discretion and flexibility to states party to the Convention. Irish national policy (non-statutory) on the protection of the archaeological heritage in the context of development advocates full recording of archaeological deposits and features being removed to facilitate development (DAHGI 1999a, 25). Clearly, this has to be implemented through the available statutory, i.e. legislative, frameworks regulating particular types of development and depending on that, and the nature of the environment in question, exceptions may arise, the most notable of which would be the partial excavation strategy applied to timber trackways in peatlands being milled by the state-owned company Bord na Móna. This is implemented under a Code of Practice agreed between that company, the Minister for Arts, Heritage and the Gaeltacht and the National Museum of Ireland (DAHG 2012). However, the authors would see the full recording policy as representing, in the normal course, a fair and consistent standard for developers to comply with and one which, equally, can be applied by the relevant authorities on a fair and consistent basis and which assures best protection for archaeological heritage (including a strong incentive towards the avoidance of unnecessary impacts). As set out previously by one of the authors (Keane 2015, 79), it can be argued that the record of recent public controversies in Ireland regarding impact of major road construction on archaeological heritage indicates that for the concerned public the minimum which will be generally acceptable is full excavation and recording and that some may not find even this acceptable.

In any event, given what has just been outlined in regard to what the two conventions actually say, the debate as between full or partial recording in advance of development is really not a debate as between Valletta and Faro. To present it as such is to create the kind of false dichotomy between the two conventions which the authors would warn against.

Valletta and the value of archaeological heritage for society

As noted above, the full title of the Faro Convention is the Convention on the Value of Cultural Heritage for Society. A wider and more detailed treatment of this issue may add to what was in the Valletta Convention as regards archaeological heritage, but it should not be thought that Valletta gave no attention to this, albeit being to a large extent focused on reconciling the interests of archaeology and development. Article 9 of Valletta requires each state which has ratified it to 'conduct educational actions with a view to rousing an awareness in public opinion of the value of the archaeological heritage for understanding the past and of the threats to this heritage'. Article 8.ii requires states which have ratified Valletta to 'promote the pooling of information on archaeological research and excavations in progress and to contribute to the organisation of international research programmes'.

It may well be the case that more needs to be done in many countries (including Ireland) to promote public interest in archaeology and to support research (including research to maximise the benefits from development-led archaeological work), but it would be a mistake to think that recognition of this only arose with Faro; all those states which have ratified Valletta have in fact already committed themselves to action in that regard.

The remainder of this paper will focus on a number of programmes and measures being undertaken in Ireland which could readily be said to enhance or promote the value of archaeological heritage for society and to go towards meeting existing obligations under the Valletta Convention, as just noted. Before focusing on particular measures, it is important to note that in 1995 (i.e. before ratification of Valletta in 1997) Ireland in fact established a statutory body with a specific remit in regard to promotion of public interest in cultural and natural heritage as well as to propose policies and priorities in regard to heritage: the Heritage Council, as established under the Heritage Act 1995. Information about the Heritage Council and its work will be found at www.heritagecouncil.ie. As will be seen from that website, a particular focus of the Heritage Council during the 20 years of its existence has been working with, and supporting, local communities and community-based actions.

Archaeology in the Classroom – It's About Time!

In Ireland a strong sense of place and pride in the parish is reflected in public interest and appreciation for heritage generally, local heritage most particularly. These values are at the core of the aspirations of the Faro Convention (Council of Europe 2005). In 2005 the Limerick Education Centre and the National Monuments Service of the Department of Environment, Heritage and Local Government launched an innovative, hands-on and imaginative programme called Archaeology in the Classroom – It's about Time! This is a teaching resource pack designed to give young children knowledge of the lives of people in the past and to introduce them to the processes which historians and archaeologists use to interrogate material culture and documentary sources.

Archaeology in the Classroom was designed and tested on a summer course in *Scoil* Dean Cussen School in Bruff, County Limerick as early as 2003. It was providing expression to the objectives of Faro as the convention was being drafted:

'Article 12 – Access to cultural heritage and democratic participation

The Parties undertake to:

(d) take steps to improve access to the heritage, especially among young people and the disadvantaged, in order to raise awareness about its value, the need to maintain and preserve it, and the benefits which may be derived from it.'

It is available for the primary (junior) school sector, to teachers and pupils alike, in hard copy or online with links to interactive historically-based games. The resource sits within the Social Environmental and Scientific Education (SESE) part of the current Irish curriculum encompassing history, geography and science, but as a specifically designed programme it fulfils age-appropriate requirements linked to other parts of the curriculum including mathematics, language and the arts. In relation to the primary resource, a hard copy issued to all primary schools (3,300) in 2005 was subsequently made available in CD format and via the website (Link 1). The Irish language (*Gaeilge*) version was translated in 2007 as there is a significant and growing cohort of *Gaelscoileanna* or Irish language based schools within the Irish education system. The resource was revised in 2013. There has been an enthusiastic response to both the quality of the resource and its functionality in relation to the SESE curriculum. The resource is also used specifically in preparing new teachers in the SESE programme as part of the overall teacher education programme.

The collaboration was the brain-child of four individuals – three archaeologists and an educator: Matt Kelleher and Denis Power from the National Monuments Service; Mary Sleeman, current Cork County Council archaeologist, and Dr Joe O'Connell, the Director of Limerick Education Centre. Matt was engaged in carrying out the Archaeological Survey of County Limerick in the early 2000s and dropped a leaflet about the work of the Archaeological Survey of Ireland into the Limerick Education Centre in 2002. Dr Joe O'Connell, who has a doctoral degree in Education, was interested in the work of the survey and approached Matt initially in relation to the school curriculum and archaeology. A dream team assembled: Matt with a background in sociology and archaeology and his Master of Arts degree in Analytical Aerial Archaeology; Mary, who in addition to her education as an archaeologist was a fully qualified secondary school teacher (post-primary level: early teens to late teens), and Denis with a joint honours degree including archaeology and a career in archaeological survey and conservation, all of whom had a passion for imparting their interest, appreciation and expertise in an imaginative, fresh, engaging way to children and their teachers.

Modules

The pack is divided into 12 modules entitled: Archaeology of the Classroom, what will survive?; Timeline Ireland; Excavation-in-a-box; Stone Age Hunters; Pots and Pottery; Making Monuments; Recording Old Buildings; Let's Look at Old Photographs; Streetscape; Exploring Old Maps; Fieldtrip; the Outdoor Classroom; My Own Place (Figure 18.1).

The modules follow sequentially but there are in-built flexibilities. Each module can be taught alone as an individual class plan. The modules are broken down into

Figure 18.1: Module 6 Resource Pack: Making Monuments – Curriculum Linkages and Class Plan (© Archaeology in the Classroom).

three stages; the first stage is the class plan itself, which is laid out in a sequential manner through to a closing activity. The second stage contains the various activity sheets for the module. These can be photocopied and distributed to the children during the activity stage. The final stage contains the teacher guidelines. The guidelines are divided into: (a) managing the module, which gives instructive information on applying the module to both junior and senior classes; (b) relevant background information, where appropriate, and (c) the skills and strands which detail how the module complements the revised Primary School Curriculum. The pack is very attractively designed, in bright engaging colours with original drawings taking inspiration from Irish art and archaeology by Rhoda Cronin, a qualified archaeologist and illustrator.

The first three modules focus on 'archaeology' in general terms of process. Module 1 is an introduction to the concept of archaeology. Module 2 uses a timeline to look at the main periods of Irish archaeology in terms of date range, typology and classifications and introduces some of the monuments and artefacts from each period. Module 3 deals with archaeological excavation as a practical exercise where the tangible evidence of a birthday party – used candles, used lollipop sticks, and even bottle tops – find themselves buried in a sand box to be re-excavated by the children as an exercise called Excavation in a Box. Themes of intercultural exchange, diversity and distinctiveness are raised in the description of how different countries celebrate birthdays. The next three modules focus on the Stone Age and feature practical experiments that explore the lifestyle of prehistoric people (Figure 18.2).

Modules 7 to 9 look at historical buildings and are designed to give the pupils some basic skills to describe and appreciate their built heritage. As the curriculum places great emphasis on local studies, the last three modules focus on this. It should be noted that module 12 is different from the others. It takes the form of a suggested project whereby the skills and abilities developed in the other modules, particularly modules 7 to 10, are applied in terms of the area immediately surrounding the school.

Due to the success of Archaeology in the Classroom, a second bespoke resource pack called Time in Transition was devised, focusing on second level students at Transition Year. Transition Year is a one-year school-based programme between Junior Cycle and Senior Cycle of Secondary School (Link 2). It is designed to act as a bridge between cycles for teenagers, who partake in this 'gap' year, where the focus from state examinations shifts to focus on gaining maturity and independence in terms of learning skills, work-place experience and career interest. In this series there are three overarching themes: 1. Worship and Commemoration, 2. Lifestyle and Living, 3. Archaeology at Work, each of which are subdivided into units. Worship and Commemoration covers Worship, Monasticism, Pilgrimage and Commemoration and Memorials. Lifestyle and Living

Figure 18.2: Unit 1, Worship: The Magic Ring – Forming the Stone Circle and Orientation of the Stone Circle (© Archaeology in the Classroom).

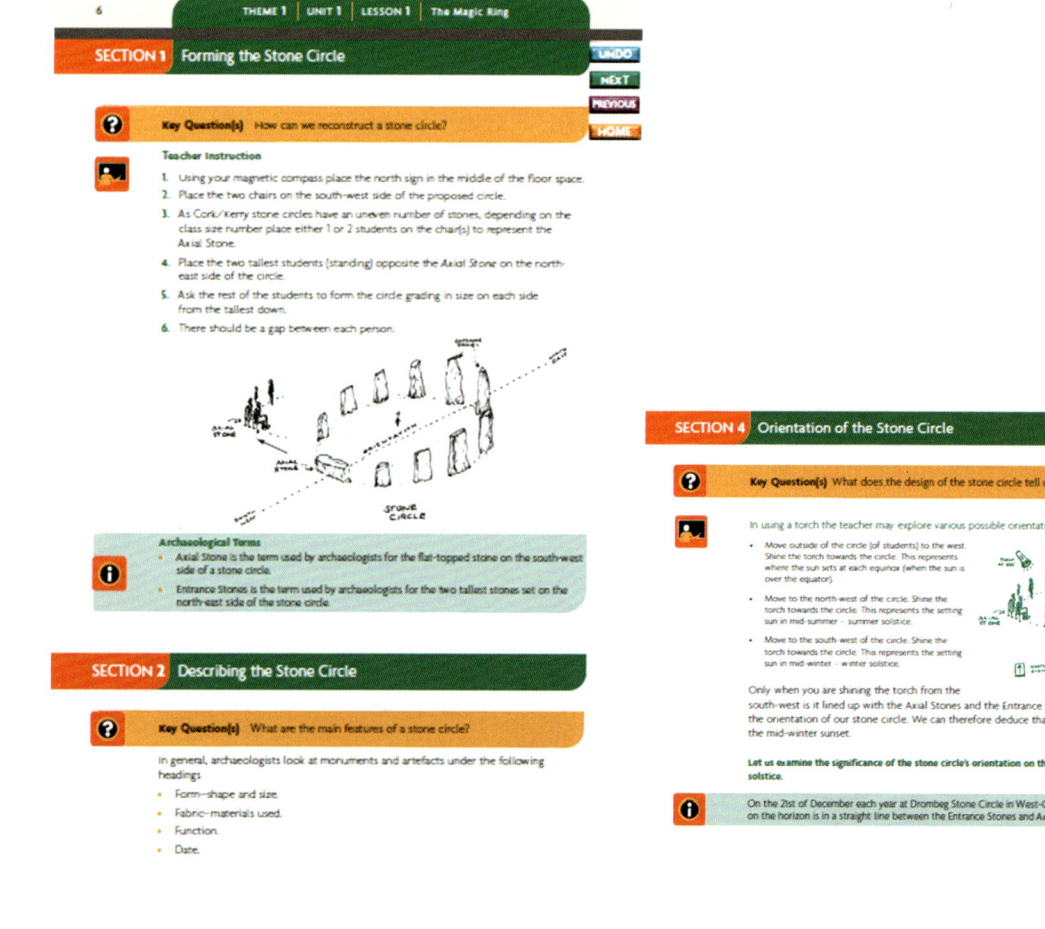

covers Housing, Defence, Towns and Lifestyle, whilst Archaeology at Work looks at how archaeologists research and examine the past, and it describes the processes of archaeological excavation in Excavation, Post-Excavation and Keepers of the Past.

Worship and Commemoration has a unit on worship which focuses on a lesson called The Magic Ring. This unit is the basis of a lesson, or lessons, which allows students to create a Bronze Age Stone Circle. The lesson explores the formation of a stone circle, how to describe this monument class, how to interpret the circle form, the associated numerology, symbolism and orientations and the results of archaeological excavations of this type of monument and the learning outcomes based on the results of excavation.

Time in Transition

Time in Transition (launched 2009/revised edition republished 2012) has also been very well received at post-primary level and was endorsed by the History Teachers' Association as a valued resource.

This is also available in translation into Irish (Link 3).

It was distributed in hard copy to all schools which offer the Transition Year option. The Professional Development Service for Teachers use the resource for when they are working with Transition Year coordinators.

The pack allows for interactive engagement with baseline resources. For instance, in Unit 2 Monasticism, there are links provided to an online database of Ogham stones in 3D. Ogham consists of cut stones bearing inscriptions in the unique Ogham alphabet using a series of scored horizontal and diagonal lines inscribed around the edges of the stones to represent the sounds of an early form of the Irish language. The inscriptions and stones commemorate the names of prominent people, sometimes providing information on lineage or tribal affiliations. They are the earliest record of the Irish language and are Ireland's earliest written record, dating back to the 5th century A.D.

A series of articles in the Irish Examiner newspaper

As an addition to the It's About Time! programme, a series of articles relating to its themes were written by Mairead Weaver and Matt Kelleher of the National Monuments Service, in collaboration with the Irish Examiner Newspaper, and were published in 2013/2014. The Irish Examiner has a broad coverage in Ireland, with daily circulation figures of around 40,000 newspapers. The articles were called: *Taking time to learn about our past, Pilgrimages' progress traced from the past to the present day; Our identity is reflected in how we celebrate our heritage; How Ireland defended itself and its native communities;* and *The mysteries of archaeology unearthed*. The articles link back to the resource pack in a witty manner, engaging the reader through comparisons with current life practices and bright colourful layouts (Figures 18.3–18.4).

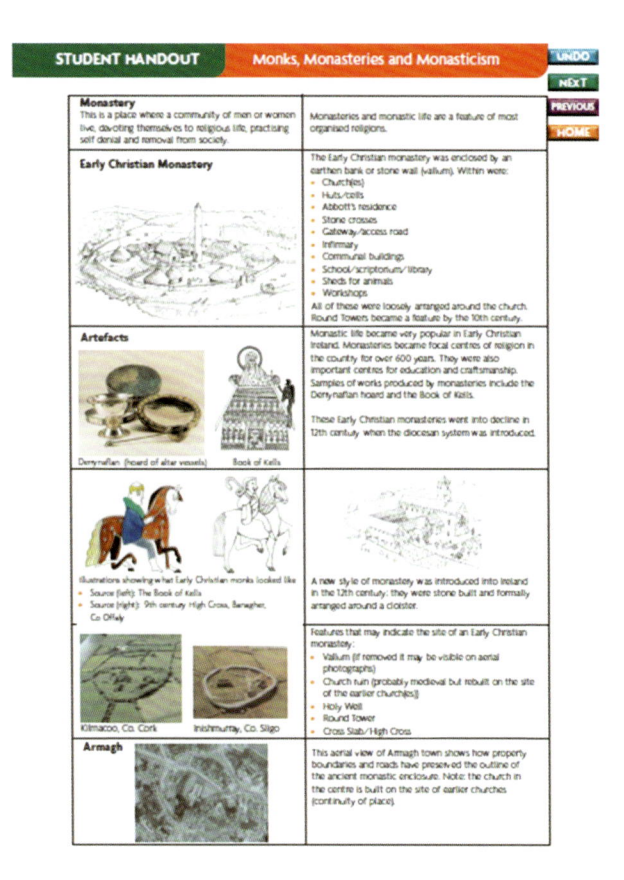

Figure 18.3: Unit 2, Monasticism; Monks Monasteries and Monasticism Student Handout (©Archaeology in the Classroom).

The Minister for Arts, Heritage and the Gaeltacht, Jimmy Deenihan, was quoted as saying: "My hope is that the Irish Examiner articles, together with 'Time in Transition', will not only foster an increased appreciation of our rich and diverse heritage, but also help to promote the preservation of that heritage through education and awareness" (Weaver & Kelleher 2013).

A final resource currently under development – another Limerick Education Centre initiative – is directed to the primary sector again and will be disseminated

Figure 18.4: Ogham Stone, Emlagh East (Imleach Dhún Séann), County Kerry: BRUCCOS MAQQI CALIACI 'of Bruscas son of Cailech' (© Nora White).

to schools in the mid-west. It will be called AHA! (Archaeology and History through the Arts) and it will be launched in December 2015.

Relationship to Faro

In its widest sense Archaeology in the Classroom – It's about Time! and Time in Transition, and Archaeology and History through the Arts, the programmes themselves, the revisions thereof, the CPD associated with it for the educational sector and all the subsequent outreach, such as the off-shoot series of articles published in the Irish Examiner during 2013, fulfil the provisions of Article 13 – Cultural heritage and knowledge, in which the Parties undertake to:

'(a) facilitate the inclusion of the cultural heritage dimension at all levels of education , not necessarily as a subject of study in its own right, but as a fertile source of studies in other subjects;
(b) strengthen the link between cultural heritage education and vocational training'.

That this programme anticipated the development of the Convention shows how events and thinking at the periphery of Europe can reflect the concerns at the heart of Europe.

The Faro Convention outlines a framework for considering the role of citizens in the definition, decision-making and management processes related to the cultural environment in which communities operate and evolve.

INSTAR

During the 1990s and into the 2000s there was a huge increase in privately and publicly funded rescue excavations (Keane 2015). It was widely recognised by Irish archaeologists that several levels of disconnect had developed in the profession in response to the increased numbers and complexity of excavations being commissioned (Cooney et al. 2006; University College Dublin 2006).There was a chasm between the amounts of data and archaeological objects being retrieved on foot of licensed excavation in Ireland and the knowledge creation accruing to the discipline. The vast and growing archaeological archives of the state were rarely accessed or utilised by academia. The commercial/consultant sector felt that their professionalism was under scrutiny by their colleagues in the academic and state sectors without a real understanding of the challenges of working under the time-bound constraints of rescue archaeology (Cooney et al. 2006; University College Dublin 2006).

'In Ireland, there is insufficient cooperation between the university Departments/Schools of

Archaeology, between them and the State institutions with responsibility for archaeology, and with the archaeological consultancy sector itself.' (University College Dublin 2006)

At the core of these issues was an imbalance between archaeological enquiry as research for its own sake and archaeological practice as a service sector within the construction industry.

In January 2006 the then Minister of the Department of Environment, Heritage and Local Government, Dick Roche, commissioned a report to examine research needs in Irish archaeology. In dialogue with a forum of members representative of the archaeological profession, research themes were selected and a structure devised with the additional aims of building research capacity and providing access to previously unpublished excavations. The report was published (The Heritage Council 2007) and the Irish National Strategic Archaeological Research fund, a collaborative cross-sectoral mechanism to evaluate grant applications for archaeological research, was founded. Funded directly from the budget of the National Monuments Service, with projects selected by international peer review, the programme has been administered since inception through the Heritage Council.

Seven research themes were identified:

1. Cultural Identity, Territory and Boundaries,
2. Resources, Technology and Craft,
3. Exchange and Trade,
4. Religion and Ritual,
5. Environment and Climate Change,
6. Landscapes and Settlement,
7. Archaeology and Contemporary Society.

A broad range of projects from the Palaeolithic through to the recent past, 37 different proposals in fact, have been successful in attracting funding. Some have brought together the work of original excavators and new collaborators together with additional funding used to progress specialist reporting towards completion of previously unpublished significant research excavations. In this way the pioneering work of Seamus Caulfield, Gretta Byrne, Noel Dunne, Martin Downes and others looking at the sub-peat Neolithic and Bronze Age landscapes of North Mayo at the Céide Fields and beyond has been progressed towards publication (Caulfield et al. 2009).

Another project, called Wodan, has developed an integrated wood and charcoal database for researchers in Ireland (Stuijts et al. 2009). Research frameworks have been designed: Burren Landscape and Settlement (Jones & Comber 2008) and The Archaeological Remains of Viking and Medieval Dublin (Simpson et al. 2010). An international collaborative project, based in the University of Reading, using source material from Bord Na Móna bogs, sought to develop techniques (Branch et al. 2008) that will provide a precise chronological and palaeoclimatic context for the archaeological remains in the midland bogs. Another of the most successful INSTAR projects to date: EMAP – Early Medieval Archaeological Project, has successfully synthesised the results of the excavation of hundreds of early medieval monuments during the last 20 years (O'Sullivan et al. 2014). Yet another – Mapping Death – traces populations and individuals across international boundaries and carries out ground-breaking science in ancient DNA, isotopic analysis and osteoarchaelogy, seeking to examine burial practices as indicators of social practices.

What has been crucial to this programme is that in order to qualify for research funding each application must fulfil mandatory criteria in relation to collaboration and *must* include at least two partners from the academic (national and international), commercial and state sectors. The work is primarily research, *not* excavation, but looks to utilise the results of development-led excavation work. It is carried out by archaeologists working towards advanced degrees including post-doctoral work. Thus the product outcomes are not simply knowledge creation and publications but increased collaboration and synergies across the profession, enhanced learning by individuals and a gain in academic qualifications and standing. Some unanticipated outcomes have included increased opportunities to attract external funding, the development of more nuanced methodologies being applied to rescue excavation, and the development of new research tools from existing databases. While analysing the environmental results of a range of excavations at sites dated to the Neolithic period as part of the Cultivating Societies – assessing the evidence for agriculture in Neolithic Ireland INSTAR project, gaps in the sampling methodologies were observed which led to complications and challenges at the final analysis stage. These problems, associated with the selection, processing and reporting on environmental work, have been addressed by the development of new guidelines relating to such work (McClatchie et al. 2014). It is hoped that the widespread use of these guidelines and embedding of such practices in environmental reporting will improve the standards of work and the outcomes of environmental research and analysis.

Challenges ahead

One of the most challenging matters for archaeologists and heritage managers into the future has to be to consider Article 7.b of Faro, which stipulates that signatories undertake to:

'b. establish processes for conciliation to deal equitably with situations where contradictory values are placed on the same cultural heritage by different communities'.

In Ireland the politicians who lead the relevant government departments, and regulatory bodies who administer current law and policy, are regularly called to task by heritage communities. Democracy in Ireland, where proportional representation is the voting system, is very much a local prerogative, with the public sometimes placing local concerns above the national when exercising their democratic will. Naturally, this leaves politicians very aware of needing a loyal public to provide for their re-election. In some recent controversies heritage communities, lauding the importance of the monuments and sites within their local area, do not agree with the decisions of local and central government in permissions to allow for monuments to be removed by excavation, even under strict scientific excavation by-hand methodologies (Healey 2015). Thus heritage communities as protesters raise their concerns in local press, in social media and directly with their public representatives. Other heritage communities, such as professional archaeological consultants commissioned to carry out excavations or the professional archaeological advisors in local and central government, place a different value on the archaeological material, and the challenge is to reconcile different voices. As professional archaeologists, because of our professional interest in the process and results of excavation, we may not fully share the attachment to place and to continuity of preservation *in situ* which other heritage communities have.

While the planning system in Ireland provides an opportunity for the public to voice their concerns in relation to proposed development, the current National Monuments Act does not provide for direct public involvement in the decision making process, which is the system for licensing excavations. For consultants and their employees, excavation work provides their livelihood. For regulatory authorities, like planning authorities, there may be concerns other than strictly archaeological, such as providing for local employment and providing for sustainable development, which colour their decision making. National archaeological policy allows for general concerns to be addressed and for public involvement; however, there is no current process for that engagement.

'Applicants for archaeological excavation licences will have to satisfy the Department of Arts, Heritage, Gaeltacht and the Islands with regard to the following factors:

(a) That the proposed archaeological excavation is justified or necessary (DAHGI 1999b, 10)'

and

'The Minster for Arts, Heritage, Gaeltacht and the Islands has specific responsibility for the protection of the archaeological heritage, but the general public and all public and private bodies also have a key role to play.' (DAHGI 1999a, 12)

It is these questions – the provision of clear and transparent decision making processes and the interaction between the regulatory authorities, the public and politics – which are the real challenges of Faro, as set out in the Amersfoort Agenda.

Acknowledgements

The authors wish to acknowledge the assistance of colleagues Matt Kelleher, Denis Power and Mairead Weaver in the drafting of this article and for the creativity and persistence and professionalism of those colleagues, along with Mary Sleeman and Dr Joe O'Connell in bringing Archaeology and the Classroom and its successors, Time in Transition and AHA!, to fruition.

References

Branch, N. P., Young, D., Elias, S., Mansell, J., Denton, K., Swindle, G. E., Matthews, I. & Whitaker, J.

2008: Examining the relationships between environmental change, raised bog development and Bronze Age human activities: recent multi-proxy investigations at Clonad Bog and Kinnegad Bog, Ireland, *Proceedings of the 13th International Peatland Congress*, 52–47.

Caulfield, S., Byrne, G., Downes, M., Dunne, N., Warren, G., McIlreavy, D., Rathbone, S. & Walsh, P. 2009: *Neolithic and Bronze Age Landscapes of North Mayo, Summary Report 2009,* https://www.ucd.ie/archaeology/documentstore/instarreports/northmayoproject/NBNM2009_SUMMARY_REPORT.pdf (accessed 18.12.2015).

Cooney, G., Downey, L. & O'Sullivan., M. 2006: *Archaeology in Ireland: A vision for the future,* Royal Irish Academy, Dublin.

Council of Europe 2005: *Framework Convention on the Value of Cultural Heritage for Society*. European Treaty Series 199. Council of Europe, Strasbourg.

Council of Europe 2014: *Action for a Changing Society Framework Convention on the Value of Cultural Heritage for Society*. https://www.coe.int/t/dg4/cultureheritage/heritage/identities/Faro-brochure_en.PDF (accessed 07.12.2015).

Department of Arts, Heritage and the Gaeltacht 2012: *Code of Practice between the Department of Arts, Heritage and the Gaeltacht, the National Museum of Ireland and Bord na Móna*, Dublin.

Department of Arts, Heritage, Gaeltacht and the Islands 1999a: *Framework and Principles for the Protection of the Archaeological heritage,* Dublin: Stationery Office.

Department of Arts, Heritage, Gaeltacht and the Islands 1999b: *Policy and Guidelines on Archaeological Excavation*. Dublin: Stationery Office.

Healy, C. 2015: 'Waterford historians unimpressed by US company's plan to clear ancient settlement', *thejournal.ie* , 25 July, 2015, www.thejournal.ie/waterford-knockhouse-ring-fort-west-pharmaceutical-2228226-Jul2015 (accessed 18.12.2015).

Jones, C. & Comber, M. 2008: *Burren Landscape and Settlement: Developing a Research Framework Report 2008,* The Heritage Council, http://www.heritagecouncil.ie/fileadmin/user_upload/INSTAR_Database/Burren_Landscape_and_Settlement_Final_Report_08.pdf (accessed 17.12.2015).

Keane, A. 2015: An Interpretation of Valetta from the Críoch Fuinidh (the remote or end country) in: P.A.C Schut, D. Scharff & L. de Wit (eds): *Setting the Agenda: Giving New Meaning to the European Archaeological Heritage,* EAC Occasional Paper no. 10, Amersfoort, 75–89.

McClatchie, M., OCarroll, E. & Reilly E. 2014: *NRA Palaeo-environmental Sampling Guidelines Retrieval, analysis and reporting of plant macro-remains, wood, charcoal, insects and pollen from archaeological excavations,* (revision 4), National Roads Authority, http://www.tollcompare.ie/archaeology/seanda-2014/june-2015/NRA_Palaeo-environmental-Sampling_Guidelines_rev-4.pdf (accessed 17.12.2015).

O'Sullivan, A., McCormick, F., Kerr, T.R., & Harney, L. 2014: *Early Medieval Ireland, AD 400–1100. The evidence from archaeological excavations*, The Royal Irish Academy, Dublin.

Schut, P.A.C., Scharff, D. & De Wit, L.C. (eds) 2015: *Setting the Agenda: Giving New Meaning to the European Archaeological Heritage,* EAC Occasional Paper No. 10, Amersfoort.

Simpson, L., Gowen, M. & Co. Ltd. 2010: *The Archaeological Remains of Viking and Medieval Dublin: a Research Framework*, Irish National Strategic Archaeological Research (INSTAR), The Heritage Council and Dublin City, http://www.heritagecouncil.ie/fileadmin/user_upload/INSTAR_Database/Medieval_Dublin_Archaeological_Research_Framework_Final_Report_08.pdf (accessed 17.12.2015).

Stuijts, I., Bunce, A. & O'Donnell, L. 2009: *The development of a wood database and a charcoal database for Ireland (WODAN)* http://www.wodan.ie/documents/FINAL_REPORT_2009.pdf (accessed 17.12.2015).

he Heritage Council 2007: *A Review of Research Needs in Irish Archaeology,* The Heritage Council, http://www.heritagecouncil.ie/fileadmin/user_upload/Publications/Archaeology/Research_Needs_in_Irish_Archaeology.pdf (accessed 18.12.2015).

University College Dublin 2006: Archaeology 2020. Repositioning Irish Archaeology in the Knowledge Society, School of Archaeology, UCD, Dublin, http://www.heritagecouncil.ie/fileadmin/user_upload/Publications/Archaeology/Archaelogy_20_20.pdf (accessed 18.12.2015).

Weaver, M. & Kelleher, M. 2013 'Taking time to learn about our past', *Irish Examiner,* 16 May 2013.

Links

Link 1 www.itsabouttime.ie (accessed 07.12.2015).

Link 2 www.itsabouttime.ie/transition-year.aspx (accessed 07.12.2015).

Link 3 http://www.cogg.ie/?s=seandalaíocht (accessed 07.12.2015).

19 | Assuring quality: archaeological works on Irish national road schemes

Rónán Swan

Abstract: This paper is from the perspective of the client, namely Ireland's National Roads Authority (NRA). The NRA is a State Agency (now operating as Transport Infrastructure Ireland [TII] since its merger with the Railway Procurement Agency in August 2015) that has responsibility for the provision of a safe and efficient network of primary and secondary roads. This amounts to approximately 5,000 km of road, and in the past 15 years the NRA has upgraded and improved nearly 1,500 km of road, from minor improvements to the construction of approximately 400 km of motorway. But why is the NRA interested in archaeology at all? Why does it care? There are three answers. Firstly, legislation: Irish law requires archaeology to be treated appropriately. Secondly, risk: if archaeology is not managed effectively it can be extremely costly in terms of delays and claims from the main works contractor, particularly if archaeology is only identified during construction. Thirdly, public trust: the NRA is a public body that takes its responsibility to the taxpayer very seriously and therefore seeks to ensure that not only do we achieve compliance, but that that compliance is purposeful and meaningful. In this context the NRA has spent more than €300 million on archaeology in the past 15 years and therefore has a keen interest in assuring quality.

Keywords: risk, infrastructure, legislation, management and public engagement

Introduction

Ireland has a rich and diverse archaeological heritage; while the Island has only been settled for the past 10,000 years, the known archaeological remains number more than 150,000 recorded monuments. The experience of preventive archaeology in Ireland shows, however, that the true number of archaeological sites is a multiple of this figure.

At EAC 2015, delegates were invited to consider preventive archaeology as it is practised in Europe today, in the early 21st century. This paper considers this topic from the particular point of view of the client or developer, in this case the National Roads Authority (NRA), which merged with the Railway Procurement Agency to create Transport Infrastructure Ireland in August 2015. Firstly, the scene will be set and the context provided as to why the NRA is concerned with archaeology. Subsequently, the following issues, which speakers at EAC 2015 were asked to consider, will be addressed:

- Finding the right expertise
- Monitoring quality
- Sharing results and ensuring lasting public benefit

Setting the scene

In considering the role of the NRA in archaeology, one can trace three broad phases in the practice and treatment of archaeology. The first was from 1994 to 2000, wherein the NRA had no in-house archaeological expertise and the emphasis was on site identification through monitoring during construction. The second phase was from 2001 to 2006; during this period project archaeologists were employed by local authorities to oversee archaeological works on behalf of, and funded by, the NRA. Additionally, the NRA appointed a Head of Archaeology and another project archaeologist at its head office in Dublin. The third phase was from 2007 onwards, wherein all the project archaeologists were directly employed by the NRA. In both the second and third phases the emphasis was on site identification and mitigation in advance of construction. The three principle factors underpinning the agency's concern for archaeology are legislation, risk and public trust, and their relationship to one another (Figure 19.1). During the initial years of the NRA, its response to archaeology was very much one of achieving legal compliance. In

Figure 19.1: The principle factors underpinning the NRA's concern for archaeology.

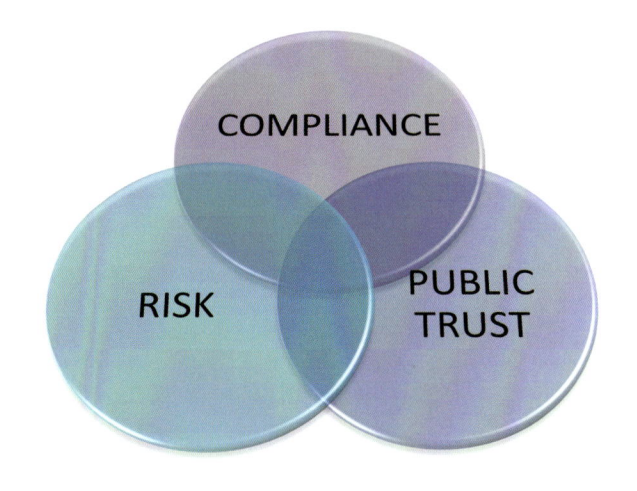

the subsequent years it became far more aware of the need to manage risk and to establish and build public trust (Swan 2014).

The protection for archaeology and heritage in Ireland derives from international conventions, EU directives and national legislation and regulations. The Council of Europe's 1992 Valletta Convention (of which Ireland is a signatory) is the primary inspiration for archaeological protection and requires that there are appropriate systems in place for the management and conduct of archaeological works. The 2005 Faro Convention (yet to be ratified by Ireland) seeks greater public participation in archaeology and heritage, and there is an onus on signatories to provide greater information on archaeology and heritage (mirroring in many ways the 1998 Aarhus Convention on providing access to environmental information).

While none of the EU directives address archaeology explicitly, the 1985 Environmental Impact Assessment (EIA) Directive and subsequent amendments address cultural heritage, including archaeology and architecture. In addition, recent EU statements have identified cultural heritage and data thereof as a strategic resource for Europe and, in particular, have called on member states to enhance the role of cultural heritage in sustainable development (Florjanowicz, this volume). At European level, we also have to be mindful of the Procurement Directive which governs the procurement of all works and services by member states, setting new requirements in terms of greater market engagement and also lifecycle pricing.

At a national level, the Roads Acts require cultural heritage (which includes archaeology and architectural heritage) to be accounted for in the preparation of the relevant EIA and Environmental Impact Statement (EIS), while the archaeological works themselves are governed principally by the National Monuments Acts 1930–2004 and the Planning Acts. Since the late 1990s, national archaeological policy has required that archaeological heritage be protected in one of three ways: firstly, consideration should be given to avoidance; if that is not feasible then every effort must be made to preserve archaeology *in situ*; and if that is not possible then the site must be preserved by record through excavation, leading to an archaeological report and the accessioning of all artefacts to the National Museum of Ireland (Department of Arts, Heritage, Gaeltacht and the Islands, 1999).

Interestingly, the National Monuments Amendment Act 2004 was introduced to resolve issues in relation to the transfer of functions and powers which had become apparent in the course of a series of legal challenges to the completion of the M50 motorway around Dublin at the site of Carrickmines Castle (Keane 2015, 79). This amendment also introduced new procedures for the treatment of archaeological works on national road schemes that were subject to an EIA and were approved by the Irish planning body An Bord Pleanála. For such schemes, the roads authority (i.e. a local authority or the NRA) is obliged to apply for specific Ministerial Directions to cover all archaeological works from the Department of Arts, Heritage and the Gaeltacht (DAHG formerly the Department of Environment, Heritage and Local Government and prior to that the Department of Arts, Heritage, Gaeltacht and the Islands). The Minister will issue Directions following consultation with the Director of the National Museum of Ireland, thus placing the onus and responsibility for the satisfactory completion of the archaeological works with the road authority rather than with the excavation director, is the case for excavations on non-approved road schemes. The National Monuments Service oversees the implementation of the Minister's Directions. From the perspective of the NRA, this approach is welcome, as it removes any ambiguity as to its obligations and provides the roads authority with the opportunity to present an overarching strategy to the Minister, setting out how it will discharge its archaeological obligations for the project as a whole.

However, inconsistencies still remain and the approach to archaeological works on roads is not as streamlined as it might be. For example, archaeological works on non-approved road schemes (i.e. projects that do not require an EIS) are governed by a series of discrete licence applications and approvals for different phases of the works. Also, rather than the roads authority being responsible and accountable for these licences, they are the responsibility of individual excavation directors, usually private sector archaeologists. Another shade of complexity is introduced as Ministerial Directions can only be introduced when the EIS for a scheme has been approved, thus pre-approval investigations on the same scheme are carried out under licences rather than Directions. Despite these dual approval mechanisms, the system of Ministerial Directions as established under the 2004 National Monuments Amendment Act is an extremely effective model, particularly as it brings clarity and understanding to both the roads authority and the National Monuments Service as to the specific archaeological mitigations that will be delivered (Department of Environment, Heritage and Local Government 2006).

Issue 1: Finding the right expertise

The NRA's response to the first issue set for discussion can be considered in two distinct ways. Firstly, as mentioned at the outset, the NRA employed its own archaeologists to manage the impact of archaeology on national road schemes, and this came about following the agreement of a code of practice in 2000 between the NRA and the Minister for Arts, Heritage, Gaeltacht and the Islands (Department of Arts, Heritage, Gaeltacht and the Islands 2000). This code of practice set out a framework for the treatment of archaeological work on national road schemes (Figure 19.2). The code was developed in anticipation of the major infrastructural developments planned in the 2000s and following the authority's experience of archaeology during the 1990s. In particular, the practice of waiting until construction to identify unknown archaeology through monitoring represented poor risk management, as on occasion it led to claims and delays on major construction contracts. Equally, from an archaeological point of view, it was not particularly satisfactory, as archaeological excavations took place within the context of construction works.

Figure 19.2: Code of Practice agreed between the Minister for Arts, Heritage, Gaeltacht and the Islands and the NRA.

The archaeological contracts developed by the NRA formed the basis for the model forms of archaeological contract on public service works in Ireland that have been in use for the past five years. These contracts have proved highly flexible and adaptable for dealing with archaeological works, whether on small-scale projects or major motorways, thus encouraging and reinforcing a consistent approach. The underlying philosophy of these contracts is to provide detailed specifications for all aspects of archaeological work from initial site identification to full publication. As the emphasis is on the whole project lifecycle, it means that costs for dissemination and publication can be built into the budget at an earlier stage. In contrast, previous archaeological contracts did not provide for dissemination, which meant that such works might be treated as an unexpected or unanticipated extra.

The contract documents that are used to procure archaeological consultants set out very specific requirements; if these requirements are not satisfied then the contract can be terminated. These contracts also specify the personnel required and provide measurable criteria for each grade; thus, site directors must be university graduates with at least three years post-graduate experience, must have passed the National Monuments Service's licence eligibility interview and must have directed at least five excavations. Furthermore, we assess tenders to determine if their tender bids are abnormally low, which is grounds for rejection in itself. Indeed, since the introduction of these contracts, this clause has been invoked successfully on several occasions.

Under the agreed code of practice, perhaps the most significant outcome was the appointment for the first time of project archaeologists directly to the engineering design teams. This ensured that archaeological concerns and issues could be raised and heard throughout the design process, from initial concept through to project completion. Thus, there were now experts on the client's side who could:

- prepare archaeological strategies,
- provide archaeological advice and comment,
- liaise with the statutory authorities (the National Monuments Service or the National Museum of Ireland),
- engage with the archaeological consultants and
- manage the archaeological works from inception to completion.

Importantly, the majority of project archaeologists who were appointed were from a private sector background and therefore had a specific awareness of and perspective on these particular issues.

A key aspect of the second part of finding the right expertise is engaging the archaeological consultant. As a public authority, all our work takes place within the context of European and national legislation, in particular procurement regulations. Since the early 2000s, archaeological projects are treated as services, and therefore all archaeological contracts with an estimated value in excess of €200,000 must be advertised through the Official Journal of the European Union. It also means EU procurement rules must be followed when it comes to awarding contracts.

Another aspect to finding the right expertise has been the capacity and capability building within the profession (e.g. the direct funding of doctoral research through the NRA's research fellowship programmes), which has also informed NRA standards and practices. These research programmes have proved extremely beneficial and demonstrate the integrated approach adopted by the NRA in the consideration of archaeology. For instance, the University of Bradford was commissioned to undertake an overview of ten years of archaeo-geophysical surveying on national road schemes. This study provided essential empirical data to determine the efficacy of such techniques (Bonsall et al. 2014). Meanwhile, Trinity College Dublin was commissioned to undertake a study of the prehistoric woodland, and this research has directly contributed to the development of palaeo-environmental sampling guidelines and standards which are incorporated into the NRA's archaeological contracts (Figure 19.3).

Issue 2: Monitoring quality

In any quality monitoring exercise, it is essential that all parties fully understand what is expected of them through the course of the project, by way of the specifications and requirements set out in the contract documents. However, the experience of the NRA, following the excavation of more than 2,000 sites, is that

Figure 19.3: Examples of NRA guidelines.

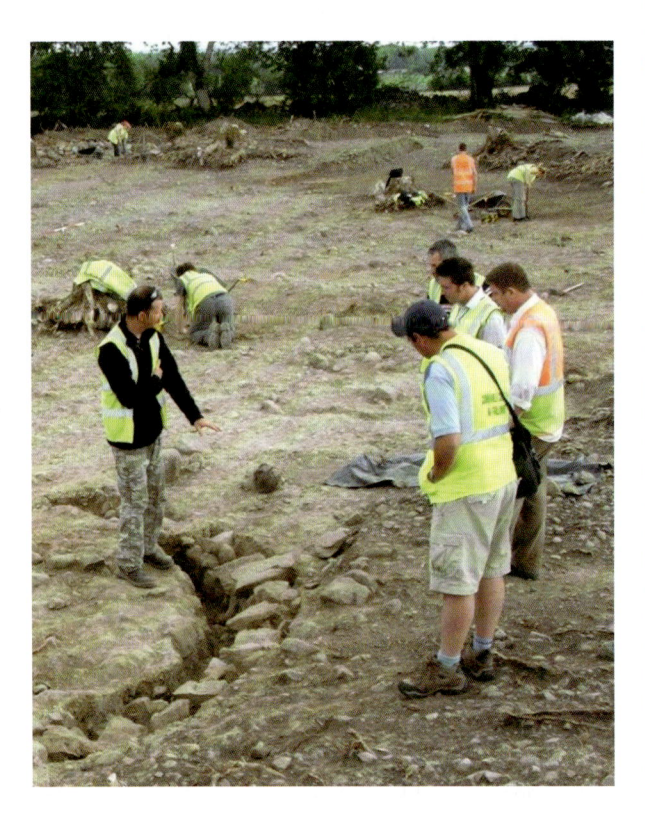

Figure 19.4: TII Archaeologist Martin Jones, second from left on site with Director Tony Bartlett of Rubicon Heritage. (Photo, © Jerry O'Sullivan)

quality is something that must be worked at constantly, throughout the life of a project, and that it cannot be taken for granted. But the most important approach to ensuring quality is to closely supervise the progress of works on site and challenge poor performance as it happens or as we become aware of it. This is supported by a process of ongoing contractor performance assessment, which is stipulated in the contract. We have a rigorous schedule of project reporting and assessments, which means that archaeological data and information is captured throughout the process, and this continues through the post-excavation stage right up to publication. In addition, we will assess each of our contractors' compliance with relevant employment and health and safety legislation. If the company does not remedy any deficiencies identified, we have the power to terminate the contract.

The NRA archaeologists will be on site with their engineering colleagues overseeing the site works from the very outset (Figure 19.4). Their primary responsibility on site is to ensure that the archaeological works undertaken are thorough and rigorous. In their 2001 study, Hey and Lacey noted that, of the various techniques adopted to identify potential archaeological remains, machine trenching under archaeological supervision was by far the most successful strategy. They also noted that 'to guarantee the degree of confidence between 5% and 10% of the site should be seen' (Hey & Lacey 2001, 54). They further advocated that a 'combination of techniques' and a 'multi-phase approach to evaluation allows a more problem-oriented investigation and strengthens the interpretation of results' (ibid, 61). Thus, at the initial site identification stage the emphasis is on identifying, insofar as possible, any previously unknown archaeological sites. A multifaceted approach is adopted to site identification, including, as appropriate, field inspection, machine investigation, archaeo-geophysical surveying and LiDAR (Figure 19.5). Under the contract, approximately 12.5% of the extent of a scheme will be sampled using mechanical excavators; this percentage excludes any testing taking place at known or potential sites that were identified during the EIA. All potential archaeological sites identified at this stage will be investigated further to establish the nature, extent and character of archaeological remains. Those sites verified as being archaeological in nature will then be fully excavated by hand (within the footprint of the road scheme).

Figure 19.5: Examples of advance archaeological surveys and works. Main photo: archaeological investigation and excavation (© Margaret Gowen & Co.); upper right: LiDAR image (© Stephen Davis); lower right: geophysical surveying (© James Bonsall).

Figure 19.6: Submission rate of archaeological reports between 2007 and 2015.

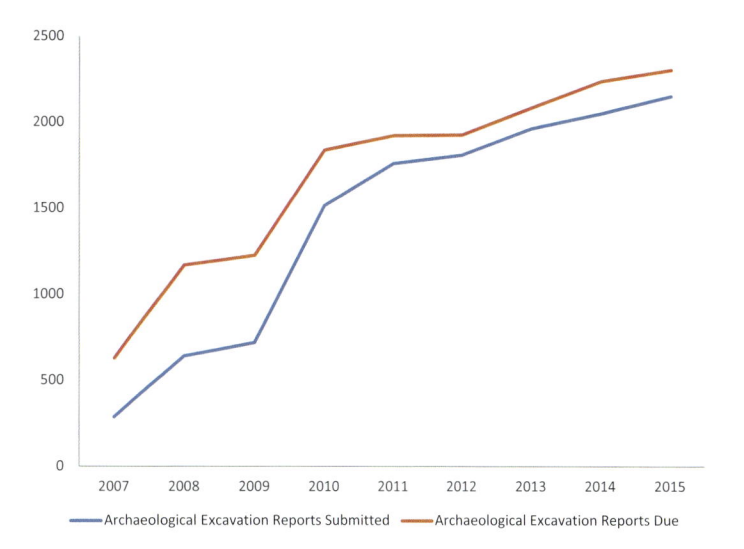

This multi-faceted and multi-phased approach to the identification and excavation of archaeological sites is now the standard on all national road schemes in Ireland, from minor safety improvements to major inter-urban motorways. In his discussion of global archaeology, Martin Carver (2011, 106) presented a case study of the M3 motorway from Clonee to Kells in County Meath, which passed close to the Hill of Tara, and he commented that 'this must constitute one of the most thorough and sensitive archaeological responses to a new road ever undertaken; for NRA [sic], mitigation meant a great deal more than the creation of a record; newly uncovered features were placed in their palaeoenvironmental and historical context and the results were widely disseminated'.

As mentioned previously, monitoring quality is something that must take place throughout a project; thus we audit excavation report submission annually, which allows us to identify archaeological consultancies who are not meeting commitments. Of the 2,200 excavations carried out since the signing of the code of practice, 2,101 of the required reports have been submitted, which equates to a 96% completion rate. The practice of auditing excavation reports commenced in 2007, when there was a completion rate of less than 40%. Figure 19.6 not only tracks the annual figures of archaeological excavation reports submitted but also helps graphically illustrate the extent of archaeological works between 2007 and 2015.

Issue 3: Sharing results/ensuring lasting public benefit

The NRA is very mindful of the public's trust and believes that the best way to respect this is to ensure that all data from the archaeological datasets (especially, the archaeological excavation reports themselves) are available to those who want them, be they local landowners, artists, members of the general public, planners, archaeologists or researchers. We are also mindful that different audiences want different products and we seek to specify these in the contract documents.

It is a truism to say that all archaeology is local and we find that there is a great interest in the discoveries made in any locality. We seek to satisfy this interest through the provision of lectures and papers to local historical societies. Working in partnership with local museums, heritage officers and community groups, the NRA has also established both permanent and temporary exhibitions throughout Ireland (Figure 19.7).

One of the most important vehicles for dissemination to the general public is the annual seminar that the NRA organises in Dublin as part of its contribution to

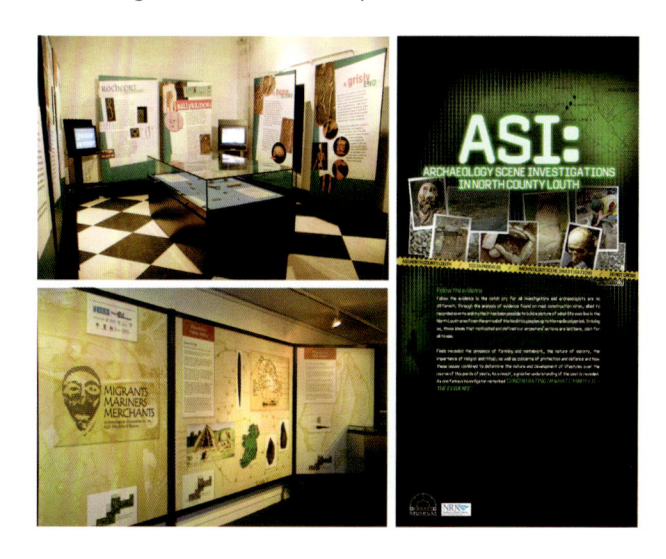

Figure 19.7: Examples of NRA exhibitions. Upper left: photo of 'Hidden Landscape: searching for the lost Kingdom of Mide' (© Studio Lab); lower left: photo of 'Migrants, Mariners, Merchants' exhibition (© Studio Lab); right: display board from 'ASI: Archaeological Scene Investigation in North Louth' (© County Museum Dundalk and NRA).

Figure 19.8: National Heritage Week events in August 2015 (© Transport Infrastructure Ireland, except where indicated). Top left: walking tour of medieval Buttevant, Co. Cork (© Marion O'Sullivan, Buttevant Heritage Group); top right: 'Romans in Ireland: fact or fiction?' public lecture and re-enactments in Kilkenny city (© Dylan Vaughan); centre: industrial heritage field trip in Connemara, Co. Galway; bottom left: archaeologist and broadcaster Julian Richards launching *Illustrating the Past* with author Sheelagh Hughes and designer Roisin McAuley; bottom right: 'The Archaeology of Roads and Light Rail' public seminar in Dublin city.

National Heritage Week. These seminars generally adopt a thematic approach and seek to address diverse topics ranging from archaeological science to culture and identity. We have contributors from across the discipline of archaeology, including excavation directors, specialists, academics and NRA archaeologists. A key practice of the NRA is to publish the proceedings of these seminars in our

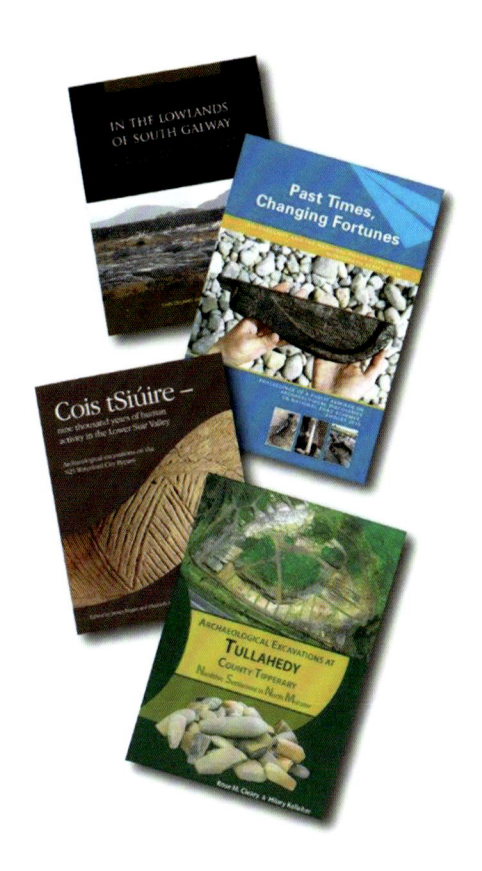

Figure 19.9: Selection of books published and / or funded by the NRA.

Archaeology and the National Roads Authority Monograph Series, and to date 11 seminar proceedings have been published. The 2015 seminar in Dublin is being supplemented by multiple regional events, including a pop-up museum, re-enactments, lectures and field tours (Figure 19.8).

The NRA also publishes a second monograph series dedicated to the archaeology discovered on specific road schemes, with 17 books published to date and at least 15 more planned (Figure 19.9). These books are generally written by the consultants and, depending on the archaeology discovered, can provide either an in-depth analysis of a single site or collection of sites, or furnish summary accounts of all the archaeological sites along a route, setting them into a broader context. One of the key challenges of these publications is to make them engaging for the public. Consequently, we request that authors write for a general audience and that our editors seek to make the books as accessible as possible by eliminating unnecessary jargon, etc. In addition to these traditional approaches to dissemination we have also a significant on-line presence, including an e-zine, downloadable audio-guides and videos.

As mentioned at the outset, the NRA merged with the Railway Procurement Agency in August 2015 to form Transport Infrastructure Ireland (TII). As a result of this merger, there will be a single monograph series (the TII Heritage series) to continue the established tradition of publishing the results of archaeological investigations on national roads, and to allow for the publication of books related to the broader topic of heritage as it relates to both light rail and roads projects. The first monograph in this series is *Illustrating the Past* (Hughes 2015), which is a compendium of some of the best reconstruction images which have been commissioned in the course of the archaeological works on Irish road schemes. This book, written by a non-archaeologist, specifically caters to a popular audience and seeks to introduce readers to the wealth of archaeology that has been discovered in the last 15 years or so (Figure 19.10).

The NRA also actively participates in major research; for example, as part of the Irish National Strategic Archaeological Research (INSTAR) programme we have been industry partners on projects such as the Early Medieval Archaeology Project, Cultivating Societies, and The People of Prehistoric Ireland. Similar synergies are planned for the future, such as a forthcoming study into the later prehistoric period in partnership with University College Cork and the University of Bradford.

Another major challenge in achieving public benefit is making the core data itself as accessible as possible. A common refrain from both members of the profession and the general public has been the difficulty experienced in accessing the primary results of archaeological works. To facilitate this need we established a database of archaeological excavation results in 2008; however, not only did this require significant resources to maintain, but users also advised us that, while they liked having summary data, what they really wanted were the individual reports so that they could follow their own research paths and reassess the primary data themselves. In response we initiated a pilot project with the Discovery Programme and the Digital Repository of Ireland (DRI) which will not only make final excavation reports available on-line but will also curate them into the future. This is a significant step forward in terms of making the primary data of reports available. In advance of the launch of this on-

Figure 19.10: Reconstruction image of Ballinvinny South moated site by Digitale Archäologie as featured in *Illustrating the Past*.

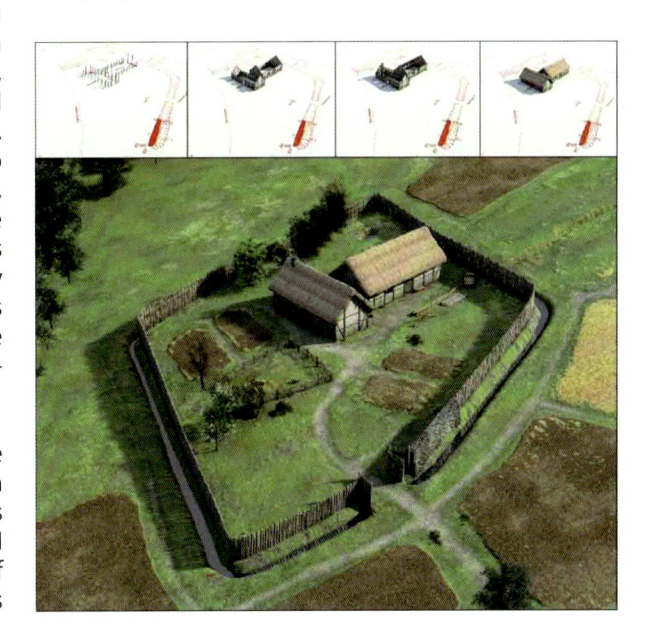

line resource, the NRA is happy to make the excavation reports available on request (Link 1), either as single reports, as road scheme 'bundles' or as an entire set (currently approximately 100 GB in size). The only condition we apply is that researchers acknowledge the authors and the NRA, and that they share their findings.

Conclusions

A discussion of quality with regard to archaeological works on national road schemes cannot only be about the process or the system; it must also be about the outcomes and results. Perhaps the most significant outcome has been the collection and collation of a substantial corpus of archaeological data, with more than 2,200 excavation reports completed, covering every era from the Mesolithic to the modern period. Another significant outcome has been the publication of 30 books to date, presenting the results of this archaeological endeavour in an extremely accessible way. The 10,000 or so radiocarbon dates that have been commissioned in tandem with this work are improving our understanding of previously enigmatic archaeological periods in Ireland, such as the Iron Age. Indeed, of all the sites now attributed to this period more than 50% were identified in the course of pre-construction works on national roads. With the almost 5,000 burials that have been excavated, there are tremendous opportunities for future research. Indeed, one such project (the Ballyhanna Research Project) has noted that the cystic fibrosis F508d mutation was not as prevalent in the medieval period as it is in Ireland today, which is allowing medical researchers to further their understanding of the development of this disease (McKenzie et al. 2015). Meanwhile, the thousands of palaeo-environmental samples collected during the course of archaeological works as a contractual requirement help to illuminate our understanding of the development of early agriculture in Ireland. A selection of these dates have featured in the debate regarding the practice of dairying versus tillage during the Neolithic (Smyth & Evershed 2014). With regard to the historic period, it has been noted that 'NRA publications in themselves, as well as the vast array of data behind them in archives and collections are likely to be transformational in our understanding of the early medieval period' (O'Sullivan et al. 2014, 26).

Ultimately, for the NRA (and now TII) the best measure of quality is to see the results of the archaeological works being used and applied, whether in local exhibitions or international research projects, by artists or archaeologists, by academics or planners. This approach has received enthusiastic support from the individual field archaeologists and excavation directors who want to see the results of their endeavours being used. As all of this work has been completed with the public's money it is essential to remember that the results belong to them, and we hope to continue to showcase the value of this expenditure for many years to come.

References

Bonsall, J., Gaffney, C. & Armit, I. 2014: A Decade of Ground Truthing: Reappraising Magnetometer Prospection Surveys on Linear Corridors in light of Excavation evidence 2001–2010 in H. Kamermans, M. Gojda, and A.G. Posluschny (eds): *A Sense of the Past: Studies in current archaeological applications of remote sensing and non-invasive prospection methods*, BAR International Series 2588, Oxford, 3–16.

Carver, M. 2011: *Making Archaeology Happen: Design versus Dogma*, Left Coast Press.

Department of Arts, Heritage, Gaeltacht and the Islands 1999: *Framework and Principles for the Protection of the Archaeological Heritage*, Stationary Office, Dublin.

Department of Arts, Heritage, Gaeltacht and the Islands 2000: Code of practice between the National Roads Authority and the Minister for Arts, Heritage, Gaeltacht and the Islands. Stationary Office. Dublin http://www.tii.ie/tii-library/archaeology/Code%20of%20Practice/code-of-practice.PDF (accessed 04.08.2015)

Department of Environment, Heritage and Local Government 2006: Guidance Notes on Procedures, National Monuments (Amendment), Act 2004, Section 14A, Directions http://www.archaeology.ie/publications-forms-legislation (accessed 30.05.2014)

Hey, G. & M. Lacey 2001: *Evaluation of Archaeological Decision-making Processes and Sampling Strategies*, Oxford Archaeological Unit, Oxford.

Hughes, S. 2015: *Illustrating the Past: archaeological discoveries on Irish road schemes*, TII Heritage 1, Transport Infrastructure Ireland, Dublin.

Keane, M. 2015: An interpretation of Valetta from the Críoch Fuinidh (the remote or end country) in P. A. C. Schut, D. Scharff & L. C. de Wit (eds): *Setting the Agenda: Giving New Meaning to the European Archaeological Heritage*, EAC Occasional Paper No. 10, 75–89.

McKenzie, C. J., Murphy, E. M. & Donnelly, C. J. (eds) 2015: *The Science of a Lost Medieval Gaelic Graveyard: the Ballyhanna Research Project*, TII Heritage 2, Transport Infrastructure Ireland, Dublin.

O'Sullivan, A., McCormick, F., Kerr., T. R. & Harney, L. 2014: *Early Medieval Ireland AD 400–1100: The Evidence from Archaeological Excavations*, RIA, Dublin.

Smyth, J. & Evershed, R. J. 2014: Milk and molecules: secrets from prehistoric pottery in B. Kelly, N. Roycroft, & M. Stanley (eds): *Fragments of Lives Past*, Archaeology and the National Roads Authority Monograph Series No. 11, NRA, Dublin, 1–13.

Swan, R. 2014: Archaeology in advance of motorway construction in Ireland in M. Carver, B. Gaydarska & S. Montón-Subías (eds): *Field Archaeology from Around the World: Ideas and approaches*, Springer, 107–10.

Website

Link 1: http://www.tii.ie/technical-services/archaeology/resources/ (accessed 16.09.2015)

20 | Archaeology as a tool for better understanding our recent history

Peep Pillak

Abstract: In 1987 the Estonian Heritage Society set as its goal the restoration of Estonian national memory, and archaeologists played their role in this process. They are engaged in activities which differ from the routine tasks of archaeology, such as clarifying the fate of victims of the Soviet regime. The exhumation and repatriation of the remains of the first President of Estonia from Russia to his homeland in 1990 consolidated Estonian people in their determination to restore the independence of their state. Exhumations related to the recent history of Estonia can be highly politicised, as was the case with the reburial of the remains of Soviet soldiers in 2007, which resulted in a polarisation of Estonian society. The timing of sensitive archaeological excavations is of paramount importance.

Keywords: importance of civil society, forensic anthropology, exhumations, victims of Soviet regime, war graves

Introduction

The Estonian Heritage Society was founded by a citizens' initiative in 1987. It was the beginning of major changes in occupied Estonia, set against the backdrop of *perestroika* and *glasnost* launched by Mikhail Gorbachev. The Heritage Society was the first legal mass movement opposing the Soviet occupation authorities, with its main aim to restore Estonian independence – the Republic of Estonia – as the core heritage of the people. By the end of the 1990s this aim, uniting Estonians, had brought the Heritage Society close to 10,000 members (the population of Estonia is 1.3 million). One of the first large-scale movements was to restore the monuments commemorating the Estonian War of Independence (1918–1920) that had been demolished by the Soviet regime (Strauss et al. 2002). The role of the Heritage Society can be compared to that of *Solidarność* in Poland in the 1980s. In 1991, when Estonian independence was restored, organisations and institutions characterising a democratic state began their work. Many leading figures of the Heritage Society became influential politicians or high-ranking civil servants.

The Estonian Heritage Society at present

Now the Estonian Heritage Society focuses on heritage protection in its classical meaning, while its membership numbers have dropped dramatically to 700. The Society, as a non-governmental organisation, collaborates with the Ministry of Culture and the National Heritage Board (i.e. the state institution that operates under the authority of the Ministry of Culture and is responsible for the management of heritage and its preservation, including archaeological heritage and heritage conservation areas; the Board also maintains the state registry of cultural monuments, issues licences for archaeological excavations and so on). In addition, the Society collaborates with local authorities, many other institutions and organisations, and includes both individual members and non-profit organisations. One of the most active of the latter is the Estonian Archaeological Society, which includes professional archaeologists. So the Heritage Society brings together professional archaeologists working in both public and private sectors. The aim of the Heritage Society is not to safeguard the interests of the state; that is the task of the National Heritage Board. Our point of departure is to assure the quality of preventive archaeology without forgetting the interests of the public and individuals, which may conflict with those of the state. In both cases there is a danger of pursuing a policy that serves the commercial interests of a narrow group that are presented as national or social. Because corruption and manipulated decisions can occur in public decisions-making processes, the presence of a strong third sector besides a national institution is very important. It is only the third sector that can argue for alternative solutions that would bring benefit not only to the economy and business community but also to the socio-cultural environment at large. Not every citizen, however, can easily understand the difference between the National Heritage Board and the Heritage Society – the first representing the state and the second the third sector.

The research of Estonian archaeologists is annually propagated by the Estonian Heritage Society within the framework of the Heritage Month (from 18 April to 18 May) and the European Heritage Days (one week in September). The Heritage Society, with the financial support of the Ministry of Culture, has organised the European Heritage Days in Estonia for years; in 2015, however, the Ministry unexpectedly decided to delegate the task to the National Heritage Board. The reason behind this decision could be that the European Heritage Days have always received the type of positive media coverage that the Heritage Board badly needs given its present social status of a state institution that is more concerned with sticks than with carrots. Besides the Heritage Month and the Days, we have been organising excursions to archaeological sites guided by professional archaeologists. This is an opportunity for archaeologists to let the public know about their work and get feedback. The excursions include site-maintenance activities where participants help to clean ancient castles, settlements, and cemeteries. Their joint work on the sites tightens relations between researchers, amateurs/volunteers, and the local population. The Estonian Heritage Society has shown initiative in different fields of heritage preservation, including archaeology.

The archaeology of terror

Traditionally, archaeology is seen as a branch of historical research that relies on material sources and studies the early periods of human society with no or scanty written records to give us a comprehensive picture. The case of the recent history of the 20th century could be different – written and printed matter, films, photos, sound recordings, etc. should be abundant. Unfortunately, this is not so because the terror regimes of the recent past have tried to eliminate every trace of their crimes against humanity. In order to reveal the nature of terror regimes one has to use archaeology and exhume the mute witnesses lying in the ground – the victims of the regime. The archaeology of terror is an extreme branch of historical research that has much in common with forensic medicine. Archaeological research into terror and the knowledge one can obtain from it is not of positive value emotionally, but the knowledge as opposed to unawareness is definitely of value. The study of mass crimes and their public evaluation is of preventive importance in ensuring their non-repetition in future (for more see Poliitilised 2008).

Projects of significance from recent history: the reburial of President Konstantin Päts

The aforesaid has to be supported by examples. The first of them is related to the fate of our highest-ranking civil servants under the Soviet regime. During 1918–1940, before the Soviet occupation, the Republic of Estonia had had 11 heads of state. One of them, August Rei, escaped and died in Sweden. He was reburied, with his wife, from Bromma Cemetery in Stockholm to Metsakalmistu (Forest Cemetery) in Tallinn in 2006. Another – Otto Strandman – committed suicide after a commission to go to the NKVD (i.e. the People's Commissariat for Internal Affairs – in Russian Народный комиссариат (Наркомат) внутренных дел, abbreviated to НКВД – that functioned under this name during 1934–1946 and was the precursor of the KGB) and was buried at Siselinna (City) Cemetery in Tallinn in February 1941. The remaining 9 were arrested and their further fate was long unknown. The NKVD/KGB archives revealed that all 9 were either killed or died in imprisonment with their graves unknown (Pillak 2015a). The only exception was the best-known Estonian politician and first Estonian President, Konstantin Päts. Several biographies of him were published in the 1930s in Estonia and also later by Estonian publishers in exile in the West. In the Soviet historiography he was always portrayed in a negative light. In the 21st volume of the *Big Soviet Encyclopaedia* one can read his bio-sketch, which states as his date of death 18 January 1956, and the place – the Kalinin oblast (Big Soviet 1975, 294). Now both the oblast and its central city once again bear their historical name: Tver. In Estonia, annexed to the Soviet Union, no more information about the President was available. It was only in 1988 that the collaborators of the KGB in the Estonian SSR, updating their practices in accordance with *perestroika* and *glasnost*, made public facts about the last years of his life and about his death. It was revealed that after his deportation in 1940, and the years of imprisonment following it, his last place of detention was the mental hospital in Burashevo near Kalinin, where he also died. Several medical experts who studied his case history found no reason to treat or keep President Päts in a psychiatric hospital. He had, however, serious health problems, as his weight was about 50 kilograms. His post-mortem identified coronary and bloodstream insufficiency, sclerosis, remnants of myocardial infarction, and nephrolithiasis, the latter being cited as the cause of his death. The burial place of the 'anonymous' patients of the hospital was said to have been a small wood about a kilometre from the hospital. The graves were not marked, and there was no plan of the burial site. In 1956 there had been 80 burials; in January, President Päts and three other persons were buried. Luckily, his physician Dr Yevgenia Gusseva (5 January 1905 – 8 August 1994), a major in the medical service who had taken part in the Second World War, was still alive, and, although she herself was not present at the burial, she knew the place used at that time. In November 1988, archaeologist Vello Lõugas and photographer Rein Kärner drove to Burashevo, where Dr Gusseva showed them the burial spot as she remembered it. They were accompanied by the grandson of the President, Matti Päts. The latter was born in 1933, and in 1940 he was deported to Russia and sent, after the arrest of his parents, to an orphanage. His younger brother Henn died there in 1944 of malnutrition; their father, Viktor Päts, died in 1952 in the Butyrka prison, Moscow. But their mother was released in 1946 and returned with Matti to Estonia.

In February 1989 Matti Päts handed in a formal application to the board of the Estonian Heritage Society asking for assistance in finding the grave of Konstantin Päts and reburying his remains in his homeland. The leader of the expedition organised by the Society was the well-known Estonian archaeologist Vello Lõugas (6 April 1937 – 21 May 1998), who assembled an expedition team including historians, archaeologists, and archivists (Lõugas 1991). As there were no anthropologists of sufficient expertise in Estonia, Lõugas contacted his Lithuanian colleagues, who agreed to participate in the endeavour. They were: Dr Gintas Česnys, a biologist; Dr Vytautas Urbanavičius, a historian; and Dr Rimantas Jankauskas, a physician, all of whom – in 2005 – received the 5th class of the Order of the Cross of Terra Mariana, as proposed by the Estonian Heritage Society.

The expedition, in which I also took part, departed for Burashevo on 14 May 1989. Although the tide was turning in the Soviet Union, the system was still fairly steady. In order not to attract undue attention, the participants were officially on their summer vacation. Everyone was optimistic and expected to complete the mission within a maximum of one week because the burial place had been identified by the doctor of medicine. We reached Burashevo by the midday of 15 May and identified the place to dig, as shown by Dr Gusseva. But then there came two elderly ladies from the neighbourhood who had worked in the hospital as nurses, and, having heard what we were looking for, they confidently directed us to a totally different location. We pinpointed this site in the landscape and began our work. By the next morning we had found and examined three skeletons, but had to admit that these belonged to younger persons. Now we went to the place that Dr Gusseva had indicated, but found nothing at the spot she had suggested. Probably the landscape had changed significantly over the decades – her orientation had relied on the positions of paths and surrounding trees.

Local people took a great interest in our activities – to the extent that we had to circle ourselves with a safety barrier to keep them off the area we were examining. Many of them said that they remembered the President well and could show us his grave. Having already excavated places we had been sent to with great confidence, we were sceptical by now. Our initial optimism was gradually turning into doubts about our ability to succeed. In the days to follow, our hope was inversely proportional to the cubic metres we were excavating: from among the number of graves

we had, it seemed impossible to attempt to find the one we were looking for by the method of trial and error. Soon we learned that the agency keeping its eye on everything had lost its patience: although we had a permit for the excavations, we were told now that our presence had caused too much disturbance in the neighbourhood. On 18 May, when a militiaman was sent to keep watch at our site, we had to stop our work. Next day we backfilled the graves we had cut and drove back to Estonia.

In summer 1990, we had our second expedition to Burashevo. Meantime we had analysed our results and obtained more information. As the memory of the failure of our previous attempt was still vivid, many participants refused to experience it anew, and it was difficult to find others to replace them. Nonetheless, we left Estonia early in the morning on 18 June, and by noon of the next day we were there. We had accepted the fact that our chances of locating the right grave were in the hands of destiny. But the probability was increasing with every grave we opened and with every spadeful of soil we dug up. The locals, seeing the amount of earth we had moved manually, offered us their help in the form of tractors or bulldozers. We were grateful to them for their kind offer but continued to work with spades, shovels, and brushes. The time allotted for the expedition was running out, and this

Figure 20.1a: Uncovering skeleton No 46 in Burashevo. In the foreground Vello Lõugas, the leader of the expedition (© Peep Pillak).

Figure 20.1b: The earthly remains of Konstantin Päts, the President of the Republic of Estonia, in Burashevo (© Peep Pillak).

Figure 20.2: Dr Gintautas Česnis examining the remains of President Päts (© Peep Pillak).

Figure 20.3: Reburial of President Päts in his homeland (© Peep Pillak).

time no one had hindered us. In the first year we had excavated only 10 skeletons; this time the number had exceeded 30 already. On only three occasions the anthropologists/experts had needed more time to think and investigate. The last day of the expedition had come. Our spirits were low – we were to return again without results. We also realised that there would be no third expedition. We discussed among ourselves whether we should take with us a few handfuls of earth or find a small oak-tree to replant in Estonia. Before taking our leave, we decided to try our luck with two other graves. Soon we saw the remains of a coffin, a skeleton that had been dissected, fragments of a textile, and tennis shoes (Figure 20.1 a, b). Most of the bodies in the graves had been buried naked and without a coffin; in many graves there were several skeletons. But Dr Gusseva had told us that President Päts had been buried clothed and in a coffin. Everybody gathered round the grave, leaving their work in other places. The Lithuanians had a long discussion in Lithuanian and consulted Vello Lõugas, who was also familiar with the language. By the time that skeleton No. 46 had been entirely uncovered, Dr Gintautas Česnis announced that it was probably that of President Päts (Figure 20.2). A more thorough examination on the spot confirmed this. We packed the remains carefully and could begin our journey back home. We reached Estonia on 23 June, which is in Estonia not only Midsummer's Day but also the Victory Day that recalls one of the decisive battles in the Estonian War of Independence. After thorough analysis, the remains of President Konstantin Päts were reburied with full honours on 21 October 1990 in Metsakalmistu, Tallinn (for more details see Pillak 2007). The ceremony was organised by the Estonian Heritage Society; thousands of Estonians participated in it, and the ceremony was also broadcast on television (Figure 20.3). So on the one hand, it was an archaeological expedition of professional standards; on the other it was an act of extreme political significance in the process of the reestablishment of Estonian independence.

It is of importance to observe that in the local museum attached to the secondary school in Burashevo there is now a permanent exhibit about the life of Konstantin Päts (Figure 20.4). The local school has established contacts with schools in Estonia, and for some years pupils have exchanged visits. While in Estonia, the children from Burashevo always visit the grave of President Päts in Metsakalmistu and light candles. Moreover, on the initiative of the local government, the former grave of the Estonian President in Burashevo has been marked by a wooden cross, and on 28 February 2015 a memorial stone was unveiled there with a portrait of President Konstantin Päts on it and the following text in Russian and Estonian: 'Here was the grave of Konstantin Päts (23.02.1874–18.01.1956), the President of the Republic of Estonia, reburied in Metsakalmistu, Tallinn' (Figure 20.5). The unveiling

Figure 20.4: In the museum of domestic life at the comprehensive school in Burashevo there is a permanent exhibition about the life and work of President Päts (© Peep Pillak).

Figure 20.5. On the former grave of President Päts in Burashevo there is now a wooden cross and a memorial stone (© Ants Kraut).

ceremony in Burashevo was attended by local inhabitants as well as by people of Estonian descent now living in Tver and its surroundings, representatives of the Estonian Embassy in Moscow, and a delegation from Estonia including schoolchildren. The Estonian Heritage Society distributed at the ceremony a Russian-language brochure: *Konstantin Päts and Burashevo* (Pillak 2015b). This is a good example of how a dramatic past can be a uniting factor between people in the present.

The cases of General Laidoner and the poet Lydia Koidula

In 1995 an expedition was organised to Vladimir to find the remains of General Johan Laidoner (12 February 1884 – 13 March 1953), the Commander-in-Chief of the armed forces of the Republic of Estonia, who had died in detention in the Vladimir prison. It was known that he had died on the same day as the Polish Deputy Prime Minister Jan-Stanislaw Jankowski, and that they were buried together next day in the city graveyard next to the prison wall; a little iron gate leading from the prison territory to the graveyard is still there. German and Japanese prisoners of war, Poles, Ukrainians, and citizens of Vladimir itself had also been buried there. As the probable territory of the burial was too large, we applied for more information about the grave of General Laidoner from the Ministry of Foreign Affairs of the Russian Federation. In 1996 the Embassy of the Russian Federation in Tallinn sent us their answer. It said that, unfortunately, the exact grave cannot be specified more accurately. So it was decided to commemorate General Laidoner by a tablet on the gate of the Vladimir cemetery similar to the ones the Poles had put there for their Deputy Prime Minister Jankowski, and the Japanese for their compatriots. The tablet was unveiled on 12 February 1999 (the 115th anniversary of General Laidoner's birth), and a booklet with his biography in Estonian, Russian, and English was published to mark the occasion (Pillak 1999). Perhaps at some later date it will be possible to have an Estonian, Polish, Ukrainian, German, and Japanese joint excavation in Vladimir that might lead us to the remains of our compatriots (for more details see Pillak 2000).

Lydia Koidula (24 December 1843 – 11 August 1886), a poet of the 19th-century Estonian National Revival, lived the last years of her life in Kronstadt near St Petersburg, where she also died and was buried in the town's Lutheran cemetery. In 1884 her son, and in 1907 her husband, was also buried there. In 1946 Koidula's remains were repatriated from Kronstadt to Metsakalmistu in Tallinn against a backdrop of majestic Soviet propaganda, but the remains of her little son and her husband were 'forgotten' in Kronstadt. Since the second half of the 1990s, when the former closed city of Kronstadt could be visited again, attempts have been made to locate the graves of the husband and the son of this great poet in order to reunite the family. Unfortunately, the Lutheran graveyard has been so severely despoiled that the family resting place cannot be found anymore (for more details see Olesk & Pillak 2000).

The Bronze Night, or the memorial to the soldiers of the Red Army

A sharp social conflict arose following the development of events in April 2007 concerning the relocation of several war graves and an accompanying memorial to the soldiers of the Red Army – a bronze sculpture in Soviet military uniform. They were moved from the centre of Tallinn to the cemetery of the Defence Forces, a part of the City Cemetery. Preparations for the prior excavation and identification of the remains began in Tõnismägi, in the centre of Tallinn, on 26 April. The area was surrounded by barriers before the work was due to begin next morning. Because of the unrest that broke out that night, the excavations had to be postponed. A crowd consisting mostly of the Russian-speaking community of Tallinn gathered on the spot, and antagonism between them and the police turned into a riot that was instigated and coordinated on a professional level from abroad. Stones were thrown at policemen; protests against the state and the government rocketed; state flags were burned. The rioters – about 1,500 of them – were forced to leave the site, but continued rioting in the city centre, burning cars, smashing windows, robbing and setting fire to shops and kiosks, etc. The unrest subsided within a couple of days while also spreading to a smaller extent to the cities of North-East Estonia with a predominantly Russian population. One person was killed as a result of clashes between the rioters, about two hundred were injured, hundreds were detained by the police, and the estimated damage was about 25 million Estonian kroons (c. €1.5 million). This was the first act of this kind of vandalism in Estonian history and a major shock that revealed the polarisation of society into two hostile parts. The identity of the Russian population was and is closely related to the victory in the Great Patriotic War (1941–1945), which was commemorated at the Bronze Soldier annually on 9 May, when the monument was covered with red carnations, people sat in the park drinking vodka, playing concertinas, singing and dancing. For Estonians, a burial place in the very centre of the city, at a bus-stop, was bizarre. The monument had been erected at this site on 22 September 1947 – the third anniversary of the 'liberation' of Tallinn – and it was known as the Bronze Soldier or Alyosha – a symbol of Soviet occupation. But there were many Estonians who thought the monument could stay where it was. A public debate between those who were for and against its replacement had been going on in Estonia for years (for more details see Petersoo & Tamm 2008).

Once the riots were over in late April and early May, the excavations were conducted. They revealed the remains of 11 men and 1 woman who had been buried there in April 1945 (Figure 20.6). After the remains had been identified, attempts were made to find their relatives for DNA tests. The remains of three of the identified persons were handed over to their relatives living in Russia, where they were reburied; the remains of one person were handed over to relatives living in Ukraine, and one skeleton – the only female, Lenina (Yelena) Varshavskaya – was reburied on the Mount of Olives in Jerusalem, Israel (Grishina 2007; Zenger 2008). Others found and also identified were ceremonially reburied on 3 July 2007 in the cemetery of the Defence

Figure 20.6: Excavation of the remains of the Soviet soldiers in Tõnismägi in the centre of Tallinn (© Ants Kraut).

Figure 20.7: Reburial of the remains of Soviet soldiers removed from Tõnismägi to the cemetery of the Defence Forces (© Ants Kraut).

Forces (Figure 20.7), where the monument from the city centre was also re-erected. Some years have passed since the dramatic 'Bronze Night', and it looks as if the Russian population has accepted the new location of the monument, where they continue with their annual celebrations on 9 May (Figure 20.8). The dead have been reburied from a bus-stop to consecrated soil, and in

Figure 20.8: Victory Day on 9 May 2015 was expansively observed at the Bronze Soldier, moved from Tõnismägi to the cemetery of the Defence Forces in 2007 (© Peep Pillak).

Figure 20.9: In the Tallinn cemetery of the Defence Forces, the graves of soldiers of the Republic of Estonia buried before the Soviet occupation (their unified memorial stones can be seen in the foreground) were overlain by burials of Soviet soldiers, whose memorial stones are inscribed in Cyrillic script (© Peep Pillak).

European culture this is of importance (for more details see also Bronze Night 2007 and Cheremnykh 2007).

The Estonian Heritage Society, in collaboration with the Ministry of Defence and the Military Museum, is active in restoring and maintaining the cemetery of the Defence Forces. During the Soviet occupation, the graves of Estonian soldiers who had been killed in the War of Independence or died afterwards were destroyed and the site was reused for the interment of Soviet officers (Figure 20.9). It was like a continuous war in which even the dead soldiers participated. The monuments in the cemetery were demolished (Hallas-Murula 2008), and not only those erected for Estonian soldiers, but also the one for the German soldiers of the First World War, whose burial ground was reused for Soviet soldiers, and the one for the British marines who had died in Estonia during the Estonian War of Independence. The restoration of the cemetery of the Defence Forces has been an ongoing project since the restoration of Estonian independence, and probably it will continue for many years to come, including several cases of reburials.

So we have had reburials of expatriates who left Estonia at the onset of the Soviet occupation. In September 2014, a holder of the Estonian Cross of Liberty, Colonel Arthur von Buxhoeveden, and his wife were reburied in the cemetery of the Defence Forces. During the First World War, this Baltic-German baron from Saaremaa had fought in the tsarist army; later he took part in the Estonian War of Independence and in the expansion of the Estonian army in the 1920s. He had left Soviet-occupied Estonia for Germany in 1941 as a German *Nachumsiedler* (i.e. a person returning to his/her German homeland) and died in Karlsruhe in 1964. As the lease on his burial plot in Karlsruhe was due to expire and he had no close relatives left in Germany, the reuse of his grave for a new burial would have been an imminent prospect. The tombstone had been removed from the grave already. Therefore, the Estonian Heritage Society, with the help of the Estonian Ministry of Defence, the authorities of Karlsruhe, the Estonian Embassy in Berlin and the German Embassy in Tallinn, decided to arrange for his reburial in his native Estonia, which had been his last wish (Figure 20.10a, b). All in all the process took three years, but the result was significant (Pillak 2014a).

Figure 20.10a: The burial place in Karlsruhe of Colonel Buxhoeveden and his wife, with the tombstone already removed (© Ants Kraut).

Figure 20.10b: The grave of Colonel Buxhoeveden and his wife after their reburial at the cemetery of the Defence Forces in Tallinn (© Peep Pillak).

Identification of the graves of Forest Brothers

Archaeologists and military historians, with the help of local people and volunteers, are working to find the battle sites and burial places of the civilian partisans known as the Forest Brothers, who fought against the Soviet occupation authorities. Their burials are identified by means of excavations, and once the persons have been identified, they are reburied in cemeteries.

One of the last and biggest reburial ceremonies of Forest Brothers was in September 2013, when the remains of 13 of them who had lost their lives in combat with the Soviet occupation authorities were interred in the small Vastseliina cemetery in South Estonia. On 28 December 1945, there had been a fight between them and the Soviet internal forces in Lükka Luhasoo, South Estonia. There were 11 men in the bunker that night. As no one was on guard, the early attack at 5 a.m. came to the men as a surprise. The Forest Brothers coming out of their bunker, which had been set on fire, found themselves under a barrage, and nine of them were killed; of the two who managed to escape one also later lost his life.

The excavation in November 2011 has been described by the archaeologist Arnold Unt: 'Wet woodland is not the best place for an ideal excavation, especially given the limited daylight that was available to us. However, we had rubber-trousers and rubber-gloves, and considerable experience in work like this. The first skeleton turned out to be only partially preserved – the upper body was totally missing. It was buried on top of another skeleton – the feet of the former resting on the head of the latter. The two skeletons of the men buried with their heads westwards were also mutilated, especially their heads: these were still in place, but the skulls were smashed and many of their parts were missing. The one on the left hand side from the bunker had been covered with a dark woollen coat with buttons; the one on the right had some fragments of cotton underwear on it. The time and other circumstances of the burial can be only guessed. It was definitely not representatives of the official power who had conducted this burial: the shallow grave had been covered by fir-tree branches, which is a symbol of final respect. A copper wire at the feet of one of the victims could have been used to drag a rotten corpse to its grave, and the partially buried body could indicate that the burial of the men had taken place much later than the fight – perhaps in spring? We found in the grave also a few pieces of burnt timber from the bunker. It testifies to nothing more than the fact that at least one of the men had been close to the bunker after his death and while dragging his body to its grave, some of the charcoal had come with it' (Unt 2013).

Figure 20.11a & 11b: Excavation of killed Forest Brothers from the mass grave in Reedopalo forest, South Estonia (© Arnold Unt).

Years later, on 29 March 1953, there was a bunker battle in Puutlipalu, Võrumaa, South Estonia. The security officials had caught a man who had helped those hiding in a bunker, and he had been tortured so that he would reveal where the bunker was. Twice, in the early morning, he led the security forces to the wrong place, but

Figure 20.12a: Memorial service of killed Forest Brothers in St Catherine's Church, Vastseliina (© Peep Pillak).

Figure 20.12b: The graves of killed Forest Brothers, as reburied in the graveyard in Vastseliina (© Peep Pillak).

while undergoing a third session of torture he gave up and the bunker was found. The Forest Brothers refused to capitulate and their fatal fight lasted for 3 hours until all 8 of them were dead. In accordance with a 1946 directive, the fallen Forest Brothers had to be buried unofficially, in secret locations, so the Soviet authorities hid the bodies in the nearby Reedopalo forest (Figure 20.11a, b). They were discovered in a mass grave hidden under garbage, together with two other Forest Brothers who had been killed a few days earlier (for more details see Kaitsepolitsei 2012).

In St Catherine's Church, Vastseliina, a memorial service was conducted featuring a speech by the Minister of Defence. A guard of honour from the Defence League of Võrumaa carried the coffins to the cemetery, where they were interred with full honours (Figure 20.12a, b). Among the participants of the ceremony was a 90-year-old Forest Brother who had escaped the persecution, relatives of the fallen partisans, members of the Defence League, and of the Women's voluntary defence organisation, as well as representatives of other organisations, and local people (see Pillak 2014b).

Conclusion

The concern of the Estonian Heritage Society is not only archaeology in its classical sense, which informs us about times immemorial, but also archaeology that is about our recent past. The reburial of the remains of President Päts from Russia to his homeland was possible only within a small window of opportunity: a year or two earlier the very idea of finding the grave of a 'public enemy' and his ceremonial reburial would have been a criminal offence in the Soviet Union. In 1988 we were lucky enough to have Dr Gusseva still with us, who could walk, leaning on her stick, and show us the approximate location of the burial. In August 1991 Estonia became an independent state again, and an excavation on the territory of another state, in the Russian Federation, would have been much more problematic, if not impossible; in the present political situation it would be unimaginable. In 1990 President Päts was the symbol of independence for Estonians;

by now historians have evaluated his activities from a highly critical perspective (Ilmjärv 2004a, 2004b).

As regards the reburial of the remains of the Soviet soldiers and the removal of the Bronze Soldier from the bus-stop in the centre of Tallinn to the cemetery of the Defence Forces, it can be said that the right time for that had passed. Had it been done immediately after the restoration of Estonia's independence, the antagonism would probably have been much milder, and the social polarisation could have been avoided. After the collapse of the Soviet Union, the identity of Russians has increasingly been tied to their victory over fascism in the Great Patriotic War. Russian media presented the removal of the monument as a plan to demolish it and stigmatised this act as a manifestation of fascism in Estonia, which incited a part of the Russian-speaking population of Estonia that is prone to be manipulated by Russian propaganda. By now the situation has calmed down, but at the cost of a hard lesson.

Colonel Buxhoeveden was an unknown refugee in Germany but an important and colourful person in the history of Estonia, where he had been forgotten for decades. Now the cooperation of Estonian and German institutions and the reburial of his remains has restored his significance. It was a lucky chance that the expedition of the Estonian Heritage Society reached Karlsruhe cemetery at the right moment, when his grave had not yet been reused.

It is high time to try and find the graves and battlefields of Forest Brothers because a few members of the generation that still remember the events of the 1940s and the 1950s are still alive and can take us to those remote places. For the Forest Brothers and their family members still living, the reburial is of immense value as it demonstrates that the hopeless fight of the partisans was not pointless.

In this light it can be said that the principle guiding the work of the Estonian Heritage Society is: the future of our past is in our hands.

Acknowledgements

I would like to thank the organisers of the EAC meeting in Lisbon for inviting me to give a paper and especially Paulina Florjanowicz for giving useful comments on my draft. While writing the article I was grateful to my friend and mentor in the field of archaeology – Vello Lõugas, and to all the other participants of the expeditions. I would like to express my thanks to Arnold Unt, who has long studied war graves, and especially to my colleagues of various expeditions: Ants Kraut and Helle Solnask from the Heritage Society, who also encouraged me to deliver this paper.

References

Big Soviet 1975: Большая советская энциклопедия, Vol. 21, Moscow.

Bronze Night 2007: Bronze Night, https://en.wikipedia.org/wiki/Bronze_Night; Aftermath of the Bronze Night, https://en.wikipedia.org/wiki/Aftermath_of_the_Bronze_Night (accessed 14.09.2015).

Cheremnykh, K. 2007: The lessons of Tallinn. The Russian-Estonian crisis in the global context, http://www.globoscope.ru/eng/content/articles/77/ (accessed 13.09.2015).

Grishina, M. 2007: Гришина, М. „Время новостей": Покой на Святой земле. *Израильтяне похоронили советскую медсестру по всем канонам еврейской традиции,* http://www.portal-credo.ru/site/?act=monitor&id=10754 (accessed 13.05.2015).

Hallas-Murula, K. 2008: *Edgar Johan Kuusiku Vabadussõja monumendid.* Eesti Arhitektuurimuuseum, Tallinn.

Ilmjärv, M. 2004a: *Hääletu alistumine. Eesti, Läti ja Leedu välispoliitilise orientatsiooni kujunemine ja iseseisvuse kaotus 1920. aastate keskpaigast anneksioonini.* Argo, Tallinn.

Ilmjärv, M. 2004b: Silent submission. Formation of foreign policy of Estonia, Latvia and Lithuania. Period from mid-1920-s to annexation in 1940. *Acta Universitatis Stockholmiensis. Studia Baltica Stockholmiensia 24,* Stockholm University Department of History.

Kaitsepolitsei 2012: = Aegumatute rahvusvaheliste kuritegude uurimine, in *Kaitsepolitseiamet. Aastaraamat,* Tallinn, 28–32.

Lõugas, V. 1991: Archeology of terror. Archeological excavations of mass-murder sites in the Soviet Union. *Fennoscandia archaelogica VIII,* 77–84.

Olesk, S. & Pillak, P. 2000: *Lydia Koidula 24.12.1843–11.08.1886.* Tallinn.

Petersoo, P. & Tamm, M. 2008: *Monumentaalne konflikt. Mälu, poliitika ja identiteet* tänapäeva Eestis. Varrak, Tallinn.

Pillak, P. 1999 (ed.): *Johan Laidoner 12.02.1884–13.03.1953.* Tallinn.

Pillak, P. 2000: The search for General Johan Laidoner. *Global Estonian 2,* 94–9.

Pillak, P. 2007: Buraševo ekspeditsioonid 1989 ja 1990, in A. Velliste (ed.): *Alasi ja haamri vahel. Artikleid ja mälestusi Konstantin Pätsist.* MTÜ Konstantin Pätsi Muuseum, Tallinn, 196–207.

Pillak, P. 2014a (ed.): *Kolonel Arthur von Buxhoevedeni (VR I/2) ja tema abikaasa Kira ümbermatmine Tallinna Kaitseväe kalmistule 12. septembril 2014 / Umbettung von Oberst Arthur Buxhoeveden (Freiheitskreutz I/2) und seiner Ehefrau Kira auf den Friedhof der Streitkräfte in Tallinn.* Eesti Muinsuskaitse Selts, Tallinn.

Pillak, P. 2014b: Langenud metsavennad maeti Vastseliina kalmistule, in *Eesti Kodu-uurimise Selts, Eesti Muinsuskaitse Selts, Eesti Genealoogia Selts: Aastaraamat,* Tallinn, 84–5.

Pillak, P. (ed.) 2015a: *Eesti riigipead 1918–1992 / Estonian Heads of State 1918–1992.* Estonian Heritage Society, Tallinn.

Pillak, P. 2015b: = Пиллак, П. *Константин Пятс и Бурашево.* Эстонское общество охраны памятников старины, Таллинн.

Poliitilised 2008: = *Poliitilised repressioonid 1940. ja 1950. aastatel.* Eesti Muinsuskaitse Selts, Tallinn.

Strauss, M., Krillo, A., Pihlak, J. & Viljat, R. 2002: *Vabadussõja mälestusmärgid I,* Keila.

Unt, A. 2013: Paper at a memorial conference of forest brothers in Vastseliina on March 27, http://www.spordilinn.ee/failid/vastseliina_ettekanne_Unt.pdf (accessed 30.06.2015).

Zenger, M. 2008: Зенгер, М. Памятник Ленине Моисеевне Варшавской, http://toldot.ru/tora/articles/articles_2663.html (accessed 13.09.2015).

21 | Archaeological sites: the need for management and legislation improvements (some thoughts on the Albanian reality)

Ols Lafe

Abstract: This paper presents a discussion of archaeological heritage and its management in Albania, seen from the legal and practical perspective. The Cultural Heritage Law and education system are analysed as part of the discussion regarding heritage management. There has been ongoing debate for years on these issues, and this has sparked the revision of the Cultural Heritage Law, which is still under preparation. With many archaeological sites still in need of management plans and experiencing an increase in visitor numbers, the debate within the archaeological community is intense.

Keywords: Albanian archaeology, cultural heritage law, education, management plans

When we talk about archaeological sites (and I mean we as archaeologists), often in my opinion we forget to think about the communities who live at those sites. By this strong statement I do not want to spark needless debate on whether this is true or false, but rather want to share my personal experience of practising archaeology.

Thus, when we dig or survey and look at a site or a given area which in our opinion offers clues to the past, sometimes we simply forget the very important role that locals have in preserving, protecting and otherwise safeguarding these cultural heritage resources. Although many of us have had community representatives working with us in our trenches, preparing our food, or even driving us around, sometimes we have not communicated further with them about the values of this heritage.

What I am tackling is just a small part of what we can and have to do in order to manage archaeological sites, and that is involve the local community (Feilden & Jokilehto 2010). Education and communication of heritage values are thus key issues which need to be addressed when we undertake actions towards the management of archaeological sites.

In most cases, the countries of Europe, despite their outward similarities, have totally different approaches to archaeological heritage management. These differences are evident both in the institutional arrangements for the management of archaeology and in the way that archaeology is perceived.

Before 1991, archaeological issues in Albania were determined by the needs of the totalitarian state (Kamberi 1993). The strategy was based upon a strong Marxist ideology which compelled archaeologists to undertake research in three areas: 1) the ethnogenesis of the Illyrians and their evolution as an autonomous community; 2) the cultural, economic and social development of the Illyrians; 3) establishing an archaeological connection between the Illyrians and the first Albanians. During this period, often with limited financial support from the government, Albanian archaeologists – especially after the 1960s – tended to concentrate upon tackling these issues by undertaking excavations as opposed to survey (Anamali 1969).

The National Law on Cultural Heritage

The legal framework in the field of cultural heritage is governed by the Law on Cultural Heritage, No. 9048 of 7 April 2003 (with later amendments). This law grants the status of 'cultural heritage' for different categories of monuments and aims to protect cultural heritage in the territory of the Republic of Albania. It focuses on cultural heritage values, providing the legal framework to protect them.

The law also defines the main duties and responsibilities of the principal state institutions and bodies operating in the field of cultural heritage. The law was amended twice – in 2006 and 2008 – with the purpose of improving the overall legal and administrative framework, and also to provide solutions to issues facing urban and territorial planning with regards to the integrated development of cultural heritage assets. This law has reflected the major conventions of UNESCO (Figures 21.1–21.3) and the Council of Europe regarding tangible and intangible heritage.

The administrative management of tangible cultural heritage is mostly coordinated by the Ministry of Culture and is implemented by a series of institutions and agencies. Basic expenses for inventorying, restoring, preserving and administering cultural heritage properties are mostly covered by the state budget and are dispensed by the institutions and entities directly responsible for administration of such properties.

Figure 21.1: Butrint (© O. Lafe, 2011).

Financing from the central state budget is the most important source, but contributions from foreign foundations or NGOs are also important. There are also some Albanian NGOs, mostly aided by foreign foundations or organisations engaged in supporting culture and cultural heritage, but they need more coordination and support locally. Income derived from the use of cultural heritage properties should normally be an important source of finance for the restoration and maintenance of cultural properties, but, for the time being, such income is at a very low level, and in most cases it is not enough to cover the relevant expenses. In December 2011 the National Park of Butrinti was granted the right to manage 90% of its self-generated income (Law 2011).

Analysis of the cultural heritage management system shows that its functioning is impaired by several factors, such as limited specialised staff officially involved in various aspects of the management of cultural properties, gaps in capabilities within institutions and from one institution to another, a low degree of employee autonomy, lack of capabilities in project management, non-existent financial appraisal of project feasibility studies, lack of fundraising skills, low levels of financial support, a lack of financial support from local government, and insufficient institutional overlapping.

The system also needs a more integrated approach when dealing with territorial and spatial planning with the purpose of conserving the natural and cultural values of sites. In terms of human resources, in some cases positions that require a high level of scientific as well as administrative professionalism are covered by inadequately qualified people. There is a clear necessity to increase the budget in order to have more resources for archaeological parks and sites and monuments in general.

I will lay forward in brief the Albanian experience in managing archaeology, starting from the creation of the Archaeological Service Agency in 2008, which, in my opinion, has changed radically the way that both archaeologists and the wider population of non-archaeologists see the management of archaeology.

The Archaeological Service Agency (ASA – established in 2008) – an institution operating under the aegis of the Ministry of Culture – is responsible for:

Figure 21.2: Gjirokastra (© A. Islami, 2006).

- archaeological excavations in the context of urban and spatial planning in the Republic of Albania,
- the establishment of criteria and observance of supervision of excavations,
- reorganisation in museum-related aspects of the archaeological heritage,
- and for the administration and enrichment of museums with artefacts from archaeological excavations.

The work of the ASA, is supplemented and overseen by the National Archaeological Council (NAC) – a collegial decision-making body, which meets periodically at the ministry responsible for cultural heritage, and is chaired by the minister.

The NAC approves in principle the research criteria, documentation and archiving of data and archaeological materials, approves the archaeological research integration strategy (driven by developments and fundamental issues of archaeological research), defines the criteria for the exercise of the archaeological profession, approves permits for private entities and individuals involved in archaeological excavations and archaeological activities in general, and also approves all projects of intervention in archaeological areas, in accordance with article 30 of Law No. 9048 On Cultural Heritage, dated 7 April 2003 (as amended).

The NAC approves the storage, integration and final displacement of archaeological remains, following excavations held anywhere within Albania, and also endorses archaeological projects of a scientific research nature, in cooperation with the Institute of Archaeology. The Council also determines the criteria for presentation of archaeological findings, as well as approving the storage and maintenance modalities.

How has this system worked in the last 7–8 years? Is it successful? Although the majority of specialists who work in the cultural heritage sector have the appropriate technical experience and capacity to carry out their duties, vocational training and maintaining up-to-date professional standards remains a challenge. To improve this situation, heritage institutions have recently been cooperating with standard-setting international organisations to facilitate staff training and capacity building.

In 2008 the Archaeological Service Agency (ASA) began the process of mapping boundaries of archaeological sites using GPS devices. In reality, the inventory system has not yet adopted the Council of Europe Core Data Standards (for architecture and archaeology).

Identifying heritage assets in the context of other management systems, such as cadastral records or in spatial and land-use planning, is still in its initial stages. The inventory system is relatively up-to-date; however, it lacks procedures designed to update it on a systematic basis. The way that information related to heritage is used by other planning authorities is as follows: planning authorities require the ministry

Figure 21.3: Berat (© A. Islami, 2008).

responsible for cultural heritage to provide information about the presence of monuments or otherwise important archaeological and historical objects in a given area where development is planned (at present the planning authorities do not have digital access and have to formulate the request of information). In turn this information is inserted into the area plans and respective actions follow.

Education and cultural heritage

The education of various sectors of society regarding the cultural heritage and its national importance and significance has progressively started to develop and gain special attention, as well as the engagement of both central and local institutions responsible for education, culture, and information culture. Government institutions are paying greater attention to the importance of public awareness about the cultural heritage and national values and the need to prevent their destruction by natural and human factors, as this could adversely affect the integrity of monuments and national culture in general. The Ministry of Culture and its subordinate institutions are obliged to raise public awareness of the need for the maintenance, conservation, rehabilitation, and restoration of monuments and archaeological sites; they are also required to attract and promote cooperation among partners and include in this effort national and foreign experts.

The education of the young generation concerning the values and importance of cultural heritage is also the responsibility of another central institution: the Ministry of Education and Sports. The Law on Pre-University Education (69/2012) has emphasised that one of the main purposes of this law is to know, accept, respect, and protect the national identity and further develop the heritage and our cultural diversity.

'Our cultural heritage' is already a specific subject in the curriculum of the pre-university education system (10th grade – c. 16 years old). This attention is also extended to the national minorities living in Albania as well as children of Albanian nationality living abroad (Law 2012, articles 10 & 11).

To conclude, we can say for certain that Albania is on the right path towards implementing the newest and best standards in the management of its archaeological sites, while more work is needed in upgrading the management system's infrastructure and empowering staff throughout the country with regards to decision-making. The future is good, and collaboration with other international partners should and always has to be on the agenda.

References

Anamali, S. 1969: Arkeologjia shqiptare gjatë 25 vjetëve të çlirimit, *Studime Historike* 4, 91–102.

Feilden, M.B. & Jokilehto, J. 2010: *Udhëzues për menaxhimin e trashëgimisë kulturore botërore*, ICCROM, Tiranë: Gent Grafik.

Kamberi, Z. 1993: Archaeological research and researchers in Albania, *University College London, Institute of Archaeology Bulletin* 30, 1–27.

Law 2011: Vendim 928 dt.28.12.2011 Për përdorimin e të ardhurave të Parkut Kombëtar të Butrintit.

Law 2012: [Law on Pre-University Education] Për sistemin arsimor parauniversitar në Republikën e Shqipërisë, 69/2012.

Résumés

1 | Itinéraires vers un contrat du « savoir producteur »

Kristian Kristiansen

Au cours de cette présentation, j'expose quelques récentes modifications dans l'évolution du contrat archéologique en Europe, et les débats qui s'en suivent. Au moins la crise économique après 2008 a mis en évidence la vulnérabilité de certaines formes de contrats archéologiques, qui ont augmenté les demandes pour une organisation plus durable sur le terrain. La hausse aussi de « Big Data » et des demandes d'accès libre déterminent de nouveaux défis pour une meilleure intégration européenne des données et de la recherche archéologique. Par conséquent le temps est venu de moderniser les conventions et codes internationaux de conduite à l'égard des associations professionnelles afin de s'adapter à ces nouvelles réalités.

Mots-clés : contrat européen d'archéologie, production d'un savoir durable, qualité de gestion, Big data, convention de la Valette

2 | Attitudes stimulantes – s'assurer la faveur du public

Adrian Olivier

Beaucoup d'interventions entreprises par les archéologues aujourd'hui puisent dans les fonds et financements publics et se poursuivent en son nom. Les dernières décennies ont vu une réelle augmentation de la prise de conscience du public, et de l'intérêt envers l'archéologie, cependant, une grande partie de cette communication se fait de haut en bas et à sens unique. L'intérêt du public est facile à prétendre, mais il est beaucoup plus difficile à définir ou à démontrer dans la pratique. Les approches à assumer envers l'intérêt public changent, mais il demeure peu de compréhension, ou d'articulation avec ce que le public (ou les publics) souhaitent de la part des archéologues. Si l'archéologie veut survivre et prospérer, les archéologues doivent apprendre à mieux remplir un rôle public en s'attirant les communautés en tant que co-créateurs - plaçant le passé au service du public afin qu'il soit pertinent et utile dans le contexte de leur vie quotidienne.

Mots-clés : intérêt du public, archeology préventive, valeurs patrimoniales.

3 | De La Valette à Faro avec une escale à Bruxelles. Contextes légal et réglementaire internationaux en archéologie ou simplement la compréhension du patrimoine au niveau européen

Paulina Florjanowicz

L'archéologie contemporaine est davantage liée à la « vie réelle » que tout autre aspect du patrimoine culturel. Aménagement du territoire, infrastructure du transport, protection environnementale, agriculture – toutes ces zones ont un impact direct sur le patrimoine archéologique et le menace. Afin de neutraliser ces risques, différentes mesures légales et réglementations ont été introduites à la fois aux plans national et européen. Ceci représente une tentative d'exposer par un archéologue la dernière perspective.

La plus largement connue est naturellement la Convention européenne pour la protection du patrimoine archéologique (Conseil de l'Europe 1992) ; cependant la Convention de Faro sur la valeur du patrimoine culturel pour la société (Conseil de l'Europe 2005) est tout aussi importante. Ces deux conventions du Conseil de l'Europe sont assez différentes, illustrant ainsi l'évolution de l'approche patrimoniale. Mais même si elles diffèrent de façon significative, elles demeurent encore plus complémentaires que contradictoires.

Lors de l'examen de la politique et de la législation internationales relatives à l'archéologie, il ne faut pas oublier l'Union européenne. En vertu de l'article 3.3 du Traité de Lisbonne, [l'Union] doit respecter ses diversités culturelle et linguistique riches et s'assurer que le patrimoine culturel de l'Europe soit sauvegardé et amélioré. De même le traité stipule que la culture représente un champ de réglementations au sein duquel l'Union supporte seulement les états membres, ce qui exclut toute harmonisation des lois nationales.

Cependant, l'archéologie étant liée à tant d'autres champs d'activités de la population, elle est constamment affectée par l'action de l'Union européenne. Jusqu'à présent, le potentiel social et économique de cette catégorie du patrimoine a été ignoré, augmentant ainsi les menaces. Voilà pourquoi diverses tentatives ont été effectuées au cours des dernières années afin de changer la compréhension du patrimoine culturel par l'Union européenne et son rôle en Europe. Les récentes réalisations, telles que l'adoption de deux conclusions du Conseil de l'Europe en 2014 et le la résolution du parlement européen du 8 septembre 2015, reconnaissant directement les aspects positifs du patrimoine culturel pour la communauté européenne, préparent le terrain vers des changements qui pourraient avoir d'énormes conséquences pour l'archéologie aussi – qu'elles soient bonnes ou mauvaises dépend largement des archéologues eux-mêmes. C'est pourquoi il est tellement important de comprendre le processus qui se met en place maintenant.

Mots- clés : Conseil de l'Europe, Union européenne, Parlement européen, cadre légal international, intervenants, approche intégrée

4 | Un aperçu de la gestion patrimoniale en Allemagne, en particulier en Saxe-Anhalt

Harald Meller et *Konstanze Geppert*

La Convention de Malte de 1992 a été intégrée dans la loi fédérale d'Allemagne (Art. 36, §4 de la Constitution en Saxe-Anhalt).

En Allemagne, depuis que le secteur culturel est soumis aux états autonomes, chacun des 16 états de l'Allemagne fédérale (appelée Bundesländer) possède sa propre loi en matière de patrimoine culturel, mais tous présentent des similarités.

Le cadre légal et organisationnel de l'archéologie préventive en Saxe-Anhalt a pris comme modèle la loi sur la protection des monuments historiques, qui sera illustré ici en référence aux découvertes emblématiques en Saxe-Anhalt.

Le principe premier est la préservation des sites dans le paysage archéologique exceptionnellement riche en Saxe-Anhalt. Les sites archéologiques en tant que traces de l'histoire humaine sont des sources non renouvelables ce qui signifie que chaque fouille participe en fait à un processus de destruction.

L'administration en charge de la gestion du patrimoine et de l'archéologie, et spécialement le département de la conservation archéologique, remplit les obligations contenues dans la loi sur la protection des monuments historiques concernant les sites archéologiques. Ses tâches primordiales se concentrent sur la préservation et la protection des éléments matériels des sites archéologiques, en les enregistrant, en les documentant scientifiquement et en les étudiant.

Afin d'accomplir ces tâches, différentes méthodes (entre autres, prospections terrestres, enquêtes préliminaires sur les constructions projetées, photographies aériennes, prospections géophysiques, lidar) sont utilisées pour enregistrer systématiquement les traces matérielles sur les sites. La conservation archéologique par l'état a, à notre sens, plusieurs avantages, en comparaison avec d'autres modèles de gestion patrimoniale.

Le travail perfectionné de l'état en matière de conservation archéologique constitue à plusieurs niveaux la première étape de l'étude scientifique et de l'évaluation des découvertes et sites archéologiques, tout en établissant les bases de communication et d'explication envers le public.

La charge financière de la documentation est régie par la règle du pollueur-payeur. Cela signifie que la documentation d'un site archéologique est financée par l'aménageur qui provoque sa destruction, jusqu'à un maximum de 15% de l'investissement global.

Un des objectifs principaux du travail consiste à avoir une assistance d'experts dans le processus de permis de bâtir sur toutes ses formes et, développant à partir de ça, l'organisation, la supervision et l'exécution des fouilles préventives. Cela implique directement des experts dans différentes sciences naturelles, incluant archéobotanistes, archéozoologues, pédologues, et spécialistes de différentes époques appartenant au département lui-même. C'est la seule manière d'obtenir une compréhension plus étendue des questions en matière d'archéologie environnementale. En engageant ce type d'experts, poursuivant l'étude des sites archéologiques, c'est aussi accorder une attention accrue. C'est fréquent dans une collaboration avec des partenaires externes ou internationaux.

Un large réseau de représentants bénévoles constitue une composante indispensable au travail du département. Ils remplissent, en accord avec l'office de conservation, des tâches précises bien définies en matière de conservation archéologique.

Mots-clés : Convention de la Valette, gestion patrimoniale en Allemagne, loi sur le patrimoine en Saxe-Anhalt, principe de l'aménageur-payeur, propriétaire des découvertes.

5 | L'organisation de l'archéologie tchèque - Système légal socialiste appliqué à l'économie de marché

Jan Mařík

Les premières mesures légales pour la protection des découvertes archéologiques en Bohème et en Moravie (régions historiques de la république tchèque) furent prises dès la première moitié du 19e siècle. Cependant la véritable réglementation ne commence qu'en 1941 avec le décret officiel. Ce décret spécifiait les principes de base à utiliser pour la conduite des fouilles archéologiques ainsi que pour l'entretien du patrimoine archéologique. Ils sont encore plus ou moins d'application, même dans les lois en vigueur.

La direction des fouilles archéologiques fut confiée au à l'Institut archéologique national (prédécesseur de l'Institut d'archéologie de l'Académie des Sciences de la République tchèque) et quiconque souhaitait entreprendre une opération archéologique ne pouvait la faire sans l'accord de l'Institut. Au cours de l'évolution de la prise de décision concernant les interventions archéologiques les autorités officielles devaient confronter leur décisions avec le l'Institut archéologique national. De plus, les découvertes archéologiques étaient considérées comme propriétés nationales.

La loi actuellement existante est déjà entrée en vigueur en 1987. Bien que la loi ait été créée dans un environnement d'état socialiste, elle fut conçue dans un esprit très progressiste. Malgré le fait que la loi n'a pas été révisée de manière significative depuis qu'elle est appliquée elle s'acquitte à la majorité des obligations auxquelles la République tchèque a consenti en adhérant en 2000 à la convention de la Valette.

Cependant, en 1987, les législateurs ne pouvaient envisager les modifications fondamentales politiques aussi bien que sociale qui se produiraient en république tchèque deux ans plus tard, en 1989. Le passage vers une économie de marché ainsi qu'une croissance significative de la construction allait augmenter très fort le nombre de demandes en matière d'interventions en archéologie préventive.

Ce progrès a entraîné, parmi d'autres choses, étant donné l'augmentation des demandes, la création de nouvelles autorisations donnant droit à une intervention archéologique. Outre les musées et les universités, des firmes privées sont apparues.

À l'heure actuelle, 110 firmes aussi bien publiques que privées possèdent la licence pour entreprendre une recherche archéologique. La mise en place du principe du pollueur-payeur a déclenché le fait que les organisations agréées utilisent les opérations d'archéologie préventive comme une de leur principale source de financement.

L'Institut d'archéologie de l'Académie des Sciences de la République tchèque a réussi à conserver sa position privilégiée au-delà même de 1989. L'Institut est la seule organisation autorisée, en vertu de la loi, à effectuer directement une opération archéologique. En outre, il a un droit d'influence significatif en matière de délivrance d'une nouvelle agréation pour une opération archéologique (le veto). Il recueille les informations concernant les opérations archéologiques en cours Il archive les rapports de fouilles et, à un certain degré, contrôle leur qualité. Cependant, la loi existante ne précise aucune norme évidente de la recherche archéologique et donc, sa qualité varie de manière très significative en république tchèque.

Le processus législatif concerné représente des problèmes fondamentaux en République tchèque : spécification vague des règles, droits aussi bien qu'obligations, non seulement du côté des organismes agréés mais aussi des aménageurs, des propriétaires et de l'administration. L'état actuel des choses ne pourra être résolu qu'en passant à une nouvelle loi.

Mots-clés : décrets légaux, République tchèque, protection du patrimoine archéologique, modification politique

6 | Recherche archéologique en République slovaque – le positif et le négatif

Matej Ruttkay, Peter Bednár, Ivan Cheben et *Branislav Kovár*

Jusqu'en 2002, la recherche archéologique en République slovaque, était très peu réglementée par la loi. Toutefois en 2002, la situation a considérablement changé après l'introduction du décret no 49, amendé plus tard en 2010 et 2014. Le décret a apporté des changements positifs mais aussi de nombreux résultats contre productifs.

Dans cet article, nous essayons d'évaluer sa contribution à la recherche archéologique slovaque. Nous esquissons quelques aspects problématiques du décret, à savoir l'introduction de licences archéologiques, l'ouverture de l'archéologie aux firmes privées, le problème urgent des pillards et les détecteurs de métaux sur les sites archéologiques.

Mots-clés : loi, recherche archéologique, firmes archéologiques privées, pillage des sites archéologiques

7 | L'archéologie préventive française : organisation administrative, rôle des intervenants et procédures de contrôle

Bernard Randoin

Au cours des années 1999, la loi de 1941 concernant l'archéologie française a dû être modifiée en raisons des évolutions modernes tant au niveau de la société que de la discipline archéologique. Le nouveau système législatif a été très débattu pendant une longue durée au Parlement et le résultat des choix variés n'ont pas été réalisés par les archéologues mais par les représentants de la société française.

Cet article a pour but de décrire le système qui couvre l'organisation administrative, les différents rôles des acteurs variés dans la prise décision, le travail de terrain et la qualité du contrôle.

Mots-clés : organisation de l'archéologie, Code du patrimoine, contrôle de qualité, opérateurs

8 | Un regard de la Turquie sur les conventions de La Valette et de Faro : efficacité, problèmes et état des lieux

Mehmet Özdogan et *Zeynep Eres*

La Turquie est fière de son propre patrimoine archéologique riche et varié. Bien que chaque année il y a de nombreuses fouilles archéologiques à haute échelle scientifique, la quantité d'archéologie préventive est faible comparée à l'allure qu'a pris le secteur de la construction. Selon la législation turque, l'état assure l'autorité légale et la responsabilité pour tout le patrimoine archéologique cependant, pour un site sous protection, il doit être enregistré.

L'enregistrement d'un site passe par une procédure extrêmement bureaucratique, le nombre total de sites recensés en Turquie en 2015 s'élevait seulement à 12.757. Les problèmes rencontrés en Turquie pour la préservation du patrimoine archéologique sont de loin plus importants en matière d'échelle et plus complexes que dans la plupart des autres pays européens : il n'y a pas d'inventaire des sites presque complet, l'instance supérieure ne s'est pas efficacement adaptée aux incessantes opérations de sauvetage et les sites présentent de prodigieuses dimensions.

Mots-clés : Turquie, patrimoine archéologique, fouilles de sauvetage, recensement des sites, sensibilisation du public

9 | Tout ce que vous avez toujours souhaité connaître à propos de l'archéologie commerciale aux Pays Bas

Marten Verbruggen

L'archéologie commerciale a été introduite de manière informelle depuis 1995 aux Pays-Bas, en adéquation avec la mise en œuvre d'un certain nombre de principes de la convention de La Valette, tels que le principe du pollueur-payeur et une interaction directe entre l'archéologie et l'aménagement du territoire. Le nouveau système, intégré seulement en 2007, consistait à réagir envers le système défaillant de la gestion antérieure du patrimoine archéologique.

En 2011, l'application du modèle de La Valette a été évaluée positivement par un bureau d'études : la politique de préservation in situ a été fructueuse et le rythme des publications de fouilles est très élevé. Cependant, récemment, une faille du système a été mise en évidence. Depuis que le prix de la recherche archéologique est devenu le seul critère différentiel pour l'aménageur, cela – ainsi que la crise économique – a entraîné une substantielle baisse des prix. Ce qui a conduit à une chute dans l'équité des groupes commerciaux, qui par la suite engendrera un déclin de la qualité de la recherche.

Parce que le gouvernement national est responsable d'un bon fonctionnement du système de gestion du patrimoine archéologique, il est aussi de son devoir de garantir la qualité au niveau souhaité.

Le gouvernement n'a pas à le faire lui-même, mais cela peut être sous-traité par des groupes privés.

Mots-clés : archéologie commerciale, Pays-Bas, Convention de La Valette

10 | Écosse et un « Dialogue national »

Rebecca H. Jones

En 2014, l'Écosse installait un dialogue national au sujet de sa place dans le Royaume-Uni, avec un referendum à propos de l'indépendance qui connu une majorité en faveur de sa demeure en l'union politique. L'année 2015 témoigne d'une année importante pour l'archéologie en Écosse avec une célébration tout au cours de l'année de l'archéologie, le meeting annuel de l'Association des archéologues européens et le lancement de la première stratégie archéologique écossaise.

Mots Clés : Écosse, archéologie, stratégie, mesure, promesse

11 | La Direction générale du patrimoine culturel, compétences dans un contexte de protection et information concernant le patrimoine archéologique portugais

Maria Catarina Coelho

La Direction générale du patrimoine culturel, fondée en 2012, a pour objectif la protection du patrimoine archéologique du Portugal continental. Sa tâche consiste en son étude, sa gestion, sa protection, sa préservation et sa diffusion. Sa stratégie au niveau de la gestion et de la protection du patrimoine l'archéologique national tend à favoriser le contact et le dialogue entre les différents acteurs de la société engagés dans la protection du patrimoine archéologique : agréments de partenariat avec les plus hautes institutions d'enseignement, aussi bien qu'avec les institutions de nature locale et régionale, indispensables à l'équilibre de la prise de conscience du public où qu'il se trouve.

Mots-clés : Portugal, patrimoine archéologique, gestion, diffusion du patrimoine, public local

12 | Travailler pour une clientèle commerciale : méthodes du Royaume-Uni face au développement mené en archéologie

Dominic Perring

Cet article décrit les procédés actuels au niveau de l'évolution menée en archéologie dans le Royaume-Uni. La clé des problèmes au sujet des forces et des faiblesses de l'offre du marché est examinée avec attention. Cela répond à des préoccupations sur la façon dont la croissance de l'archéologie comme une entreprise n'a pas été accompagnée par une croissance équivalente dans les prestations publiques de nos activités. Cela se voit, en partie, dans la manière dont les politiques de conservation ont été appliquées, aggravant un clivage entre l'archéologie de la gestion des ressources culturelles et un secteur universitaire autrement organisé.

Mots-clés : évolution menée en archéologie, gestion des ressources culturelles, appel à la concurrence, projet de recherche, règlement

13 | Equilibre entre intervenants aux Pays-Bas. Un appel pour une haute qualité de l'archéologie municipale

Dieke Wesselingh

La mise en œuvre de la Convention de La Valette aux Pays-Bas a intégré pleinement l'archéologie au sein de l'aménagement territorial. Les instances locales ont pris la majorité des décisions étant donné que ce sont eux qui élaborent des plans de zonage et délivrent des permis appropriés. L'archéologie néerlandaise dite de « Malte » est une tentative scientifique aussi bien qu'un service préalable à la construction.

L'un ne doit pas exclure l'autre comme en témoigne la démarche en usage à Rotterdam. L'aménagement territorial sans destruction du patrimoine archéologique précieux et, pas moins important, sans fouilles inutiles, est crucial pour acquérir et retenir le soutien politique et social. Les archéologues doivent être sélectifs et soucieux d'expliquer leurs choix, de manière à rencontrer les attentes de tout autres intervenants.

Mots-clés : archéologie préventive, les Pays-Bas, archéologie municipale, intervenants, travail d'évaluation

14 | Les bases légales et l'organisation de l'archéologie préventive en Pologne

Michał Grabowski

Au début des années nonante, la Pologne a subi non seulement une transformation de son système politique, précédée par la chute du communisme, mais aussi a ultérieurement assisté à une période de développement sans précédent de ses infrastructures nationales et de constructions industrielles. Simultanément, un débat substantiel a débuté au sujet du rôle de l'archéologie préventive dans le progrès de la science. Le débat est encore en cours.

Récemment des changements ont été introduits concernant la réglementation du travail archéologique entrepris sur des sites patrimoniaux, qui ont réduit la recherche archéologique à un simple service subordonné à la construction d'industries. Cela montre que l'introduction de règlementations excessivement libérales a apporté des changements assez négatifs et fait de la protection et de la gestion du patrimoine archéologique une tâche très difficile dans un secteur qui, par sa nature même, exige un contrôle et une gestion attentifs. Et même bien que la plupart de ces changements ont été heureusement révoqués seulement quelques mois plus tard, la situation a démontré qu'il y a un manque de concept au niveau gouvernemental pour une politique de conservation cohérente qui va définir des normes pour les travaux archéologiques et l'étude ultérieure ainsi que le dépôt des découvertes.

Mots-clés : normes professionnelles, archéologie préventive, projets archéologiques à grande échelle

15 | L'archéologie préventive en Wallonie : les perspecives

Alain Guillot-Pingue

Dans cet article, l'auteur met en évidence l'évolution des vingt-cinq années de l'archéologie préventive en Wallonie qui ont suivi la régionalisation de la Belgique en 1989. Il y expose l'intégration de l'archéologie wallonne dans la Direction générale de l'aménagement du territoire, la publication d'un décret, l'organisation structurelle, etc

L'auteur évoque aussi ce qui se prépare pour le futur et les outils légaux, structurels et techniques mis progressivement en place afin d'améliorer le dialogue entre les intervenants, d'opérer des choix rationnels mais aussi de répondre à la demande du citoyen.

Mots-clés : planification, nouveaux codes, fonds, changements structuraux, outils opérationnels

16 | Tout le monde est-il heureux ? La satisfaction de l'utilisateur après dix ans de gestion de la qualité de l'archéologie commerciale menée en Europe

Monique van den Dries

Le sujet de la dernière session du symposium annuel de l'*Europae Archaeologiae Consilium* (EAC) à Lisbonne en 2015 était d'assurer la qualité de l'archéologie liée à l'aménagement ou de l'archéologie préventive. Dès la communication de départ, il a été exposé, qu'un des plus grands défis de l'archéologie liée à la construction ou de l'archéologie préventive est de déterminer comment contrôler la qualité – la qualité à la fois du processus de la recherche archéologique et la valorisation des résultats.

La dernière inclut le processus du choix entre différents publics-cibles (chercheurs et public) et en assurant un profit public durable. La suggestion dont je voudrais débattre dans cet article est d'examiner cela dans la perspective des praticiens ou clients de l'archéologie contractuelle et d'essayer de « mesurer » leur degré de satisfaction.

Mots-clés : archéologie liée à l'aménagement, qualité de gestion, publics-cibles, intervenants, satisfaction de l'utilisateur (client)

17 | Défis et opportunités pour la diffusion de l'archéologie au Portugal : différents scénarios, différents problèmes

Ana Catarina Sousa

De La Valette à Faro, beaucoup a changé dans l'archéologie portugaise : la législation, les archéologues, l'administration du patrimoine et la communication envers la société. Plusieurs intervenants archéologiques reconnaissent que la diffusion demeure encore un des fossés majeurs de l'archéologie portugaise post « La Valette ».

Cet article veut analyser séparément les principaux problèmes et opportunités qui concernent la diffusion de la connaissance au Portugal, utilisant des cas d'études et des croisements de données avec quelques perspectives personnelles. Pour différents acteurs et contextes, il y existe différents défis et opportunités, un grand nombre perdus, d'autres redécouverts.

Les scénarios seront rétrospectivement analysés : 1. L'archéologie urbaine (Lisbonne), 2. L'archéologie préventive dans la plupart des projets (EDIA – Alqueva Compagnie de développement et d'Infrastructure). 2. L'archéologie municipale (Mafra). 3. L'archéologie dans les universités et les centres de recherches (UNIARQ, Centre d'archéologie de l'université de Lisbonne). 4. L'archéologie sous l'autorité du patrimoine culturel (IPA, IPPAR, IGESPAR, DGPC). 7. L'archéologie communautaire et associative.

Ce bilan a pour objectif de couvrir la période entre 1997–2014, commençant avec la date de la ratification de la convention de La Valette par le Portugal.

Mots-clés : Portugal, archéologie, diffusion, archéologie publique, La Valette

18 | De La Valette à Faro – éviter une fausse dichotomie et travailler dans le sens de la mise en œuvre de Faro en regard du patrimoine archéologique (réflexions d'une perspective irlandaise)

Margaret Keane et Sean Kirwan

Même si la convention de Faro n'est pas ratifiée (Conseil de l'Europe 2005), les aspects clés de la gestion du patrimoine en Irlande reflètent déjà ses valeurs et principes. Cela exprime le fait qu'il n'y a pas de conflit entre Faro et La Valette. Faro est une convention-cadre qui soutient le secteur spécifique des conventions en matière de patrimoine culturel telle que la Valette. Présenter les choses autrement serait créer une fausse dichotomie.

Des débats à propos de questions telles que la fouille partielle ou totale en réponse aux impacts des aménagements sont très nécessaires, mais ne doivent pas être présentés comme conflictuels entre Faro et La Valette. Dans cet article les auteurs suggèrent que Faro s'unisse et soutienne La Valette dans le développement continu de la gestion du patrimoine archéologique en Europe.

Cette relation complémentaire plutôt qu'évolutive entre les conventions de La Valette et de Faro est démontrée dans quelques programmes particuliers qui ont été mis en œuvre aux de la dernière décennie en Irlande. L'archéologie en classe est un programme adapté qui permet aux enfants entre cinq et douze ans de l'étudier et d'apprécier leur patrimoine.

Ceci sert comme un mécanisme pour la protection et la conservation de ce patrimoine dans le futur, acquérant la préservation à travers l'éducation. Résultant de la mise en œuvre de la convention de La Valette en Irlande, un programme collaboratif subventionné – Recherche stratégique nationale irlandaise (INSTAR) - fut établi afin d'encourager le double but de faire progresser les vastes quantités de données au niveau de la connaissance et d'assurer une coopération parmi les groupes archéologiques professionnels incluant les établissements commerciaux, universitaires et publics.

Mots-clés : fausse dichotomie, préservation par l'éducation, conventions complémentaires, collaboration

19 | Assurer la qualité : projets des opérations archéologique sur les routes nationales irlandaises
Rónán Swan

Cet article se place du point de vue du client, à savoir l'Autorité des routes nationales irlandaises (*Ireland's National Roads Authority* - NRA). La NRA est une agence nationale (travaillant maintenant comme Infrastructure de transport d'Irlande (*Transport Infrastructure Ireland* - TII) depuis sa fusion avec le *Railway Procurement Agency* en août 2015 et qui a la responsabilité d'assurer la sécurité et l'efficacité du réseau des routes primaires et secondaires.

Cela atteint environ 5.000 km de routes et, il y a quinze ans, la NRA a ajouté et amélioré près de 1.500 km de routes à partir de d'améliorations mineures en construisant approximativement 400 km d'autoroute. Mais pourquoi la NRA est-elle intéressée par l'archéologie ? Pourquoi s'en préoccupe-t-elle ? Il y a trois réponses.

Premièrement, la législation, la loi irlandaise requiert que l'archéologie soit traitée de manière appropriée.

Deuxièmement, le risque, si l'archéologie n'est pas gérée efficacement cela peut être extrêmement coûteux en termes de retards et de réclamations des principaux travaux de l'entrepreneur, particulièrement si l'archéologie est seulement repérée durant la construction.

Troisièmement, la confiance du public, la NRA est un organisme public qui prend ses responsabilités très sérieusement envers le contribuable et, par conséquent, tente d'assurer que non seulement nous sommes arrivés à être en conformité, mais que cette conformité est résolue et significative. Dans ce contexte, au cours des quinze dernières années, la NRA a investi plus de 300 millions d'euros en archéologie et, par conséquent, porte un intérêt aigu envers la qualité.

Mots-clés : risque, infrastructure, législation, gestion et engagement public

20 | Archéologie en tant qu'outil d'une meilleure compréhension de notre histoire récente
Peep Pillak

En 1987, la Société estonienne du patrimoine prit comme objectif la restauration de la mémoire nationale estonienne, et les archéologues jouèrent leur rôle dans ce processus. Ils se sont engagés dans des activités qui diffèrent de leurs tâches archéologiques routinières, comme clarifiant le destin des victimes du régime soviétique. En 1990, l'exhumation et le rapatriement, de Russie vers sa patrie, des restes du premier président d'Estonie ont conforté le peuple estonien dans leur détermination de restaurer l'indépendance de leur état.

Les exhumations liées à l'histoire récente de l'Estonie peuvent être hautement politisées, comme ce fut le cas avec la ré-inhumation des restes de soldats soviétiques en 2007, qui a eu pour résultat la radicalisation de la société estonienne. L'opportunité de fouilles délicates est d'une suprême importance.

Mots-clés : importance de la société civile, expertise anthropologique, exhumations, victimes du régime soviétique, tombes de la guerre

21 | Sites archéologiques : la nécessité d'améliorations de gestion (quelques réflexions au sujet de la réalité albanaise)
Ols Lafe

La discussion du patrimoine archéologique et sa gestion en Albanie, vue à partir les perspectives légale et pratique. La loi du patrimoine culturel et du système d'éducation sont analysés comme faisant partie de la discussion concernant leur gestion.

Depuis des années, il existe un débat permanent à propos de ces questions et cela a provoqué la révision de la loi du patrimoine culturel, qui est encore en cours de préparation. Avec de nombreux sites archéologiques nécessitant de plans de gestion et une augmentation du nombre de visiteurs, le débat au sein de la communauté archéologique est intense.

Mots-clés : archéologie albanaise, loi du patrimoine culturel, éducation, plans de gestion

Contributors

Peter BEDNÁR
Institute of Archaeology of Slovak Academy of Sciences
Akademicka 2
94921 Nitra
Slovakia
peter.bednar@savba.sk

Ivan CHEBEN
Institute of Archaeology of Slovak Academy of Sciences
Akademicka 2
94921 Nitra
Slovakia
ivan.cheben@savba.sk

Maria Catarina COELHO
Direção-Geral do Património Cultural
Palácio Nacional da Ajuda
1349-021 Lisbon
Portugal
mccoelho@dgpc.pt

Zeynep ERES
Faculty of Architecture,
Istanbul Technical University
34437 Taşkışla Taksim İstanbul
Turkey
zeyneperes@yahoo.com

Paulina FLORJANOWICZ
PhD Candidate at Institute of Archaeology and Ethnology
Polish Academy of Sciences
Al. Solidarności 105
00-140 Warsaw
Poland
paulina.florjanowicz@gmail.com

Konstanze GEPPERT
State Office for Heritage Management and Archaeology/
State Museum of Prehistory
Richard-Wagner-Strasse 9
D-06114 Halle (Saale)
Germany
kgeppert@lda.mk.sachsen-anhalt.de

Michał GRABOWSKI
PhD Candidate at Institute of Archaeology
Rzeszów University
Poland
micgrabowski@wp.pl

Alain GUILLOT-PINGUE
Public Service of Wallonia
Rue des Brigades d'Irlande 1
5100 Namur
Belgium
alain.guillotpingue@spw.wallonie.be

Rebecca H. JONES
Head of Archaeology Strategy
Historic Environment Scotland
Longmore House, Salisbury Place
Edinburgh EH9 1SH
Scotland, UK
Rebecca.Jones@gov.scot

Branislav KOVÁR
Institute of Archaeology of
Slovak Academy of Sciences
Akademicka 2
94921 Nitra
Slovakia
branislav.kovar@savba.sk

Kristian KRISTIANSEN
Department of Historical Studies
University of Gothenburg
Box 200, 40530 Gothenburg
Sweden
k.kristiansen@archaeology.gu.se

Jan MAŘÍK
Institute of Archaeology of the Czech Academy of Sciences, Prague, v. v. i.
Letenská 4, 118 01 Prague
Czech Republic
marik@arup.cas.cz

Harald MELLER
State Office for Heritage Management and Archaeology/
State Museum of Prehistory
Richard-Wagner-Strasse 9
D-06114 Halle (Saale)
Germany
hmeller@lda.mk.sachsen-anhalt.de

Adrian OLIVIER
London
UK
adrian.olivier@btinternet.com

Mehmet ÖZDOĞAN
Prehistory Section, Faculty of Letters
Istanbul University
34132 Laleli İstanbul
Turkey
c.mozdo@gmail.com

Dominic PERRING
Director, Centre for Applied Archaeology
(incorporating Archaeology South-East)
Institute of Archaeology, University College London
31-34 Gordon Square
London WC1H 0PY
UK
d.perring@ucl.ac.uk

Peep PILLAK
Chairman
Estonian Heritage Society
Pikk Str. 46
10133 Tallinn
Estonia
Peep.Pillak@gmail.com

Bernard RANDOIN
Ministère de la Culture et de la Communication
Direction générale des patrimoines
Sous-direction de l'archéologie
182 rue Saint-Honoré
75033 - Paris cedex 01
France
bernard.randoin@culture.gouv.fr

Matej RUTTKAY
Institute of Archaeology of Slovak Academy of Sciences
Akademicka 2
94921 Nitra
Slovakia
matej.ruttkay@savba.sk

Rónán SWAN
Head of Archaeology and Heritage
Transport Infrastructure Ireland
Parkgate Business Centre
Parkgate Street
Dublin 8
D08 YFF1
Ireland
ronan.swan@tii.ie

Monique H. VAN DEN DRIES
Archaeological Heritage Management
Faculty of Archaeology, Leiden University
P.O. Box 9514, 2300 RA Leiden
The Netherlands
m.h.van.den.dries@arch.leidenuniv.nl

Marten VERBRUGGEN
RAAP Archeologisch Adviesbureau
Postbus 5069, 1380 GB Weesp.
The Netherlands
m.verbruggen@raap.nl

Dieke WESSELINGH
City of Rotterdam Archaeological Service (BOOR)
Ceintuurbaan 213B
3015 XG Rotterdam
The Netherlands
da.wesselingh@rotterdam.nl

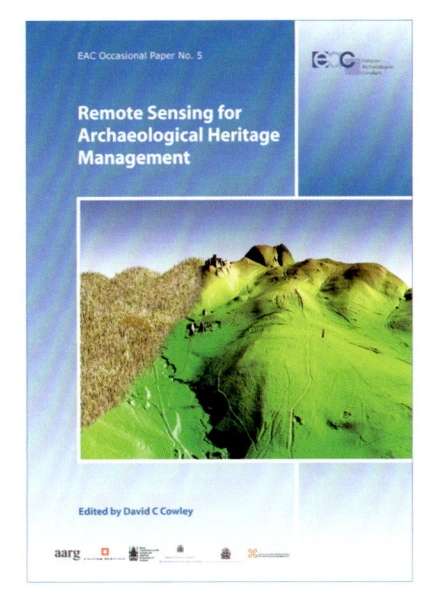

EAC Occasional Paper No. 5

Remote Sensing for Archaeological Heritage Management

Edited by David C Cowley

Remote sensing is one of the main foundations of archaeological data, underpinning knowledge and understanding of the historic environment. The volume, arising from a symposium organised by the Europae Archaeologiae Consilium (EAC) and the Aerial Archaeology Research Group (AARG), provides up to date expert statements on the methodologies, achievements and potential of remote sensing with a particular focus on archaeological heritage management. Well-established approaches and techniques are set alongside new technologies and data-sources, with discussion covering relative merits and applicability, and the need for integrated approaches to understanding and managing the landscape.

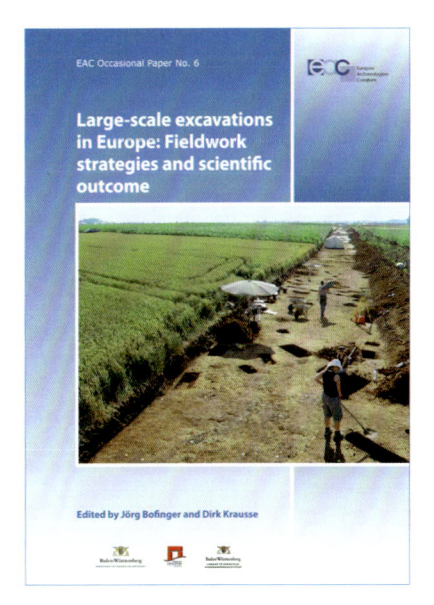

EAC Occasional Paper No. 6

Large-scale excavations in Europe: Fieldwork strategies and scientific outcome

Edited by Jörg Bofinger and Dirk Krausse

During the last decades, the number of large-scale excavations has increased significantly. This kind of fieldwork offers not only new data, finds and additional archaeological sites, but also gives new insights into the interpretation of archaeological landscapes as a whole. New patterns concerning human "offsite activities", e.g. field systems, or types of sites which were previously underrepresented, can only be detected by large-scale excavations. Linear projects especially, such as pipelines and motorways, offer the possibility to extrapolate and propose models of land use and environment on the regional and macro-regional scale.

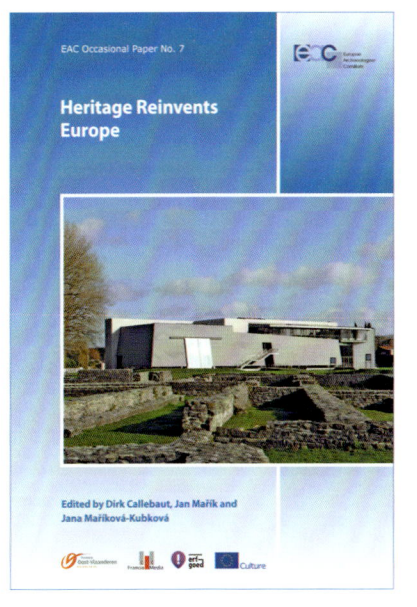

EAC Occasional Paper No. 7

Heritage Reinvents Europe

Edited by Dirk Callebaut, Jan Mařík and Jana Maříková-Kubková

Unity in Diversity, the motto of the European Union, has, since World War II, seldom been as relevant as it is today. In these difficult economic times Europe is more and more confronted with the phenomenon that citizens openly stand up for the defence of their national and regional interests. This has put enormous pressure on the process of European integration and the concept of a shared European identity based on the cultures of individual EU member states. Thus, understanding the diversity of European cultural heritage and its presentation to the broadest audience represents a challenge that can be answered by diversified group of scientists, including archaeologists, historians, culturologists, museologists etc.

By choosing "Heritage reinvents Europe" as the theme for the 12th EAC colloquium that was held between the 17th–19th March 2011, in the Provincial Heritage Centre in Ename, Belgium, the board of the Europae Archaeologiae Consilium made its contribution to the understanding of the key concept of a shared European identity.

EAC Occasional Paper No. 8

Who cares? Perspectives on Public Awareness, Participation and Protection in Archaeological Heritage Management

Edited by Agneta Lagerlöf

The increasing numbers of reports on tampering with ancient monuments and archaeological materials may reflect more acts of plunder. But it could also reflect a higher incidence of reporting of such acts to competent authorities or a combination of them both. A third solution is of course that acts of plunder are currently deemed more newsworthy than before in our part of the world. And if this is the case, we must ask why has this become important now, and also, how does this influence our understanding of what is happening? The complexity of this problem and the ethical issues it raises require us to examine our view of the archaeological source material and archaeology as a profession in relation to society at large. An international conference took place in Paris 2012 with participants from different European countries. The purpose of the conference was to discuss the kind of measures that need to be taken and what the societal consequences of these may be.

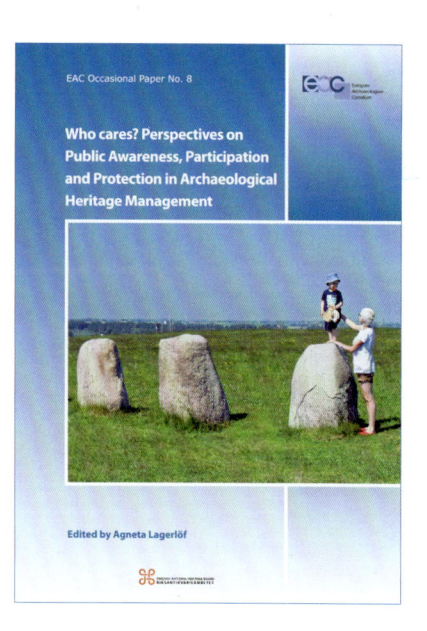

EAC Occasional Paper No. 9

The Valletta Convention:
Twenty Years After – Benefits, Problems, Challenges

Edited by Victoria M. van der Haas and Peter A.C. Schut

The Valletta Convention (1992) was the result of a process which started with the Convention of London (1969) where the foundation for contemporary archaeological preservation was laid. The inclusion of archaeology in the process of spatial planning was one of the most important milestones. In most European countries it meant a strong growth of archaeological research, and now, in 2014, we can say that Valletta has become visible in all parts of archaeology. Not only are new residential quarters, industrial and infrastructural works archaeologically investigated, also within the field of public information and cultural tourism there are important achievements. The implications for education are great. In this publication the main topics are addressed. Not only the successes, but also the challenges and possible solutions are addressed. Due to articles written by experts from different parts of Europe, this publication provides the reader with a good view of the state of affairs in various countries.

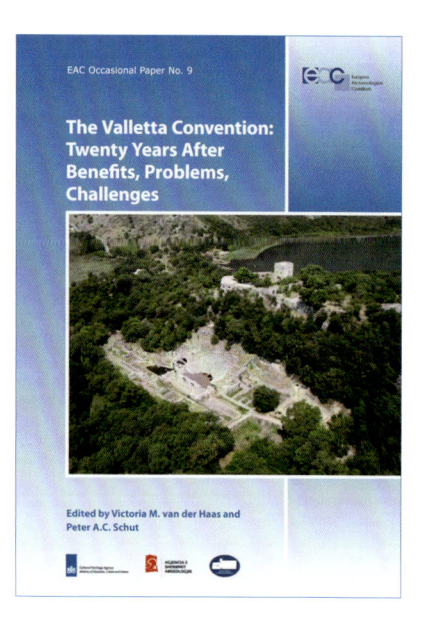

EAC Occasional Paper No. 10

Setting the Agenda:
Giving New Meaning to the European Archaeological Heritage

Edited by Peter A.C. Schut, Djurra Scharff and Leonard C. de Wit

More than two decades after the signing of the Valletta Convention the time is ripe to draw up a new agenda for how Europe should manage its archaeological heritage. With this purpose in mind, the EAC organised two symposiums that were attended by heritage managers from 25 European countries. The first symposium was held in Saranda, Albania, and the second in Amersfoort, the Netherlands, which took the form of a working conference. The results are published in this volume, which largely comprises the Amersfoort Agenda for managing the archaeological heritage in Europe. This agenda ties in with the ideas of the Council of Europe's Faro Convention on the Value of Cultural Heritage for Society (2005) among others. The zeitgeist calls for an acknowledgement of the multiple values of archaeological heritage for society and recognises the potential role of archaeological heritage in sustainable development. The various articles in this book explore this topic in greater depth. Reports of the break-out sessions have also been included so that readers can follow the discussions that have led to the Amersfoort Agenda.

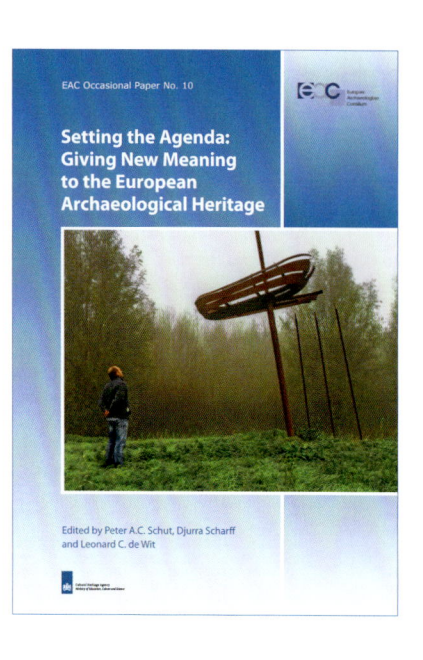